MW01520137

HOSPITALS

Planning, Design and Management

HOSPITALS

Planning, Design and Management

PRINCIPAL AUTHOR

G D KUNDERS
Formerly CEO, Cauvery Medical Center
and
Administrator, St. Martha's Hospital
Bangalore

ASSOCIATE AUTHORS

S GOPINATH
and
ASOKA KATAKAM
Principals
Gopinath & Katakam
Architects and Interior Designers
Bangalore

Tata McGraw-Hill Publishing Company Limited

NEW DELHI

McGraw-Hill Offices

New Delhi New York St Louis San Francisco Auckland Bogotá
Lisbon London Madrid Mexico City Milan Montreal Caracas
San Juan Singapore Sydney Tokyo Toronto

Tata McGraw-Hill
A Division of The **McGraw-Hill** *Companies*

Information contained in this work has been obtained by Tata McGraw-Hill, from sources believed to be reliable. However, neither Tata McGraw-Hill nor its authors guarantee the accuracy or completeness of any information published herein, and neither Tata McGraw-Hill nor its authors shall be responsible for any errors, omissions, or damages arising out of use of this information. This work is published with the understanding that Tata McGraw-Hill and its authors are supplying information but are not attempting to render engineering or other professional services.

ISBN   0-07-462211-0

Published by Tata McGraw-Hill Publishing Company Limited, 7 West Patel Nagar, New Delhi 110 008, typeset in Garamond at Scribe Consultants B4/30 Sufdarjung Enclave, New Delhi 110 016 and printed at Replika Press Pvt. Ltd., Plot No A 229 DSIDC Narela Industrial Park, Narela, Delhi 110 040

*This book is lovingly dedicated to the
memory of a very dear friend*

C. A. Keshava Murthy

*who laboured with us but did
not live to enjoy the fruits*

About the Authors

G D Kunders, one of the nation's few professional hospital administrators, has an unusually rich background and experience in hospital administration, spanning over twenty-five years. Mr Kunders holds a master's degree in English language and literature from the Madras University. Later he received his master's degree in business administration from the Atlanta University in the USA, distinguishing himself by earning recognition from Delta Mu Delta National Honor Society in Business Administration (USA) for high scholastic attainment in MBA. Formerly of the staff of Christian Medical College and Hospital, Vellore, Grady Memorial Hospital, Atlanta, and MMM Hospital, Kolenchery, he served for several years as administrator of St. Martha's Hospital, a more than a century-old, 600-bed hospital in Bangalore, and as chief executive officer of the Cauvery Medical Center, also in Bangalore. He was consultant to the Bangalore-based Manipal Hospital, a superspecialty hospital, during its developmental period. In the late 1980s, Mr Kunders had the rare distinction of being invited as a visiting hospital administrator in Chicago, Illinois, USA.

Mr Kunders has written a number of articles on management of hospitals, which have been published in health care journals. He is the author of two books entitled: *Managerial Leadership* and *Are You Executive Material?* The publication of his two other books, *Are You Leader Material?* and *How to Market Your Hospital Without Selling Your Philosophy*, is currently in the works. His name figures in Who's Who in the British Commonwealth, and his biography is included in the International Who's Who of Intellectuals and Men of Achievement, all of which are the publications of International Biographical Centre, Cambridge, England.

S Gopinath, a well-known architect based in Bangalore, graduated from Abhinava Kala Vidyalaya of Architecture in Pune. After working for some of the leading architects in India and Iran, he teamed up with his architect wife, Tara, and set-up his practice with other principals in a firm called Gopinath and Katakam.

Some of the major hospital projects he has designed are Manipal Hospital, a 600-bed superspecialty hospital, Bangalore; KMC Hospital, a 400-bed teaching hospital in Mangalore; and the Ravi Kirloskar Memorial Hospital, Bangalore. Mr Gopinath is currently associate architect to Cauvery Medical Center, a 350-bed high-tech corporate hospital in Bangalore. Among the major hotels he has designed are the Valley View International Hotel, Manipal; Balu and Velu Hotel, Ooty; Comfort Inn Ramanashree, Bangalore; and Malabar Hotel Renovation, Cochin. In addition, he has designed or has been consulting or associate architect for a number of health care facilities and for other projects like hotels, tourist resorts and villas, both in India and Iran.

Mr Gopinath is an associate of the Indian Institute of Architects, Indian Institute of Interior Designers and is also a registered member of Council of Architecture. Gopinath is an active Rotarian involved in a variety of social service programmes.

Asoka Katakam has had the unique opportunity of receiving his architectural training in the United Kingdom, where he received his bachelor's and master's degrees in architecture from the Cambridge University, in addition to a diploma in architecture from London University. He then worked for the department of environment and Jarvis K. Sons. His singular experience in hospital architecture came when he worked as an architect for the renowned Llewelyn-Davies & Weekes, London, the reputed firm of hospital architects of international fame. In 1982, he started his practice with other principals under the name of Katakam & Associates, and is based in Secunderabad. He has also designed some of the hospitals, hotels and other projects listed under the second author. Among the various projects currently on hand are a maternity hospital in Hyderabad; and Kamineni Hospital, a 250-bed superspecialty hospital in Hyderabad to which he is a consulting architect. His overseas projects include the PSA Hospital in Riyadh; St. Mary's Hospital in London and Queen Mary's Hospital in Hong Kong. He is a corporate member of the Royal Institute of British Architects, and a Registered Architect with the Council of Architecture, India.

Preface

The technological advances made in the twentieth century have brought scientific marvels into our hospitals, and these in turn have made an unprecedented demand on hospitals' medical services, particularly in areas such as surgery, clinical laboratories and radiological services. As a result, during the time when these spectacular changes and adjustments were taking place, the attention of hospital planners and designers was primarily focused on the proverbially acclaimed medical services or what we may call the glamorous aspect of hospital services. The less noticeable physical facilities, the administrative areas and supportive services have, like the shoemaker's progeny of old, been impatiently waiting for their new shoes. A planning guide which covers all the services in one volume has been long overdue. This book attempts to fill that void.

The primary objective of this book is to promote good planning and design in order to build efficient hospitals. Not only are the operational costs for running an inefficient hospital high; they also mean less health service and more expenses for the patients. The initial cost of the building is insignificant compared to the cost of operating, staffing and maintaining it over, say, twenty years. Studies have shown that in some cases it is 18 to 20 times the initial cost. Of significance for the planners and administrators to remember even at the initial stage of building a hospital is the axiom that if a hospital has to be successful, it should be based on a triad of good planning, good design and construction, and good administration. The absence of any one of these results in a mediocre hospital or one that is doomed to failure. By the same token, the real test of any hospital—a rugged test which many do not pass—is the quality of health care it provides. It is obvious that the system inside the hospital has to work effectively to meet that test.

The material provided in this book depicts the functional requirements of various departments, units and services. They are intended to be used as a basis for planning, for developing architectural programmes and for stimulating new ideas and techniques. To facilitate this, we have delineated the functions of each department, and discussed, among other things, location and inter- and intra-departmental relationships. With imagination and professional acumen on the part of the architect, planning and design of facilities can be as innovative and as best as they can be. However, the criteria established in this book are intended only as a guide and not to be followed exactly. They should be adapted to suit the individual requirements of each institution.

For the proper interpretation and use of these guidelines, it is important to understand that the criteria, diagrams, schematic plans and other material are presented to illustrate a possible approach and not a recommended pattern. The material, collected from various sources, was combined with the authors' ideas of what an appropriate design should be. If these ideas and plans were to be literally lifted from these pages and used without any modification, it is possible that they would not represent the optimum plan for any one particular hospital. In our view, the best perspective for considering the plans, diagrams and charts is simply to call them "graphic guidelines".

While these guidelines are in keeping with the present day practices and requirements, it is necessary that newer concepts must be continually searched and evaluated. Therefore, it cannot be over-emphasized that a high degree of flexibility should be incorporated in planning in order to meet the future needs of expansion and changes.

We make no claim that much of the material in this book represents new or original thinking on our part. The material was developed from many sources and from the work of well known authorities in the fields of health care, architecture and management and blended with many years of our learning and experience in hospital administration and architecture in some of the leading institutions both in India and abroad. It represents what we consider should be an attainable criteria to plan and design an efficient hospital and to make it work. We are indebted to all those from whom we have received what we are now able to pass on to others.

All those who are concerned with planning and designing medical facilities should find something of value here. It is our hope that the book will contribute in some measure to the growing awareness of the need for efficiency in building health care facilities and for creating an environment that is more comfortable for patients, visitors and personnel, besides making these facilities viable and economical to maintain.

Architects, planners, designers, consultants, trustees and administrators of hospitals for whom this book is intended have both an awesome responsibility and a unique opportunity to bring about constructive changes in building hospitals. It is hoped that this book will be a source of information and enlightenment to all who are interested in the subject.

One can predict that the coming decade or so will see more and more high-tech hospitals being set up. This publication presents guidelines for planning and designing that class of hospitals. At the same time, it is developed to satisfactorily meet the requirements of all types and sizes of hospitals.

We have thought of the criticism that may be raised in certain quarters that the facilities and criteria we have discussed in this book are too elaborate for most hospitals and may probably be unattainable. But then we thought of the awesome responsibility that was placed on us as authors to delineate what is needed in an ideal set-up where the planners have also the option to choose what they need or what they can afford to have. We would rather not face the criticism that the criteria and facilities we have provided are inadequate for a good set-up. It is for the individual hospitals as informed planners to intelligently choose from these guidelines to build an optimum facility for themselves according to their budget.

We express our sincere appreciation of the help rendered by our secretaries and assistants without whose pleasant cooperation the book would not have been possible. Inspiration and partnership were provided by Ms P.L. Padma, the principal author's secretary. The manuscript preparation and typing was a monumental task which she did cheerfully, draft after draft, often putting up with unreasonable demands and work schedules, thus assuming a major responsibility for the preparation of the book. Padma's commitment and dedication are truly commendable.

We also acknowledge the work of the secretary and assistants of the second author — Ms Sunanda Menon, Ms J. Chithra and Mr Ram Prasad — whose enthusiasm in the work of the book by way of typing, development of diagrams and drawings is much appreciated. A special mention is made of Chithra who was often subjected to exacting demands in the preparation of plans and diagrams.

The text of this book has been written entirely by the principal author while the diagrams, plans and drawings have been provided by the associate authors.

G D Kunders
S Gopinath
Asoka Katakam

Acknowledgements

We owe a great debt of gratitude to the celebrated professionals listed below who have distinguished themselves in the fields of health care, architecture and management, for having so generously given their knowledge, time and expertise for making a pre-publication review of the book.

1. Dr V I Mathan
    Director, Christian Medical College and Hospital
    Vellore

2. Dr R M Varma
    Former Director and
    Emeritus Professor of Neurosurgery
    NIMHANS
    Bangalore

3. Rev. Dr Percival Fernandez
    Director
    St. John's Medical College and Hospital
    Bangalore

4. Dr Nalla G. Palaniswami
    Chairman and Managing Director
    Kovai Medical Center and Hospital
    Coimbatore

5. Mr H C Thimmaiah
    President
    Indian Institute of Architects
    Bangalore

6. Dr A S Fenn
    Former Principal
    Christian Medical College
    Vellore
    Currently Director
    Lutheran Hospital
    Ambur

7. Mr Jai Rattan Bhalla
    President
    Council of Architecture, India
    New Delhi

8. Dr Daleep Mukarji
   Former General Secretary
   Christian Medical Association of India
   Currently Executive Secretary
   World Council of Churches
   Geneva

9. Dr G R Ravikumar
   Chairman
   Cauvery Medical Center and Cauvery Medical International
   California

10. Professor Shireesh Deshpande
    Immediate Past President
    Indian Institute of Architects
    Nagpur

11. Dr C Dayakar Reddy
    Chairman
    C.D.R. Hospitals
    Hyderabad

12. Dr Cherian Thomas
    Former Director
    Miraj Medical Centre
    Miraj
    Currently General Secretary
    Christian Medical Association of India
    New Delhi

13. Dr R J Amruthraj
    Former Dean of Karnataka Medical
    College and Hospital
    Hubli
    Former Medical Superintendent
    St. Martha's Hospital
    Bangalore

14. Mr K S N Murthy
    Former Chief Secretary
    Government of Karnataka
    Bangalore

Special Thanks

A special and appreciative "thank you" is accorded to the following friends and experts, some of international fame, who have helped us form our ideas about the book. They shared with us their knowledge and expertise by doing a critique of chapters of their respective specialties in which they are the acknowledged authorities. With their assistance they made this a better work than what it was in the early drafts.

1. Dr A M Nisar Syed, California
2. Dr A R Suresh, Bangalore
3. Dr H B Shivaprasad, Bangalore
4. Dr Jinka Subramanya, Bangalore
5. Dr Hemanth K. Kalyan, Bangalore
6. Dr Joga Rao, Bangalore
7. Dr M H Shariff, Bangalore
8. Dr Sojan Iype, Kolenchery
9. Dr (Mrs) Shaloo Iype, Kolenchery
10. Dr V M D Namboodiri, Kolenchery
11. Dr (Mrs) Prabhavathy Kunders, Bangalore
12. Dr Malathi Rao, Bangalore
13. Dr Cyril Jayachandran, Vellore
14. Mrs Violet Jayachandran, Vellore
15. Mr C A Keshava Murthy, Bangalore
16. Mr B N S Murthy, Bangalore
17. Mr G M Vijayasingh Yesudian, Salem
18. Mr A C Varghese, Bangalore
19. Mrs Joy Varghese, Bangalore
20. Dr B S Prakash, Bangalore
21. Mrs Shobha Prakash, Bangalore
22. Mr M Rajendran, Bangalore
23. Mr R Srinivasan, Bombay
24. Mr A Vijay Kumar, Bangalore
25. Mr B Ramnadh, Bangalore
26. Mrs S G Gayathri, Bangalore

Contents

CHAPTER **4**

PLANNING AND DESIGNING MEDICAL SERVICES 131

CHAPTER **5**

PLANNING AND DESIGNING NURSING SERVICES

211

CHAPTER **6**

PLANNING AND DESIGNING SUPPORTIVE SERVICES   251

CHAPTER 9

COLOUR PLATES

CHAPTER 1

Planning and Building a New Hospital: From Designing to Commissioning

CONTENTS

- ☐ **Planning the Hospital**
  - ▪ Guiding Principles in Planning Hospital Facilities and Services
- ☐ **Preliminary Survey**
  - ▪ Study of Existing Hospital Facilities
  - ▪ Study of Required Staff and Services
- ☐ **Financial Planning**
- ☐ **Equipment Planning**
- ☐ **Permanent Hospital Organization**
- ☐ **Functional Plans for Hospital Construction**
- ☐ **The Design Team**
  - ▪ Hospital Consultant
  - ▪ Architect
  - ▪ Engineers
  - ▪ Hospital Administrator
- ☐ **Functional Programme and Design Stage**
  - ▪ Production Documents
  - ▪ Tender Documents
  - ▪ The Hospital Site
  - ▪ Bed Distribution
  - ▪ Space Requirements
- ☐ **Planning the Hospital Building**
  - ▪ General Principles
  - ▪ General Features
- ☐ **Planning Operational Units**
- ☐ **Building Contract and Contract Documents**
- ☐ **Furnishing and Equipping the Hospital**
- ☐ **Ready to Operate Stage**
- ☐ **New Building Announcement**
- ☐ **Groundbreaking Ceremonies**
- ☐ **Commissioning and Inauguration**

PLANNING THE HOSPITAL

In the establishment of a new hospital, the first step is always an idea born in the mind of some individual. If the idea is appealing and is based on sound reason, the originator is able to gather the support of other people — first a small group which gradually enlists the support of other people in the community who are sold on the idea of having a hospital in their town. A committee is then formed and is given the authority to undertake the preliminary work such as a feasibility study or survey and to raise funds to meet the expenses involved in the survey or study. If the committee is lucky, it may receive the support of a few influential philanthropic citizens and organizations. Thus one man's idea becomes an accomplished fact.

All successful hospitals, without exception, are based upon a triad of good planning, good design and construction, and good administration. The success of a hospital is generally measured in terms of patient care, efficiency and community service. The absence of any one of these closely related components means a mediocre hospital, or one doomed to failure.

To be successful, a hospital requires a great deal of preliminary study and planning. It must be designed to meet the needs of the people it is going to serve and be of a size which the promoters can afford to build and operate. It must be staffed with competent and adequate number of doctors, nurses and other professionals to render efficient service. If mediocre hospitals abound in our country or if some of them are unable to provide adequate and efficient service to the community, the principal reason is most likely the lack of proper planning, the other being paucity of funds. Fig. 1.1 shows the stages in promoting and building a new hospital.

➤ GUIDING PRINCIPLES IN PLANNING HOSPITAL FACILITIES AND SERVICES

Some important principles of planning hospital facilities and services are stated below.

High Quality Patient Care
This can be achieved by the following:
- appointing competent and adequate number of medical, nursing and other professional staff and providing necessary facilities, equipment and support services;
- establishing an organizational structure in which clearly defined responsibility and authority are assigned to each job, particularly in patient care. There should be proper accountability;

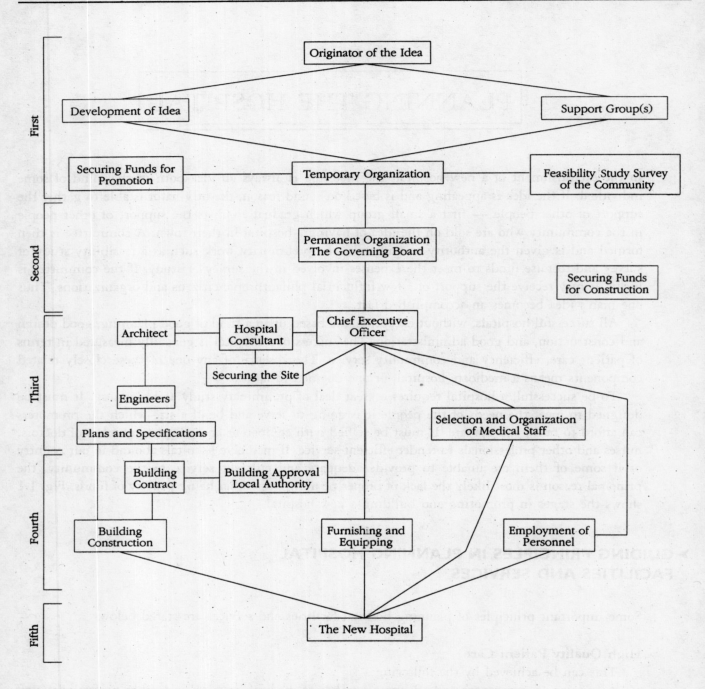

➢ **FIG. 1.1** Flowchart Representing Stages in Promoting and Building a New Hospital

* medical staff working as a team and interacting with each other and with other health care professionals;
* instituting a mechanism or procedure for continuous review of patient care provided by physicians, nurses and other professionals;
* providing continuing medical and other educational programmes to all professionals to enable them keep abreast with the latest medical and technological knowledge aimed at improving patient care;
* establishing and enforcing standards in patient care.

Effective Community Orientation

This can be done by the following:
* a governing board which is made up of individuals who are known and respected leaders of the community;
* extending programmes and services of the hospital to the community;
* ensuring the hospital's participation in community programmes in preventive care, teaching of good health care and practices, school health programmes, etc;
* the hospital administrator, other key personnel and doctors providing assistance in planning and implementing community health care programmes;
* the hospital exercising responsibility to ensure that it gains support from the community; and
* providing a public information programme to keep the community informed of the services provided by the hospital as well as its goals, plans and objectives, and encouraging members of the community to personally participate in them.

Economic Viability

This can be realized by
* accepting responsibility and accountability for a strong and viable financial position that will command the respect and confidence of the community, donors and investors;
* making available adequate operating finances for personnel and equipment necessary for providing quality patient care;
* a programme to attract and retain competent and dedicated physicians, nurses and other health care professionals to maintain high occupancy and full utilization of inpatient and outpatient facilities;
* planning new services and expansion programmes based solely on community needs;
* a planned programme for financing replacement of equipment and improvement of facilities; and
* an annual budget that will provide for maintaining services at a high level, for equipment, salary and wages, payment of interest, loan, etc., depreciation funding and capital for replacement and development. The objective is to help the hospital remain in the forefront of medical technology and knowledge.

Sound Architectural Plan

This can be achieved by

- engaging, early in the planning stage, an architect who is experienced in hospital design and construction;
- selecting a site that is readily accessible to public transport, water, sewerage lines, population concentration, etc., and is large enough to meet the requirements for parking, access road, future expansion, etc.;
- determining the size of the hospital that is adequate for various services, administrative and functional needs of departments, and patient care and treatment;
- recognising the importance of establishing traffic patterns for movement of physicians, hospital personnel, patients, visitors, and efficient transportation of food, linen, drugs and other supplies;
- a design that will avoid duplication of services but at the same time provide flexibility and inter-changeability of patient rooms for clinical departments with fluctuating census; and
- paying attention to special services like outpatient, intensive care, psychiatry, obstetrics, operating rooms, medical and surgical specialties, and to such concepts as infection control, disaster planning, etc.

PRELIMINARY SURVEY

One of the first tasks the preliminary organization must set itself to perform is a survey of the service area of the proposed hospital. This may be carried out in different ways. One is to engage a consultancy firm which has experts on its staff to carry out this kind of a survey. Some banks and chartered accountants' firms also provide consultancy services. Occasionally, the hospital organization may put together a team from among its members, including doctors, to carry out the survey. But this kind of team, consisting of amateurs as it were, may not always be successful. The biggest problem will be lack of unanimity in the committee's report, particularly if the team has some strong, self-opinionated members who are biased one way or the other.

One caution that should be exercised in case the survey is carried out by financial companies is to ensure that the focus is not entirely on financial projection and viability to the exclusion of the study of the service area — its character, needs and possibilities. Insofar as major decisions will be based on the results of the survey, it must be done in a professional manner. In this chapter we provide some guidelines on how to conduct a survey.

Hospital organizations will do well to engage a hospital consultant. As we shall see later, it will be advantageous if such an expert is appointed for the entire project. There are many aspects of the project in which his expertise can be used. The survey is only one of them.

The consultant must be familiar with hospital construction, and know hospital administration by having had experience as a hospital administrator. Above all, he must be objective, honest, and free from bias, and be capable of drawing conclusions from the facts which will be presented to him. Finally, he must have the courage to report his honest convictions whether or not they match with the preconceived ideas of the committee.

There is great merit in appointing an outside expert as a consultant. A local man is almost certain to have a preconceived opinion either for or against the project. Unconsciously, he will go about proving his point. Moreover his recommendations may not carry as much weight with the committee as those of an outside expert.

The first consideration in the survey is to determine the character, needs and possibilities of the community which the hospital is going to serve. It is on the basis of this knowledge that a decision is going to be taken as to whether or not a hospital is to be built, and if so, its type and size. If the community is preponderantly wealthy, or made up of moderate wage earners, the industrial workers or the indigent, that fact will be the deciding factor as to the kind of hospital that is going to be built. For example, if the community constitutes largely of wealthy individuals, one may plan to build a luxurious hospital with private rooms and sophisticated diagnostic and therapeutic equipment; if it is largely meant for the indigent patients, a charitable hospital is needed.

It would be a mistake to build a hospital for the richer class, some of whom would have promoted the hospital, only to find later that they patronize the hospital only for minor ailments. They actually go to the distant hospitals for treatment of serious illness which would be due to the lack of confidence in local physicians. The authors have seen this happen in quite a few places.

Apart from levels of income, other characteristics such as occupation, age distribution, etc. must be studied. They determine the amount and kind of hospital care those people need and the amount they are willing to pay. For example, if there is a large population of senior citizens, more of geriatric services may have to be provided.

The general attitude of the people is also important. What makes patients choose a particular hospital? Here are some answers:

 * Availability of specialists
 * Range of services offered under one roof
 * Availability of latest technology and sophisticated equipment
 * Personalized care and courtesy of the hospital staff
 * Overall hospital reputation
 * Travel time (not necessarily the distance) to reach there
 * Other things being equal or relative, proximity of the hospital
 * Status of roads and transportation facilities.

Contrary to popular belief, hospital charges and proximity are not high in the list. When it comes to the question of their health, people want the best and are willing to pay for it.

There are certain characteristics which make people turn away from a hospital if they have a choice. Some of them are:

 * Building not constructed as a hospital

- Hospital not clean and tidy
- Hospital is inaccessible or is in poor location because of security risk, nuisance factors or lack of parking facilities
- Inadequate size or obsolete construction
- Inadequate medical care, staff and equipment, limited services, admission restrictions, non-availability of 24-hour service, etc.

➤ STUDY OF EXISTING HOSPITAL FACILITIES

The next phase of the survey is a study of existing hospital facilities on an area-wise basis. This study should be comprehensive and involve both short and long term needs and objectives.

The most important part of the study is an inventory of facilities, beds and services of every hospital. It should cover the following areas:

- Bed capacity of the institution
- Physical condition of facilities
- Hospital occupancy
- Bed ratio
- Volume and kind of hospital services
- Quality of facilities and services.

➤ STUDY OF REQUIRED STAFF AND SERVICES

The consultant or the committee must then make a study of human resources — doctors, nurses and other professional staff — required for the proposed hospital, and the hospital's ability to initially provide and subsequently support them. This study must be extended to cover the hospital's services.

Finding and retaining qualified and competent specialist doctors and nurses in adequate numbers is not easy even at the best of times. This difficulty is particularly acute in smaller towns and remote areas. It is not unusual to find hospitals which otherwise have everything to boast of unable to provide services to the community due to lack of staff although some of them make temporary makeshift arrangements. Local politics, so rampant in small towns, often aggravates the problem by keeping local professionals away.

How many and what kind of specialists are needed for a hospital to give adequate care to the community? It is generally agreed that in addition to the traditional services such as internal medicine, general surgery, paediatrics, obstetrics and gynaecology, specialists in the disciplines of eye, ENT, dermatology, radiology, pathology, urology, etc. must be provided. Not all small and medium—sized hospitals can support specialty and superspecialty services.

The same exercise must be carried out for nurses and other professional staff. Getting trained nurses is a serious problem in our country. While qualified laboratory technicians and pharmacists are freely available, the same is not true of physical therapists and some of the rare technical staff who are difficult to find. New hospitals must be aware of these problems.

Last, but not the least, is the Chief Executive of the hospital. He should be qualified, preferably with a master's degree in hospital or business administration, and adequate experience besides possessing leadership qualities. Equally important are his key management team and other categories of employees. It should be recognized that in a hospital set-up, even the least of the employees has an important role.

FINANCIAL PLANNING

The importance of planning for funds for the hospital at this stage cannot be over-emphasized. Financial planning must take precedence over other considerations and must cover the following three areas that need financing:

Constructing, equipping and furnishing the hospital

After the schematic drawings have been agreed upon, the architect should be able to make a fairly accurate estimate of the cost of the project. Some changes will certainly develop and the estimate will have to be revised. There may also be cost escalation due to inflation and delay in executing the work.

Operating funds

As the hospital building approaches completion, the governing board begins to incur liability which will continue to accumulate over a period of two to three years or until the hospital is fully operational. Expenses on account of salaries and wages, repayment of loan(s), payment of interest and other operational and maintenance expenses should be met. Insurance against fire and other structural damage should be continuous from the contractors to the owners. Operational expenses during this initial stage are far higher than the revenues. The governing board must find funds to meet these commitments.

Financial assistance

Hospitals do not, by and large, make profit even when circumstances are favourable. This is because most of our hospitals are charitable hospitals. Maintaining and operating these hospitals need financial assistance perennially.

One question the promoters must address and perhaps study carefully is to what extent the patient fees, hospital charges and other facility funds will cover the construction and operational costs.

We do not propose to discuss how hospital organizations raise funds for their hospitals — whether it is by donations, membership fee, trust, loans from banks or financial institutions, or a combination of some of these. But we do want to sound a note of caution. We have seen hospitals come to grief because of lack of proper financial planning. The unsuspected pitfalls or dangers lurking around may prove lethal to many an unwary hospital. The following are some of the dangers: (i) in their enthusiasm, some hospitals embark on a project that is bigger than they can afford; (ii) they start building without a realistic estimate of the total cost or without assured resources to complete; (iii) quite often the much expected loan does not come through; and (iv) the delay in the execution of work which sends the costs spiralling upwards. As a result the project comes to a halt or is cut down drastically or left unfinished.

EQUIPMENT PLANNING

Hospital planning is not complete if careful attention is not given to the fixed and movable equipment that will be needed for the operation of the hospital. The time to attend to this is early in the design development stage. A series of meetings are arranged with the medical staff and other personnel to discuss the equipment needed for the hospital. A room by room equipment list is then compiled and reviewed by administrative, medical and departmental staff. The valuable information thus compiled and documented is used in coordinating details and sizes of rooms, utility services, lighting and work flow. This information is also needed for financial planning.

Another detailed equipment planning is necessary later at the stage of procuring the equipment and the actual furnishing and equipping the hospital. A separate section later in the book is devoted to this subject.

PERMANENT HOSPITAL ORGANIZATION

On completion of the survey and other studies, and based on the conclusions of the same, the consultant will sum up the whole situation and advise the preliminary organization about the feasibility of the project. In case the project is feasible, the consultant will advise on the kind of hospital that should be built, its size and the services that it would offer. He will also report on other aspects such as the means of funding the project, probable source and extent of support, financial viability and general principles which will underlie building, furnishing, and equipping.

As soon as it is decided that the project is to be carried through, the preliminary organization should proceed with the formation of the permanent organization. It should then decide on the type of organization that is aimed at — charitable, corporate or any other — and make arrangements to secure a charter. If the hospital is to be owned by an existing organization, the charter of the parent organization probably provides for ownership and management of property. As soon as the permanent governing board is formed, it finds itself faced with several problems which must be dealt with almost simultaneously. The most important ones are raising funds for construction, employment of an architect, consultant and hospital administrator, selection of a site, and preparation of plans and specifications. When the construction gets under way, the owners must initiate action for the selection and organization of the medical staff.

FUNCTIONAL PLANS FOR HOSPITAL CONSTRUCTION

The survey, other studies and the detailed preliminary planning which were discussed in the earlier section logically lead to the next phase of building a hospital — planning of physical facilities on a functional basis. It is at this crucial stage that the promoters of the hospital are likely to make the worst mistake which stems from lack of competent advice and guidance. Some succumb to the glamour of having an outwardly beautiful edifice, an artist's concept, which is no better than raising a structure instead of designing clinical and administrative services to patients. As a result, they are saddled with an attractive but totally inefficient building.

Designing and building a hospital is an intricate job that is best left to experts. Even minor defects in designing can make the operation inefficient, require more employees and markedly increase the cost of maintenance. Functional planning in hospitals is important, and the key to functional planning is the understanding that travel and adjacencies affect the operational cost. Viewed from this angle, the initial cost of building is often insignificant when compared to the cost of running and staffing it, say, over 20 years, sometimes 18 to 20 times the initial cost. It will be worse in the case of poorly designed hospitals.

Inefficient hospitals cost more to the patients too because they get less health care for their money.

THE DESIGN TEAM

➤ THE HOSPITAL CONSULTANT

As soon as the governing board has raised sufficient funds to start construction and has assurance of more funds coming in, it should employ a consultant who should continue to be in its employ until the hospital project is completed. The consultant, whether a physician or a layman, be experienced in hospital administration and must be familiar with the problems of planning and construction. With this background he brings to the project his knowledge of the principles of hospital administration in locating departments, rooms, utilities, equipment and services in a manner that ensures the better care of patients and the smooth functioning of administration. A physician may be outstanding in his profession but if his knowledge and experience are limited to the professional care of patients, and not in the myriad tasks and problems of hospital administration, he may not be suitable as a consultant. Since his work is of supreme importance to the success of the project, the consultant should be selected carefully.

The consultant's functions and responsibilities should be delineated in advance as far as possible. The following are some of the responsibilities that may be assigned to him:

- Develop a long range plan and give recommendations
- Survey
- Feasibility and other studies
- Study role
- Prepare a master plan for the facility
- Financial feasibility
- Equipment selection
- Inspection
- Function as owner's representative at the construction site and on work committees
- Preparation of inventories of construction materials and equipment
- Determination of need for expansion, renovation and new construction
- Preparation of operational and functional programmes as bases for architectural planning
- Screening of conflicting suggestions and recommendations
- Programming of space allotments and departmental groupings
- Selection of certain items of equipment.

➤THE ARCHITECT

The architect is the leader of the design team and a key figure in the hospital project. The plans drawn by him will affect the standard of patient care and the efficiency of the operation markedly. He must be given a written programme or brief by the promoters detailing the clinical and administrative requirements of the proposed hospital. The brief may have been prepared by the hospital consultant. Since he is not expected to have competency in the technical aspects of hospital architecture and may not have knowledge of clinical and administrative matters, the architect must listen to and evaluate the needs and wishes of all parties involved in the project. The responsibility of the architect is to translate the clinical and administrative needs into architectural and engineering realities. These include, among other things, site selection, orientation, adapting building design to construction contracts, and functional requirements like utilities, electrical and mechanical installations. Architects who have been most successful in hospital construction recognize that the problems associated with planning and designing hospitals are beyond their ordinary range of experience and that they need the services of not only engineers and technical consultants but of persons like hospital consultant and hospital administrator who combine a knowledge of planning and building of hospitals with experience in hospital administration. When the architect recognizes this need and welcomes the working association with a hospital consultant whose duties consist in collaborating with the architect on the one hand, and conferring with the promoters or the governing board on the other, a fuller understanding of his brief emerges. This results in the plans and specifications being complete in every detail and well suited to the needs of the hospital.

The early employment of the architect is one of the first requirements for planning a successful hospital. He should, therefore, be employed as soon as the governing board thinks it is safe to go ahead with the project. This will ensure that the details for which the architect is responsible in the preparation of the project will be speeded up. Completion of plans and specifications is a long and tedious task which takes time. Plans, primarily intended to be used in construction, are also needed during the time of application for loans from financial institutions and for obtaining approvals and sanctions. Plans will be of value in the promotional work and financial campaign.

The architect is generally selected by competition through an interview procedure, sometimes by comparison of sketches submitted by competing architects. Experts feel that this is a doubtful procedure in the case of a hospital. On occasion, the hospital may commission a known architect who has performed satisfactorily in other projects. A hospital is different from any other type of building, and there are not many architects who are qualified to plan and build hospitals. The governing board will do well to select an architect not only on the basis of his demonstrated competence but on his experience in designing and building hospitals. There should be no pressure from or on behalf of any local architect unless he has had wide experience in hospital construction and is chosen purely on merit.

Some Tips on the Selection of the Architect

As mentioned earlier, the architect is a key figure in the project. The choice of a good architect is, therefore, crucial to the building of a successful and efficient hospital. The following guidelines will help the hospital organizations in the selection of an architect:

- Firstly, the architect must be evaluated as a person and on the basis of his ability to work with people as a member of the hospital building team. He must be able to relate to people — the members of the governing board, the medical staff, consultant and administrator on the one hand and the engineers, technical consultants and contractors on the other. The selectors must check his references and antecedents carefully, and examine the projects he has completed.

- If the firm of architects has several architects, the selectors must check the person who will be responsible for their project and evaluate him. Since this person will run the show, the selection of the firm should be based on the merits of this architect.

- The committee must next study the other members of the architect's team — the structural, electrical, heating, ventilation, air-conditioning and plumbing engineers. Each one of them is of vital importance to the success of the project and one weak link will spell disaster. Their organization's profile, experience and the work they have done must be studied and evaluated.

- Then the committee should check if the architect and his supporting firms have had experience in building hospitals. This is a major deciding factor.

- The committee should explain to the architect the mission, philosophy and goals of the hospital, needs and other concerns of the project like the design, systems and functional requirements, and observe how he responds to these concerns.

- Finally they should discuss the fee structure vis-à-vis the cost of construction. They should remember that cheap is not always the best, many times not even good. After all one or two per cent increase in the construction cost does not make any material difference in the overall cost of the project. The loss will be much higher if the operation and maintenance become a bigger liability in the future.

Functions and Responsibilities of the Architect

The architect's contract generally includes the following basic services:

- Site evaluation
- Schematic drawings
- Cost estimate of construction
- Design development
- Construction documents
- Normal engineering services
- Specifications

- ◆ Inspection
- ◆ Perspective drawings
- ◆ Scale models
- ◆ List of fixed equipment
- ◆ Site utility study
- ◆ Parking.

➤ENGINEERS

Engineers who are specialized in various disciplines of engineering — structural, electrical and plumbing — form the design team of the hospital. Large architectural firms include in-house engineers in their outfits or have tie-up with consulting engineering firms. In such cases, the engineers are appointed by the architect. Where the architect does not have the entire team with him, the owners appoint the consulting engineers who form the design team. In addition, the owners also appoint a team of site engineers headed by a chief engineer or project manager to supervise the work and safeguard the interests of the owners at the work site.

The Structural Engineer

The role of the structural engineer is to provide optimum support for the building. To that end, he
- ◆ studies the report pertaining to the soil condition, the bearing capacity of the soil, the rock outcrop, the subsoil water level, etc. and advises the architect on the type of structure and foundation needed to support the building;
- ◆ prepares schematic structural layout and grid planning to enable the architect to proceed with detailed planning;
- ◆ prepares preliminary and later detailed estimates of cost for the hospital building based on the requirements furnished by the hospital consultant and confirmed by the architect;
- ◆ prepares detailed specifications of the entire civil works in coordination with the hospital consultant and the architect;
- ◆ prepares tender documents for the architect to call for tenders;
- ◆ prepares detailed drawings for the construction of the building. The structural engineer's design, judgement and analysis are of little value if they are not clearly communicated through neat and complete drawings and specifications. This reduces the cost of construction and the risk of errors on the part of the contractor;
- ◆ verifies shop drawings submitted by the contractors and certifies them for construction after coordinating with the architect;
- ◆ undertakes inspection and testing, and reviews construction to ensure that the requirements of the contract are being met and all instructions are carried out;

- provides direction and interpretation of the plans and specifications when required; and
- scrutinizes contractor's bills and certifies for payment as per the tender conditions.

Coordination of structural work with the architect on the one hand and with other engineers on the other is absolutely essential in hospital construction.

The Electrical Engineer

In any electrical system, the main concern of the electrical design engineer (as well as the hospital engineer) is the power distribution system which is the electrical life-line of the hospital. Design features must include elements of safety, reliability, cost, voltage quality, maintenance and flexibility. The engineer should study the conservation of energy and apply it most efficiently and economically. He must be aware of the public utility supply so that regular economical power distribution and an emergency supply are secured. Lighting should be so designed as to secure the mood and correct optical levels of the entire hospital staff. The electrical engineer's responsibilities are to:

- study the report pertaining to the site and to the availability of electrical load in the vicinity, calculate the load needed for the project and advise the architect and the owners as to the ideal place for locating the transformer and generator as well as routing for the cable from the supply source;
- prepare schematic electrical layout for the architect and proceed with detailed electrical system planning;
- prepare preliminary estimates of cost and later detailed estimates for the electrical system of the project based on the requirements furnished by the hospital consultant and medical staff and confirmed by the architect; and
- prepare detailed specifications for the entire electrical works in coordination with the architect. The electrical engineer's responsibility in assisting the owners in selecting quality equipment, wires and cables should be particularly noted. There have been instances, for example, of fire breaking out in hospitals due to short circuit caused by the use of poor quality wires. It was noticed in such cases that the owners had not been properly advised;
- prepare tender documents and call for bids from reputed contractors, and prepare comparative statements after evaluation of quotations;
- prepare detailed drawings of the electrical system and issue these drawings from time to time ensuring progress of construction;
- verify shop drawings submitted by the contractor and certify them for construction after coordinating with the architect;
- undertake inspection and review at site to ensure that all instructions in the drawings and specifications are carried out; and
- scrutinize contractor's bills and certify for payment according to agreed tender conditions.

The Plumbing Engineer

The purpose of providing a plumbing system in the hospital is to furnish potable water to all parts of the hospital building, and for removing liquid waste products and discharging them into the sewer. The plumbing system must be well engineered in order to provide necessary quantities of water under sufficient pressure. The backflow or back siphonage can be a real danger as, because of this, even the best plumbing system can be a source of contamination to the entire hospital.

To ensure supply of pure water in a newly built hospital, the new water supply system including pipes, fittings and fixtures must be first decontaminated before being placed in service. This is done by thoroughly flushing the system under maximum pressure, and then injecting a chlorine solution into the system. The plumbing engineer should be aware of this important principle.

The responsibilities of the plumbing engineer are to:

- study the report pertaining to the availability of water at or near the site and calculate the quantity of water required for the hospital;
- study the method of disposal of liquid waste, calculate the total discharge of waste, and check if the existing sewage in the vicinity of the site is adequate to receive this total discharge;
- make provision for alternate means of discharge, if necessary, and provide for treatment plant for purification of water before discharge;
- advise for an ideal location and capacity for the treatment plant, septic tank, soak pit, sump, overhead tank, etc. and also the general direction in which the waste is to be disposed depending upon the topography of the land, road and public sewer location;
- prepare schematic plumbing layout to enable the architect to proceed with detailed planning;
- prepare preliminary estimates of cost for plumbing works based on the requirements of the hospital as furnished by the hospital consultant and confirmed by the architect;
- prepare detailed specifications for the entire plumbing works in coordination with the consultant and the architect;
- prepare tender documents and call for bids from reputed contractors and prepare comparative statements after evaluation of the bids;
- verify shop drawings submitted by the contractor and certify them for payment as per tender conditions;
- undertake inspection and testing, and review work at the site to check whether the instructions in the drawings and specifications are being carried out correctly; and
- scrutinize the contractor's bills and certify for payment as per tender conditions.

➤ THE HOSPITAL ADMINISTRATOR

The question that is often asked is at what point of time should the administrator of a new hospital be engaged. In most cases, the governing board will find it advisable to employ an administrator shortly after the formation of the permanent organization. He may not be engaged until after the plans are under way but he should definitely be in place before the working drawings are completed. He will then be familiar with the details of planning and construction from the start. As the representative of the owners of the hospital, his sole concern is to watch their interests. As an administrator who is familiar with the departments, services and administration, his guidance and inputs to the drawings are valuable. He will also advise on the selection of equipment, organization of the hospital and matters relating to the initial development of the hospital. Elsewhere we provide a programme covering numerous areas of activity which the administrator must develop in order to keep the hospital in readiness for commissioning and operation. The employment of the administrator at the early stage is doubly necessary if the governing board is, for some reason or the other, unable to engage a hospital consultant, or if its members are unable to devote adequate time and attention to the affairs of the new hospital. Where a hospital project feels that it cannot afford to engage a hospital consultant, especially in smaller ones, the administrator can double as a consultant, as in the earlier stages of the project he will not be preoccupied with the rigours of day-to-day administrative responsibilities.

Hospitals and the public have increasingly come to realize that the hospital is a complex place to administer and that administrative ability in the industrial field in no sense qualifies an individual to properly direct a hospital with its complicated management, personnel, emotional, psychological and medical problems. A study of the numerous and varied duties that the administrator must perform makes it clear that, to be successful, he must be a person endowed by nature with special qualities which should be supplemented by education, training and wide experience. Despite all this, governing boards of hospitals are often found to be casual about selecting suitable administrators for their hospitals.

In so complicated an institution as a modern hospital, the job of an administrator demands a working knowledge of finance and accounting, industrial relations, systems analysis, public relations, marketing, social psychology, housekeeping, information management, high technology, law, computer science, personnel management, to name only a few. The administrator to whom falls this responsibility for leadership and coordination of all these activities must, of necessity, be well trained, qualified and experienced if the hospital is to discharge fully its obligations to the governing board, patients, public and staff. It is imperative, therefore, that the board exercise great care in his selection since he is totally responsible to carry out its established policies.

In the earlier days the hospital administrator was chosen from among retired government officials or army personnel, or if he was of a medical person because he was a good physician or surgeon, or because a person possessed an outstanding ability in some technical field. Professional excellence and managerial ability do not always go together. There is ample evidence to show that capable men with outstanding technical expertise proved disappointing when moved to administrative positions in which their functions were preponderantly managerial. Today we have universities and colleges offering not only management courses like MBA but also master's level degree courses in hospital administration. Hospitals will do well to look for professional administrators. The future for better trained professional administrators looks bright.

Let us briefly look at some of the qualities the administrator must possess.

♦ He must be amply endowed with tact, diplomacy and infinite patience.

- He must have organizing ability, and firmness tempered with consideration in handling people.
- He must be a respected leader in the community as well as in the hospital where he is administrator. He should possess leadership qualities and vision and be a good administrator.
- He must have a sense of responsibility of his position and must take his work seriously; nevertheless, this seriousness should be tempered by a sense of humour.
- He must be just and honourable, giving credit where credit is due.
- He must be a shrewd judge of human nature. This must be particularly evident in the selection of personnel.
- His appearance must always be neat and tidy without appearing finical. The hospital and personnel must look to him as a model.
- He must be an educator since much of his duties is in the field of educating, teaching and guiding subordinates and other personnel.
- He must have business ability and be a good buyer for the hospital, combining quality and price so that he secures the greatest value for money.
- He must have ability to work cordially with others — the governing board, medical staff, personnel and public.
- Lastly, the administrator must learn that the hospital has no hours of closing and that, like the hospital his office is never closed. What this means is that while he is responsible for the smooth operation of the hospital at all times, he should always be available, approachable and be ever willing to meet and listen to the staff and the public and do anything that is for the good of the hospital.

FUNCTIONAL PROGRAMME AND DESIGN STAGE

A major responsibility rests on the hospital consultant in the preparation of a second comprehensive written document, called operational programme, showing how all the necessary services can be provided throughout the hospital facility. The document, which is prepared in collaboration with the architect, lists all service units and programmes — a formidable list starting from the complex operating rooms and ICUs to the janitor's closet — and provides detailed description in respect of each of them such as location, functions, department utilization, intra- and inter-departmental relationships, space requirements, staffing pattern, equipment, communications, work flow, traffic flow, problem situations, and special requirements, if any.

There is no standard operational programme, and each has to be produced in a different way. The hospital administrator has a tremendous influence on this document, and could write it himself. On his part, the architect translates the written programme or document into a functional and efficient design.

We provide here a list of units in the administrative services for which descriptions have to be written. The following list many not be exhaustive and is given only for illustrative purpose:

Administration	Library
Admitting	Linen
Communications	Maintenance
Community relations	Orientation
Computers	Parking
Construction	Personnel
Data processing	Physical facilities
Dietary and food service	Planning
Disaster plan	Plant
Education, formal and informal	Public relations
Employee facilities	Purchasing
Employee health	Records
Engineering, clinical	Recruitment
Engineering, general	Security
Governing board	Sanitation
Finance	Site
Fire and safety	Stores
Fund raising	Training
Housekeeping	Transportation
Information	Utilities
Laundry	Volunteers
Legal matters	Wages and salaries

Concomitantly during the planning stage, the architect and his team of engineers should be studying the construction aspect of the project to determine the most economical and practical systems to be adopted for the foundation, structural framing, plumbing, electrical wiring, ventilating, air-conditioning and other engineering problems. They must particularly study the materials to be used from the point of a view of economy, function and maintenance and complete it before starting the working drawings.

Planning and building a hospital is a complicated job. There should be complete understanding of the various phases of the work among all concerned. Cooperation of the hospital board, administrator, consultant, architect, and building contractor is necessary to ensure the success of the project. Before any planning is done, an understanding should be reached among all the parties concerned as to the requirements of the hospital. It is on the basis of these requirements as mutually agreed upon that the consultant writes a detailed operational programme which clearly delineates the needs of various areas of the hospital. If this is done with care, much money and time will be saved, and a well planned and efficient hospital will emerge.

If the hospital site has not been chosen, the time to do it is now. Without it, the architect cannot proceed.

Armed with the written programme and after the selection of site, the architect will first prepare a small scale schematic drawing. He will translate the information in the programme into required building area, study the relation of various units and flow of traffic, and then determine the size and shape of the building. This plan should be studied by the governing board, administrator and others concerned, and changes, if any, should be incorporated into the plan.

After the schematic drawings have been approved, the architect makes an estimate of the cost of the project and the funds needed. The estimate should be fairly accurate and should include the cost of the following — construction of the building, fixed and movable equipment, architect's and consultant's fees, inspection and supervision, grading, roads, landscaping and contingencies.

Once the governing board approves the revised schematic drawings and the cost estimates, the architect will proceed with developing the plan on a larger scale.

The first principle is that planning should not be hurried. All those who are involved in the planning should be provided with ample opportunity to go through the architect's drawings and make necessary changes. The hospital consultant and administrator should visualize the hospital as delineated in the plans and mentally evaluate the procedure that will be carried out later when the hospital becomes operational in order to ensure that the arrangements are as perfect as they can be. If changes are required, the architect should be requested to effect them. All this requires time, and so patience and cooperation are vitally important.

It is necessary that all details are complete. It is a collective responsibility, not the architect's alone, that nothing is omitted. The contractors will eventually bid and construct according to what is contained in the plans and specifications. Anything omitted from the drawings and specifications cannot be subsequently expected from the contractors.

In any building project, more so in a hospital building project, changes are inevitable. If those changes are justified and accepted, they should be incorporated in the drawings. The architect will then revise the estimate accordingly. It is a sound policy that all changes receive the approval of the governing board especially if they have major financial implications. After receiving this approval, the architect may proceed with developing working drawings.

►PRODUCTION DOCUMENTS

Working Drawings

The working drawings convey to the contractor and his workmen details pertaining to the construction of the building. The drawings are drawn to scale. A complete set of working drawings consists of (i) architectural, (ii) structural, (iii) mechanical and (iv) electrical drawings. These are described below:

1. Architectural drawings show plan of site, location of building, existing and finished grades, roads and walks, utility connections, floor and roof plans, sections, elevations, schedules of doors and windows, finishes of all rooms, etc.
2. Structural drawings show location and size of foundations, footings, columns, beams, girders and slabs.
3. Mechanical drawings depict diagrams of all piping, details of plumbing, ventilating and air-conditioning work.
4. Electrical drawings show diagrams of electric feeders, locations of electric panels, fixtures and other electrical equipment.

Specifications

The final or working drawings prepared by the architect embody every possible detail, but there are points which cannot be shown in them. Specifications supplement the working drawings and furnish the information not shown in them. They are written instructions, describing in detail the construction work to be undertaken. They prescribe qualities of materials and workmanship that should be furnished by the contractor and define the work required under the contract. While the drawings define in graphic form the quantity and type of each component of the building — the size, shape, and location of various elements of the building — the specifications express in writing the quality, performance standards and the ultimate result that is expected from the proper assembly of materials and equipment identified in the drawings.

The selection of materials and equipment that are going to be used in the project is the responsibility of the architect, but before including them in the specifications, he should evaluate their suitability for use in the project. The contracting personnel should also become fully conversant with these materials and the methods that are applicable to the project.

The architect must have a working knowledge of the materials — their basic composition, how they are fabricated and installed, and their overall suitability. This knowledge is not only necessary for the development of specifications, but helps the owners to understand the nature of materials used in their building. Owners, like all those who are not familiar with construction work, comprehend more readily what is written in the specifications than what is shown in the drawings.

Once the suitable material or product is determined, the architect must write specifications for the product in such a way that it makes competitive bidding possible. For this, architects generally employ some common practices such as product description, product approval and product equal.

In the product description method, the architect provides full description of the product such as dimensions, colour, pattern, proportion of ingredients, density and finish. In the product approval method, he either approves one product that is readily available in the market, or arranges for the manufacture of the product as per the specifications provided by him. In the product equal practice, the architect lists out a few approved makes or brand names of a particular product out of which the contractor can choose any one while bidding.

It should be remembered that the contractor cannot be held responsible for any portion of the work or material not included in the specifications. When accepted and signed, specifications become a part of the contract documents.

The specifications should be written in simple, accurate and unambiguous language and divided into sections corresponding to various kinds of work or work performed by various sub-contractors.

To summarize, specifications are written instructions that accompany the architect's drawings for a construction project. The following are their objectives. They

- describe and supplement in writing what is shown in the drawings;
- include all legal requirements that apply to the owner and the contractor after the contract is signed;
- establish a basis for the preparation of competitive bids and the execution of the work required under the contract;
- assist the contractor in organizing his bidding, purchasing and construction procedures by classifying the various activities based on trade involvement on the site such as masonry, carpentry, electrical and mechanical work.
- function as a guide for the owner, contractor and architect in matters of construction and inspection of the work; and
- regulate payment to the contractors and provide a basis for legal interpretations.

►TENDER DOCUMENTS

Tender documents are indexed and bound in the form of a book. The book has a title page, a table of contents and generally includes the following sections:

- Notice inviting tenders
- Special instructions to tenderers
- Articles of agreement
- Tender form
- General and special conditions of the contract
- Technical specifications
- Bill of quantities
- Appendix.

The appendix consists of subsections on the date of commencement, date of completion, defects liability period, value of interim bills, penalty and bonus clause, retention amount, etc.

Separate tender documents have to be prepared for civil works, electrical works, water supply, plumbing works and air-conditioning works.

In addition to the above mentioned major contract works, the architect should assist the owners in calling for tenders for other contracts on works such as centralized medical gas system, public address system, laundry, kitchen, nurse call system, paging, fire protection, landscaping, etc.

➤ THE HOSPITAL SITE

Selection of the Site

The site for the new hospital must be selected and purchased as early as possible after the formation of the permanent organization, and before the architect starts his first sketches. The consultant who conducted the preliminary survey plays an important role in the site selection. The site selection should not, however, be made without the participation of the architect. Land is either not easily available these days, or is sold at a premium. That being the case, the promoters may be left with few choices, but if they do have choices, they should take the following factors into consideration while selecting and purchasing the hospital site.

- *Accessibility to transportation and communication lines* The site must be close to good roads and public transportation. This is important for the transportation of patients, visitors and supplies. Transportation costs are high and inaccessibility will add to the cost of operation. Besides, modern hospitals are designed to handle acute cases, casualties and emergencies which should reach the hospital speedily.
- *Availability of public utilities* The hospital requires water, sewerage, electricity, fuel, and telephone lines. The governing board must make certain that these utilities are freely available to meet the needs of the new hospital.
- *Proper elevation for good drainage and general sanitary measures* This is a principle that applies to all organizations where people are congregated, more so to a hospital.
- *Freedom from noise, smoke, vapours and other annoyances* Both the patients and the personnel need clean air and quiet surroundings. The site chosen should be free from undue noise and atmospheric pollution coming from sources such as railroads, main traffic arteries, airports, schools and children's playgrounds. Plenty of natural light should be available in all parts of the building.
- *Future expansion* Every hospital faces expansion problems in ten or fifteen years, some of them sooner. This should be kept in mind when a new building is planned.
- *Cost* Total cost including the cost incurred in making the site suitable for a hospital structure must be considered, not merely the initial cost. A site acquired cheaply or received as a gift may ultimately turn out to be far more expensive than a high priced plot.

Site Survey

After the land is selected, a survey and soil investigation must be made in order to obtain all information necessary for the design of the foundation, mechanical service connections and development of the site. A plat (a map or plan, especially of a piece of land divided into building lots) of the site should be prepared. The plat should indicate complete details and location of any building, structure, easements, rights of way or encroachments on the site, details of boundary walls, foundations, etc. adjacent to the site, trees that will be affected by the proposed building, detailed information relating to established curb, street, alley, pavement, etc., all utility services, location of piping, mains, sewers, poles, wires, etc. upon, over or under the site or adjacent to the site, complete information as to sanitation, disposal of sewage, storm water, subsoil drainage, etc. and any other pertinent data.

The plat should bear the certification by the appropriate authority, like the officials of the city corporation, municipality or panchayat so that streets, sewers, street lines, etc. are correctly given.

The architect must arrange adequate investigation to determine the subsoil conditions. The investigations should include a sufficient number of test pits or test borings to determine the true conditions. The architect should also have detailed information relating to the subsoil conditions such as soil bearing capacity, thickness, consistency, character, etc. of various strata, amount and elevation of ground water in each pit, elevation of rock, high and low water levels, whether the soil contains alkali in sufficient quantities to affect concrete foundation, so on and so forth.

►BED DISTRIBUTION

Functional design of the hospital is influenced by many factors such as types and extent of services, departmental relationships and administrative practices. The total number and distribution of beds also affect the design to some extent.

The functions of the hospital revolve around the total number of beds and their distribution within various departments and services. The yardstick that is applied when referring to the size of the hospital, its various services, occupancy rate, etc. is the number of beds the hospital has.

There is no universal rule to determine the expected distribution of beds in hospitals. Each hospital is unique and requires a special study to determine its requirements. If there is a specialty hospital in the vicinity — a maternity hospital or a cardiac centre, for example, or if the new hospital has certain specialists on its staff such as a cardiologist, a nephrologist, or a neurosurgeon, such factors will markedly affect the bed distribution by service.

A hospital bed is one that is installed for regular 24-hour use by inpatients during their period of hospitalization. This includes the cribs (those equipped with sides or guards) for use of young children, but not the newborn infant bassinets although these are installed in the hospital for regular 24-hour use by newborn infants.

Bed capacity is the maximum number of hospital beds, excluding the newborn infant bassinets, which can be established in a hospital at any given time. It is based upon the space required for those beds. The beds may not be physically installed. The expression bed complement of a hospital — not to be confused with maximum bed capacity — refers to the number of hospital beds (excluding newborn bassinets) normally set up and available for use of inpatients. It is this figure that is used in calculating bed occupancy. In an ideal hospital, bed capacity and bed complement should be identical.

Hospitals must remember that in planning a new hospital, certain beds are necessary for hospital purposes; nevertheless, they are not to be included while computing the bed complement as they are not available for full time care of inpatients.

Examples of such beds are:

- ◆ Labour beds used for a short spell of time. The patient normally keeps her regular bed in the room and pays for it
- ◆ Beds in the outpatient and emergency departments — used for observation and for a short period following treatment or minor surgery
- ◆ Beds in diagnostic or therapeutic departments such as x-ray, blood bank, etc.
- ◆ Anaesthesia recovery beds used by post-operative patients
- ◆ Beds in the infirmary attached to nurses' or other staff residences or hostels.

Types of Bed Accommodation

Three major types of bed accommodation are usually designated as private, semi-private and general ward. In high-tech luxury hospitals there may be other categories like deluxe rooms and executive suites. These terms, however, do not have the same connotation in every hospital. The private room is, as the name indicates, a room with one bed, the semi-private accommodation is provided in a room with two or three beds. The general ward is generally a multi-bed room or area.

The three types of accommodation should be grouped together so as to allow departmentalization of services. Some hospitals like to group private rooms in a separate section. This is inadvisable unless there is sufficient demand to justify a private pavilion. In general hospitals of average size, it is better to group all surgical cases on one floor, with private, semi–private and general ward male and female patients in order to utilize common equipment and facilities. The same arrangement should apply to the other disciplines.

Bed Distribution by Service

In the context of specialty and superspecialty services pervading our present day hospitals, the old rule of thumb concerning distribution of beds by services — say 30 to 40 per cent each for surgical and medical patients, 10 to 15 per cent for obstetrics, 7 to 10 per cent for paediatrics (other than newborn), 9 to 15 per cent for other patients including eye, ear, nose and throat — may no longer be applicable. Besides, as mentioned earlier, there are other factors that affect bed distribution such as the existence of specialty hospital or hospitals in the neighbourhood, or specialist doctors on the staff of the hospital. Some hospitals develop one or two services such as cardiovascular surgery or kidney transplant as their specialty centrepieces, put much of their resources into their development, and allocate additional beds to them.

In many hospitals, a percentage of the total beds is not assigned permanently to any particular service. The private and semi-private rooms are by and large used interchangeably for surgical and medical services and their related specialties. This flexibility may be necessary in many hospitals.

The interchangeability is not advisable in the case of maternity services although there is a tendency to permit clean gynaecological cases in the maternity unit. Proper flexibility makes possible a better bed utilization which is a factor of tremendous importance to bed distribution as well as to staffing requirements and efficient administration of the hospital.

Hospital planners must keep in mind two factors that influence bed distribution. A new hospital takes time to fill its beds — sometimes 2–3 years. For this reason, they do not throw open their full complement of beds on commissioning the hospital. They do it gradually, floor by floor or wing by wing synchronizing with demand and staffing. During this period departmentalization of all services is not possible.

The other factor is fluctuating census. Hospital design must provide interchangeability of patients, rooms and areas for clinical departments with fluctuating census.

Finally it should be emphasized that proper bed distribution cannot be planned for a new hospital without a thorough evaluation of the needs, services, hospital policies, staffing patterns and other related factors. The absence of this evaluation makes impossible the efficient and high quality patient care which is the ultimate goal of every hospital.

➤ SPACE REQUIREMENTS

Despite very wide variations in services and functions among hospitals, experience has taught that certain conclusions can be drawn with regard to the space required for various units and departments. Actual space requirements for any individual hospital can be decided only when services and programmes are known.

The fundamental principle while designing a hospital (that planners and builders should remember) is that hospitals should be planned for at least ten years ahead, or else the plans will be obsolete when they come from the drawing board.

Granting that hospitals do keep future expansion and changes in mind in what is called forward planning, they are still prone to commit a major mistake as they often do. They succumb to the dominant pressure for beds and add them without giving sufficient consideration to supportive and other infrastructural facilities. This will result in their moving from one crisis to another. Experts say that the hospital design should permit flexibility and expansion possibility in practically every area so that when expansion becomes necessary, it can be brought about with minimum disruption of daily activities and without major construction costs. It is better, they say, to omit a major service or function in the design than squeeze the whole facility.

With the rapid development of technological, medical and administrative sciences, the space requirement of almost every department has increased markedly. For example, it is said that laboratory tests and space requirements double every ten years. Some years ago, the space requirement of hospitals was between 500 to 600 square feet per bed. With the unprecedented advances in the medical fields, the space needs of a modern hospital may be put at anywhere between 900 to 1200 square feet per bed. Of course, with the financial and other constraints they have, hospitals are too eager to compromise and settle for less.

TABLE 1.1

Space Requirements for a Non-teaching General Hospital

Area	Gross Sq. Ft. Per Bed
Administration	30–35
Emergency	10–15
Outpatient	10–15
Social service	1
Admissions and discharge	2
Clinical laboratory, pathology	25–30
Delivery suite	12–15
Diagnostic radiology	30–40
Dietary and food service	25–30
Employee facilities	5–8
Education, auditorium	5–10
Speech and hearing	1
Housekeeping	4–5
Materials management	4–5
Central stores	25–35
Purchasing	2
Laundry	10–15
Medical records	5–8
Medical staff facilities	2–3
Engineering and maintenance	50–60
Nuclear medicine	4–5
Nursery	4–5
Personnel	3–4
Pharmacy	4–6
Public spaces	10–15
Pulmonary function	1–2
Radiation therapy	8–10
Occupational therapy	3–5
Physical therapy	10–12
Surgery	35–50
Circulation	100–150
Nursing units	250–300

Table 1.1 above presents a list of some of the more important departments in a non-teaching

general hospital and the space required for them. It should, however, be remembered that these are only guidelines and that there is no hard and fast rule for apportioning space to various departments. Besides, various constraints and special circumstances also influence the allocation of space — the most important one being that there simply is not enough space. Be that as it may, the foregoing table will help the planners to visualize the size of each department in the total design.

PLANNING THE HOSPITAL BUILDING

➤ GENERAL PRINCIPLES

After completing all preparations for building a hospital such as employment of architect, consultant and administrator, purchase of land and raising of sufficient funds, the governing board issues instructions for the development of final plans and specifications. Although a full and detailed discussion of hospital planning is outside the scope of this section, a discussion of the important principles which apply to planning of all hospitals, and the general features that should be borne in mind while planning, is necessary here.

We remind the readers, even at the risk of being repetitive, that the two general principles that are important to planning are — (1) planning should not be hurried, and (2) the details should be complete. There is this example of two hospitals — one which violated these two cardinal principles and spent an ultimate extra of nearly 30 per cent, and the other which spent two years in planning and an extra amount in the final bill of little over seven per cent.

The third principle that is even more important to remember is that hospital planning starts with circulation. In the words of Emerson Goble, "Separate all departments, yet keep them all close together; separate types of traffic, yet save steps for everybody; that is all there is to hospital planning". For the architect who is uninitiated to the modern hospital planning, the most important rule to remember is circulation. The numerous departments of the hospital should be properly integrated so that the different types of traffic traversing the building are separated as much as possible, traffic routes are kept short, and important functions are protected from prying eyes or intrusion (see Figs 1.2–1.7). The skill with which circulation is handled by the architect will determine the efficiency of the hospital during its life. Imagine nurses walking miles everyday unnecessarily and cursing the architect all their life for giving them varicose veins!

The following rules must be observed:
- Protection of the patient is the primary rule. Too much traffic will disturb the patients, affect the efficiency in patient care and increase the risk of infection, particularly in the case of surgical patients for whom aseptic condition is essential.

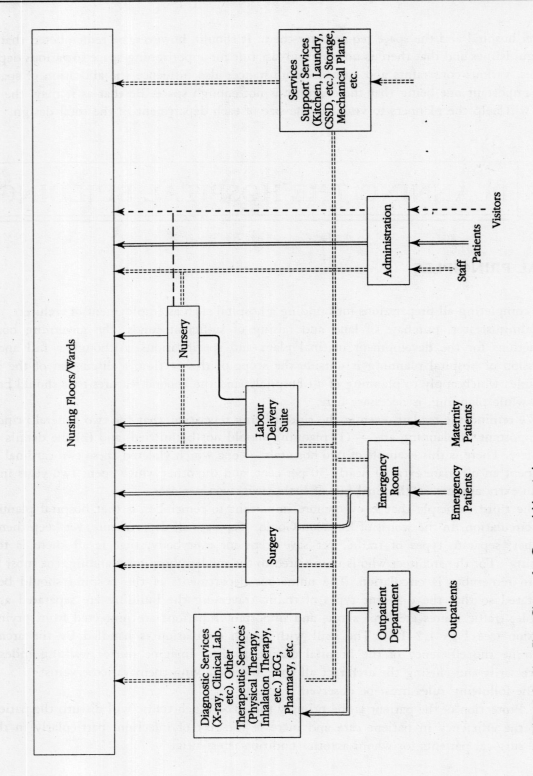

▷ **FIG. I.2** Key Flow Chart of a General Hospital

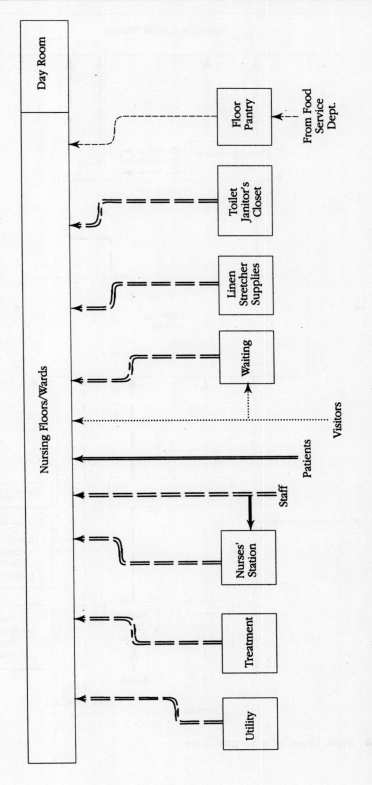

➤ **FIG. I.3** Flow Chart of Nursing Department

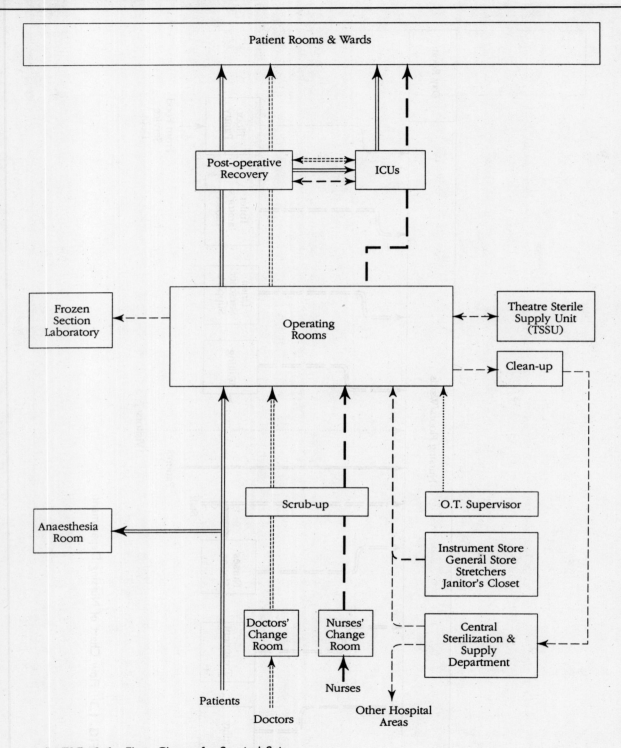

> **FIG. I.4**   Flow Chart of a Surgical Suite

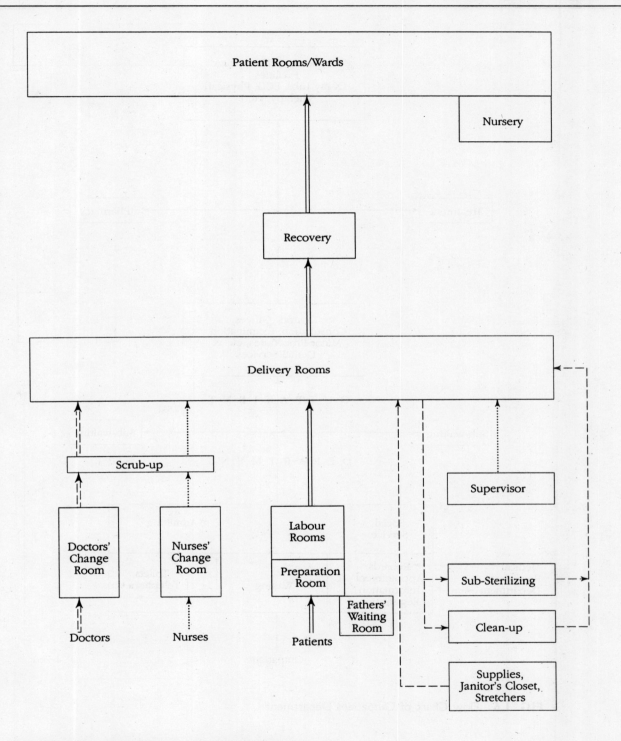

> **FIG. 1.5** Flow Chart of Obstetrics Department

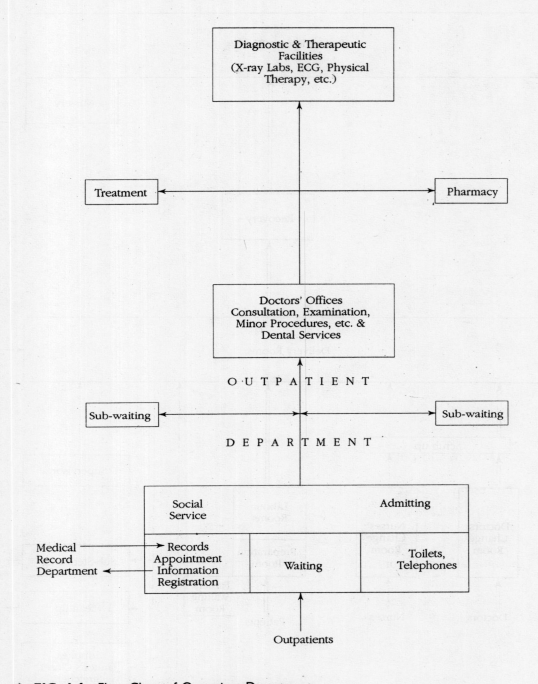

➢ **FIG. 1.6**   Flow Chart of Outpatient Department

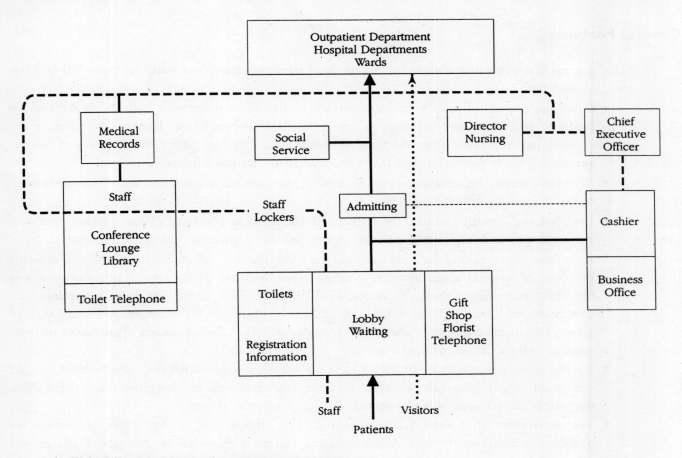

➤ **FIG. 1.7** Flow Chart of Administrative Department

- The second rule is to plan for the shortest possible traffic routes. They assist in maintaining aseptic conditions and save steps for everybody — nurses, doctors, patients and other hospital personnel. A hospital is a place where everything should be done fast. Patients' lives often depend on it. Time wasted on unwanted steps costs money besides making people suffer from fatigue at the end of every working day.

- The third rule is the separation of dissimilar activities. Examples — separation of clean and dirty operations, quiet and noisy activities, different types of patients, different types of traffic both inside and outside the building, etc.

- Control is the fourth rule to follow. A certain amount of control is inherent when dissimilar activities are separated, but that is not enough. The nurses' station should be so situated as to assist the nurses to exercise control over patients' corridors and over the visitors entering and leaving the unit. Infants must be protected from being stolen and against germs brought in by visitors and hospital personnel. Patients in the ICUs must be guarded against infection. Operating rooms should be similarly protected.

General Features

Some general features which should be incorporated while planning and building a general hospital are discussed here.

- ◆ Orientation of all or most sections of the hospital to allow maximum light to all parts of the building and maximum exposure to breeze is an important factor. In very few parts of our country patients suffer from the cold of winter. But they may suffer from the severe heat of summer. This is particularly so if the hospital is not air-conditioned.

- ◆ Windows should be screened, as far as possible, to keep out mosquitoes and flies which seem to be ubiquitous in our hospitals.

- ◆ Detailed and careful planning of traffic and entrances is necessary. These include traffic of patients, staff and employees, visitors, service deliveries, garbage and waste removal.

- ◆ As a rule there should be only one entrance for the general public. One of the difficult problems of hospital administration is the control of visitors. Thefts are not uncommon in our hospitals, particularly through unsupervised passages. Each additional entry point increases this problem. There will, of course, be additional entrances for emergency, for delivery of goods, and in some hospitals a separate entrance for staff. Added numbers of entrances increase costs of design, access roads and walks.

- ◆ By and large, patients and visitors will use the main entrance which may also be used by staff except in larger hospitals where a separate entrance may be designated for them. The outpatient department is better served if it has a separate entrance.

- ◆ The service entrance should be adjacent to the kitchen and storage areas, and wherever possible, to a service elevator. A platform and scales are provided in that area. Garbage and other solid wastes are removed from this point, so also in many hospitals, dead bodies. Access to the morgue from patient areas and for removal of dead bodies to the outside should be protected from patients' and visitors' view for psychological reasons. In some hospitals, this outside entry is combined with the emergency entrance for better control.

- ◆ The main entrance and the lobby should be attractive. While the hospital catering predominantly to the richer classes is luxuriously designed throughout, its main entrance and lobby may be furnished elaborately and elegantly in consonance with good taste. The entrance and lobby of other hospitals, like the charitable hospitals, can nevertheless be made elegant and attractive at small cost if adequate thought is given to its construction and furnishing.

- ◆ A great deal of thought should be given to planning for the development of traffic flow for patients, personnel and visitors within the building and within and between the departments. The objective is to expedite traffic, eliminate congestion and promote efficiency of operation. Flow charts, in rough draft, should be given to the architect as a part of the written planning programme, and checked in detail against his preliminary drawings. This will ensure location of departments and layout of equipment in proper relationship.

- ◆ Outpatients may be routed from the registration and medical records to the waiting area and from there to laboratory, x-ray, pharmacy, special therapeutic areas and other services. They should not be routed to pass through inpatient areas.

* Inpatients may be routed from the admitting office and medical social worker's office, if necessary, to laboratory, x-ray, and other departments and then to rooms on the patient floors according to established admitting procedure.
* Medical staff members are better routed from the staff lounge passing through the medical records departments for the convenience of completing patient records.
* Visitors' route should be controlled carefully. They should not be allowed to wander undirected through patient areas. They are best routed past information or main office. Where a system of passes is used, colour coded visitors' passes may be issued to a particular ward or floor for which the pass is applicable.
* Staff should pass from the entrance to locker rooms, and then to the place where they punch time cards or sign their attendance before entering their respective work areas.
* Floors should be constructed of materials which are determined by their future use and maintenance. They may be hard floors or resilient ones. Since the hospital staff spend many hours on their feet, efficiency will be impaired or lessened if floors are absolutely hard. Marble and mosaic floors are durable but have other disadvantages — cost and hard surface. After some years of use, the mosaic floor wears off and gets pitted with stains. Replacing these floors is very difficult in hospitals. All things considered, vinyl flooring is probably the best especially on corridors since it is soft and smooth which make walking easy and almost noiseless. It is available in a variety of bright colours and can make the floor appear bright and cheerful. The biggest advantage is that after ten years or so, the floor can be stripped and the vinyl replaced with ease.
* Corridors of seven to eight feet width with a finished ceiling height of 7 1/2 to 8 feet are the most widely accepted pattern. They should be as straight as possible. Where ramps are used, the gradient should not exceed 1 : 10.
* Walls must be smooth so as not to collect dust and dirt. They must be daddoed with tiles or polished stone up to a height of five feet as a protection against soiling and damage by wheelchairs and stretchers. Angles and sharp corners should be avoided, and in some places provided with protective metal or rubber guards to prevent stretchers, wheelchairs and carts from damaging the corners of the walls. A 5 to 7 inch skirting is useful against damage caused by casters. Walls finished in light colours are most attractive.
* Stairways may not be normally used in handling traffic when there are elevators. Nevertheless, they must be planned with care as they are essential in case of fire. There should be at least two stairways leading from the top floor to a ground level exit. They should be located in separate areas of the building. The stairway entrance should be closed with self-closing doors to eliminate noise as well as to meet fire regulations. The stairway should also be provided with lighted exit signs. A minimum width of three feet eight inches is necessary for handling stretchers in an emergency, for example, for evacuation of patients

in case of fire. The stairways should also have wide landings for the same purpose. Continuous railings on both sides at a height of approximately three feet are required for safety of patients.

♦ Elevators should be located where there is maximum concentration of traffic. Elevator doors should not open to the main lobby, but preferably to an alcove or a side corridor where patients are accommodated. Two elevators are a minimum for any multi-storeyed building, but hospitals with more than 250–300 beds require three or more. They are best utilized in a bank and not at separated locations. Where possible, separate passenger and service elevators are recommended.

♦ Hospital type elevator car size should be preferably 5 × 7 1/2 feet and the door at least four feet wide to allow handling of beds and stretchers together with attendants. Two common complaints frequently heard against elevators in hospitals are the noises emanating from the elevator machinery and their speed. As a rule, hospital elevators are too slow. There is no reason why they should not be fast as those in hotels and commercial buildings, and why with the advances in the field of engineering, they should not be made to run smoother. Elevators should preferably have dual controls. This will obviate the need for an operator during the hours when traffic is light. All safety devices including self-levelling feature, telephone, alarm and ventilation are necessary.

♦ Fire-escapes are hardly used. They are, however, mandatorily required. To be of any value, personnel should know how to operate them. By and large, personnel in our hospitals are ignorant of how to operate fire-escapes and how to handle fire-fighting equipment like extinguishers and hoses. The fire-escapes should also be sufficiently wide at the turns to allow a stretcher to be carried conveniently. There should be appropriate lighted signs to indicate their location.

PLANNING OPERATIONAL UNITS

The hospital must be planned in units of service. All areas or rooms related to any specific unit should be grouped together and arranged in such a way that coordination is possible. Thus we have units for administrative services, for accommodation of patients, for each of the diagnostic and therapeutic services, for special services such as operating rooms and intensive care units, emergency services, for supportive services such as food service, laundry, engineering, purchase and stores. Details of these units and services will be discussed under the respective departments and services in this book.

BUILDING CONTRACT AND CONTRACT DOCUMENTS

The work on drawings, specifications and legal documents must proceed simultaneously. When they are complete in every respect and are accepted, the governing board calls for bids for construction. While advertising a contract, an established form is used which has by its continued usage become almost legal in nature. For this reason it is advisable for the hospital's legal adviser to approve the advertisement before it is released to the newspapers. The plans and specifications which form the basis for bidding are secured from the architect, but the bids are delivered directly to the owners. A definite day and time is set after which no bid is accepted, and none opened before that set time.

The bids must quote definite prices for the authorized plans and specifications. If they are not made on this basis, they are not considered. A contractor may, however, propose alternatives in a separate proposal. On the basis of these bids, the contract is let. It is generally the accepted practice — can be a legal requirement in some instances — that the contract is awarded to the lowest bidder unless there are strong, sufficient and justifiable reasons to the contrary. It is not unusual for the owners to invite a few leading and responsible contractors to bid instead of floating open tenders through newspapers.

Once the contractor is selected, the governing board enters into a formal and legal agreement with him. The document should be carefully drawn by the legal adviser and should contain all the necessary ingredients. Building contracts vary from place to place, but they all have certain similar features and conditions. By and large, the contract document provides for the following.

- ◆ It binds the contractor to construct the building according to authorized plans and specifications.
- ◆ It sets a time limit by which the contractor is to complete the building and turn it over to the owners.
- ◆ It provides for penalties to be paid by the contractor if the building is not completed by the stipulated time. The owners on their part may provide for payment of incentives or rewards if the work is completed before the time limit set in the contract.
- ◆ It places the contractor directly under the supervision of the architect as far as construction according to plans and specifications is concerned.
- ◆ It provides for changes that may become necessary during construction. The costs of changes as approved are over and above the contract price. (Major changes should be formally approved by the governing board.)

- ◆ The contract provides for part payments at specified stages of construction on certification by the architect. The final payment is, however, deferred for protection against possible liability or lien.
- ◆ It specifies the amount of insurance that the contractor should carry. (The owners must ensure that the necessary insurance is carried at all times; and when it expires, the policy must be renewed in time. One way of safeguarding this is for the owners to keep the policy documents in their custody.)
- ◆ It provides scope for correction of work performed by the contractor. The cost of removing and correcting all work rejected by the architect as being defective is borne by the contractor.
- ◆ It defines the circumstances under which the owners or the contractor may legally terminate the contract before the project is completed.
- ◆ Lastly, the contract provides for the completion of the building and its being taken over by the owners.

The last clause can lead to a tricky situation which the owners must watch out. The clause should be examined and written carefully. In a new hospital, the owners will find it necessary to commence installation of equipment and furnishings before the building is completed. Some contracts provide that occupation of any portion of the building constitutes acceptance of the whole. Even if there is no such stipulation, in case of a dispute between the owners and the contractor, a serious situation may arise. There must be good understanding and some kind of agreement between the two parties to obviate this problem. The agreement should be in writing and signed by the representatives of the owners and the contractor. The furnishing and equipping of the hospital should in no way interfere with the work of the contractor.

On completion of the construction work, the building is formally taken over by the owners, and notice of completion is filed by their legal adviser or attorney. The notice of completion is necessary to protect the owners and the subcontractors. At the expiration of the lien period, the legal adviser has to examine if there are any encumbrances against the contractor. If the contract is completed to the satisfaction of the architect, he certifies to the owners that final payment may be made to the contractor.

Following the award of the contract, the successful bidder commences work on the project. He may subcontract much of the work to other contractors. However, the overall responsibility rests with the main contractor.

The "General Conditions" of the contract for construction prescribe the methods of administering the contract and the responsibilities of the owners and the contractor regarding such important matters as time schedule of payments to the contractor and bonds to be furnished by him, insurance to be carried and for what amount, protection of the owners against liens if the contractor fails to pay his bills, and other legal provisions.

While in smaller projects, the architect himself may supervise the construction through frequent visits, in larger projects it is advisable that the owners employ a superintending engineer (project executive) and several subordinate supervising engineers working whole-time on the project under the direct supervision of the architect. In view of the complexity of mechanical equipment in hospital construction, it is also desirable that a mechanical engineer is employed to be on the site full-time under the supervision of the architect to oversee installation of mechanical equipment.

FURNISHING AND EQUIPPING
THE HOSPITAL

Furnishing and equipping a hospital is an extensive undertaking — almost as extensive as planning and construction. Although the amount of money involved is not as much (except perhaps in high–tech hospitals), the degree and variety of technical knowledge required to accomplish this task are great. Almost every area is beset with problems — from the ordinary furnishings to the semi-technical equipment of the laundry, kitchen, electrical section and similar departments to the highly sophisticated and complex medical equipment. In addition there are innumerable expendable articles which should be procured and stocked in sufficient quantities.

In an existing hospital, purchasing new equipment presents no particular problem, except perhaps financial. Besides a purchasing department, there are well set mechanisms and procedures for establishing a need for an item of equipment in a department and for preparing specifications and purchasing. The hospital administrator, generally an experienced man, and his purchasing officer will easily accomplish this task. It is not so in a new hospital. The problem is compounded by the timing of delivery, warehousing, unpacking, assembling and installing of equipment. These are as important as selection and purchase. If the equipment is to be imported, the procedure will be even more complex. There are approvals and licenses to be obtained, and bureaucratic hurdles to be crossed. The lead time will be longer. Lakhs and lakhs of rupees may be wasted, operating efficiency impaired and standards of patient care severely affected by not planning and executing any one of these phases properly and adequately.

It devolves upon the hospital consultant or, in his absence, the hospital administrator who will have been engaged early in the planning stage, to determine all the items of equipment necessary for the hospital, to write or secure specifications wherever necessary, to call for and receive bids and to purchase or recommend purchases according to the policy of the hospital organization.

Equipment needed for a new hospital can be classified into the following three groups based on the usual methods of acquisition and on suggested accounting practices in regard to despreciation.

1. *Built–in Equipment* This is usually included in the construction contracts. Examples are cabinets and counters in pharmacy, laboratory and other parts of the hospital, fixed kitchen equipment, laundry (linen) chutes, elevators, dumb waiters, boilers, incinerator, cold rooms/walk–in coolers,

deep freezers, fixed sterilizing equipment and surgical lighting. The planning and design of fixed equipment built into the hospital facility is the architect's responsibility.

2. *Depreciable Equipment* Equipment that has a life of five years or more is not normally purchased through construction contracts. These large items of furniture and equipment have a reasonable fixed location in the hospital building but are capable of being moved. Examples are surgical apparatus, diagnostic and therapeutic equipment, laboratory and pharmacy equipment, office equipment, etc.

3. *Non–depreciable Equipment* Equipment having less than five years' life span is purchased through other than construction contracts. These are generally small items of low unit cost under the control of the storeroom. Examples are kitchen utensils, chinaware, tableware, surgical instruments, catheters, linen, sheets, blankets, lamps, waste baskets, etc.

The consultant must prepare a list of all the items of equipment under groups 2 and 3 given above. The first step in preparing this list is to consider each room as a separate entity in the plan and prepare a comprehensive room by room equipment list which should include additional items that may be required for the hospital. Detailed specifications must be given. This task must be undertaken during the design stage itself. Working in close association with the architect, the consultant should test the space needed for each item of equipment on the list.

The selection of technical, scientific and medical equipment requires careful analysis of the needs of each department and conscientious study that will result in the selection of equipment that will best meet the needs of the department. The present day high-tech medical equipment is so mind boggling even to medical experts that the consultant or the administrator, being an unwary layman, may be easily stumped in the selection process. Department heads and staff members should be fully satisfied with the type and quality of the equipment. For this reason, they should be consulted before purchase. It is not unusual, particularly in a new hospital, that a consultant has selected equipment for a department only to find the chief user of the department rejecting it as unsuitable. Survey should also be made to ascertain as to how a particular piece of equipment is performing in other places. Money often gets wasted by purchasing equipment that is either not utilized, or not fully utilized, or is of inferior quality. All those who are involved in the selection and purchase of equipment should exercise extreme caution and avoid procurement of equipment that will be a loss or liability to the hospital.

Consultation with the architect early in the design stage is necessary so that the facilities planned are of sufficient size to accommodate the equipment. Instances of having to break the wall, door or windows to let the equipment into the room are not uncommon. In one case, when it was found that the elevator shaft was too small for a bed elevator to be installed, it was too late to take any remedial action.

The timing of the purchase order and of delivery is exceedingly important. Delivery instructions should be keyed to building completion schedules. If delay in construction is anticipated, suppliers of equipment should be notified to defer supplies accordingly. Adequate arrangements for storage of equipment on the site should be made. These arrangements, while providing for protection against weather, theft and damage, should also not interfere with the construction work. The delivery schedules should allow ample time to unpack, check and assemble the equipment and to install it properly in the finished building. Tens of lakhs of rupees will be wasted if this schedule goes haywire.

READY TO OPERATE STAGE

When the construction work is going on, concomitantly, the administrator and his team, with the assistance of the hospital consultant, if necessary, must develop a written programme covering numerous documents, activities, policies, procedures, rules and regulations aimed at keeping the hospital in readiness to operate when it opens its doors to the general public. This will ensure smooth start and effective utilization of the hospital facilities. We list here some of the major items that should be included in the programme.

- Statement of mission and philosophy of the hospital.
- Organizational pattern of the hospital — governance and management structures — organizational chart.
- Administrative policies, rules and regulations for the proper operation of the hospital — policy manual.
- Organization of internal functions through appropriate departmentalization and delegation of duties.
- Establishment of proper relationships throughout the hospital including all major departments, and indicating them on the organizational chart.
- Medical staff organization, medical staff bye–laws, rules and regulations.
- Delineation of privileges of medical staff who are not hospital-based, full-time, salaried doctors.
- Employees' service rules, administrative standing orders.
- Qualifications and experience for all posts — medical staff, non-medical staff and employees.
- Terms and conditions of service, and policies on hours of work, overtime, holidays, absenteeism, tardiness, etc. Development of leave rules.
- Administrative responsibilities after regular working hours, on Sundays and other holidays.
- Remuneration pattern for medical staff.
- Emoluments — salary schedule and allowances for all staff members and employees.
- Statement on employee benefits — hospitalization benefits, provident fund, gratuity scheme, subsidized food and housing, free coffee and other amenities.
- Position descriptions, clear delineation of duties and responsibilities for managerial staff, and job descriptions for all other positions.
- Orientation programme, training programmes including in-service training programmes.
- Policies and procedure manuals for all departments.

- Performance evaluation programme and preparation of evaluation forms for various categories of staff.
- Management review programme for managerial staff.
- Grievance procedure.
- Disciplinary action procedure — domestic enquiry.
- Development of a manual for supervisory staff.
- An effective appointment order form.
- Arrangements for a good and effective graphics and signage system.
- Obtaining necessary licenses, permits and documents — registration, tax-exemption certificate (if applicable), pharmacy license, blood bank license, narcotic and alcohol permits, boilers, inspection certificate, motor vehicles and ambulance registration, fire and safety inspection certification, no objection certificate (if applicable), licenses, approvals for x-ray and radiotherapy departments.
- Arrangements for maintenance contracts for medical and other equipment, elevators, etc.
- Contracts for the supply of food items, milk, eggs, meat, drugs, fuel, x-ray films, linen, other hospital supplies, etc. and arrangement for laundry, garbage disposal, etc.
- With the assistance of medical staff/medical director, preparation of a hospital formulary.
- Establishment of a disaster plan and procedure to follow in case of fire and bomb threat. Organize simulated drills.
- Establish public relations with civic bodies, organizations, news media, etc.
- Establish procedures for emergency room, admitting and discharge and for outpatient and inpatient billing.
- Preparation and printing of hospital forms — medical and business forms — various outpatient, inpatient, medical record forms, prescription, doctors' order forms, and other forms.
- Preparation of policy on visiting hours, patients, attendants and children visiting inpatients.
- Arrangement for insurance — building, equipment, vehicles, etc. against fire, damage, theft, etc. and malpractice insurance.
- Schedule of rates and hospital charges for all hospital services, doctors' services, food service department, etc.
- Method of maintaining vital statistics for the preparation of records and reports.
- Establishment of accounting system and records.
- Arrangement for safe keeping of patients' valuables and hospital's cash.
- Establishment of banking procedures and a bank extension counter in the hospital.
- Institute internal audit and internal control systems.
- Planning and preparing a hospital budget showing expected receipts and expenditure.
- Development of a need based human resources position plan.
- Preparation of a good employment application form.
- Establishment of employment policies.
- Procurement of personnel files or file folders and identification cards.
- Develop external and internal newsletters.

- Organize a volunteer programme.
- Organize speakers' bureau from among physicians and other senior hospital staff and arrange for speaking engagements as a part of marketing and public relations strategy.

It is not difficult to see that the list which indicates areas of required activities is rather elaborate and that there is considerable detail in carrying out each one of these activities and that it requires time. While some of them will be handled by the chief executive officer himself, others may be directed or supervised by him or delegated to his associates.

BEFORE OPENING THE HOSPITAL

When the hospital is ready to admit patients, the governing board will take steps to make formal announcement(s) of its availability to the public. A great deal of work, however, needs to be done prior to the actual opening of the hospital.

Certain heads of departments and staff such as the medical director, director of nursing and some of the nursing supervisors and staff, chiefs of accounts, food service, housekeeping, engineering, purchase, stores and some of their staff will have been employed long before the hospital is opened to assist in getting it in operational order. Gradually, the organization of the medical staff should be completed. This should be followed by the employment of other staff and soon the hospital should be ready to function.

The formal taking over of the hospital building by the owners is a crucial stage in the realization of their dreams of opening the hospital. Apart from formalities, there are concomitant responsibilities which the owners must attend to. The engineering chief must assume responsibility for the upkeep of the building and various services. The architect must provide him a set of as-built drawings — another set should be given to the chief executive officer. The chief engineer should also have warranty details on plant, machinery, etc. It would be advantageous if the chief engineer is the same person who has been involved in the construction process so that he is familiar with all the details and aspects of the building. The other important person who must assume responsibility soon after the owners take over the building is the executive housekeeper who has to see that the hospital is kept clean.

The date of completion of the building is extremely important. Any delay as is so frequent, for whatever reason, either because of the contractor's or of the owners' fault, will spell economic doom. By the time the building construction is nearing completion, a fairly large investment would have been made, and if the money was borrowed from financial institutions, payment of interest and repayment of loan would have started. There may be escalation of cost if the construction is delayed. When the hospital is getting ready to function there will be huge expenditure on account of consumable supplies and a large number of materials required for starting the hospital. At this time and in the immediate post-commissioning period, there will be no income, and the hospital will take some time to break even. All this may land the new hospital in a financial crisis. It is, therefore, imperative that every step is taken to complete the building well in time.

No new hospital is expected to function smoothly from the very beginning of its existence. Adjustments will have to be made in order to ensure the greatest degree of perfection in the organization and its functioning during the early period. For this, certain preliminary but essential preparatory work is necessary. The following are some of the steps that should be taken into consideration:

- A skeleton staff is adequate at the outset. It is better for the hospital to grow and develop through meeting the increasing demand placed on it rather than by having complete staff at the beginning.

- The hospital should be brought into use gradually in a phased manner over a predetermined period of time. This should synchronize with the increasing patient census, occupancy and workload. The gradual increase should be in the number of beds, personnel, clinical sessions, facilities, clinical departments and services. If the building consists of several storeys, the occupancy can move vertically which will take some time. The storeys may be finished in shell form and not furnished and equipped. As we have mentioned earlier, every department should be designed in such a way that future expansion is possible following the dictates of demand and available finance.

- The time of appointment of the first complement of personnel is important too — neither too early nor too late. If this is not kept in mind, it may so happen that a whole lot of personnel are appointed and paid salary but the hospital is not ready to function, or the hospital with all the infrastructure is ready to function, but the recruitment of personnel is not complete. Either way it will be a loss to the hospital.

- Every staff member will have to be meticulously selected. If an institution is to operate efficiently, each position must be filled by a person who is not only well qualified but also the most suitable to fill it. Administrators often fail to appreciate the importance of careful investment in human resources and the enormous cost resulting from poor selection practices. Every job in the hospital is important. No amount of time and attention involved in selecting the right people can be regarded as superfluous.

- The entire staff, from top management to housekeeping, however qualified, must be given orientation — to the hospital, to the hospital's mission, philosophy and goals and to their respective departments. They should be thoroughly trained in their respective jobs. They must also be given training in public, patient and guest relations programmes. This is best done during the months before the opening of the hospital. The first batch of staff is indeed key to the success of the organization.

- It is necessary to remember that personnel do not come tailor-made to suit the hospital's needs and jobs. They come from different backgrounds, and with different training and experience. It is also necessary to remember that we do not get perfect people. Some have training and experience, others may have neither of these. Some may be completely new to their jobs and

PLATE 1

➢ **PICTURE 1**   Koval Medical Center and Hospital, Coimbatore

➢ **PICTURE 2**   B M Birla Heart Research
Centre, Calcutta

➢ **PICTURE 3**   All India Institute of Medical Sciences, New Delhi

PLATE 2

➤ **PICTURE 4**   A View of more than a century old, St. Martha's Hospital, Bangalore

➤ **PICTURE 5**   S D M Dental College Hospital, Dharwad

➤ **PICTURE 6**   P D Hinduja National Hospital, Bombay

PLATE 3

> **PICTURE 7**   Brigham and Women's Hospital, Boston, USA

> **PICTURE 8**   Vanderbilt University Hospital, Nashville, USA

> **PICTURE 9**   Provident Hospital, Chicago, USA

PLATE 4

➢ **PICTURE 10**   St. Vincent Hospital and Medical Center, Ohio, USA

➢ **PICTURE 11**   Sand Lake Doctors' Building, Florida, USA

to the health care systems. Many of them will be unfamiliar with their new environment, work, equipment, procedures and with working as a team. A great deal of harm can be done to the institution and patient care, and damage to equipment if they are not properly initiated to their work and given proper training. In each department personnel who have come from different backgrounds and training and who are familiar with a different system must be trained and made familiar with the new system. Also the personnel should be given practical "do it yourself" training to gain experience. Simulated exercises should be undertaken under the watchful eye of the supervisor with volunteers acting as patients.

- The chief executive officer must work in close association with each department's head to work out the details of the organization and functioning of each department.

- Coordination of all sections and departments and their work is most essential in the early days of the new hospital. The chief executive officer must hold frequent conferences to study the working of the organization as a whole.

- The chief executive officer will prepare a detailed plan of action to be circulated among staff, detailing step-by-step, the procedures that should be followed, particularly in crucial departments like registration, medical records, cashiering, billing, admission, etc.

- For at least one month there should be a trial run of the hospital. Before sending a ship to the sea, the navy generally requires a testing period called a "shakedown cruise". This is the time when they make certain that both the ship and the crew are in operational order. The month-long trial run of the hospital with the chief executive officer and the skeleton staff as the testing crew will provide the shakedown cruise for the new hospital. They are the first employees. It is up to them to discover the strengths and weaknesses that exist in the systems so that problems may be attended to at once. When the new hospital becomes operational, patients and staff will find the functioning smooth.

- It may be necessary to alter programmes and practices in the light of experience gained during the trial period. More changes may be necessary after the hospital becomes operational because of conditions that may develop in the new organization. For this reason, the plan of functioning should be left flexible so that ready adaptation to changes of procedure is possible.

- When a new hospital starts functioning, there is a degree of confusion all around — patients do not know where to go and whom to see. Often sick people act unnaturally, and their relatives and friends get worried and distraught. An easy to understand signage system reinforced by courteous and friendly staff to guide and help the patients will help put them at ease. The first impression of a patient is very important.

- Much of the success of the new hospital depends on the administrative ability and leadership of the CEO.

Finally, the chief executive officer and his team must recognize that the success of the new hospital depends on these six fundamental requirements: (i) organization, (ii) coordination, (iii) cooperation, (iv) efficiency, (v) economy and (vi) service.

➤NEW BUILDING ANNOUNCEMENT

The announcement of the embarkation of a new hospital is often made in the midst of a certain amount of fanfare. Important community leaders are invited and given a place of honour in the function. A press conference is convened at which the announcement is made. To make the occasion more meaningful, the architect may be asked to interpret the structure. He is best able to articulate and explain the salient features of the proposed building. He should present a photo of the rendering of the building or, still better, a model to make his presentation more effective. All news items and clippings should be preserved for future use. The handling of the news media is better left to an experienced public relations practitioner.

➤GROUND BREAKING CEREMONIES

The ground breaking is an exciting occasion to the promoters of the hospital. It really means that their dream is at last taking a concrete shape. After a great deal of hard work, presentations, conferences, deliberations, proposals and approvals, ground breaking calls for a celebration. If well organized, it will have a positive impact on the public and its leaders. Ground breaking is generally preceded by pooja or a dedication service.

Every detail of the ceremony should be meticulously planned — the venue, time, date, arrangements, placement of speakers, tables, microphone, podium and audience. The date and time should be convenient to the guests, speakers, public and press. Arrangements must be made for refreshments, music, photographer, etc. The atmosphere though professional must be warm and friendly. The ceremony should be documented by photographs or video tape. Photos should be given for press coverage immediately after the ceremony.

Imagination is needed in all the arrangements and in the design of invitation cards which should be sent to all the important people well ahead of time. If a brochure or a hand-out is to be brought out for the occasion, it should be appealing. It is useful to have an exhibition of plans, perspective plans and models for public viewing.

Speakers should be chosen carefully — those who have something important to say and can express it interestingly. Above all, they should be brief. This is no time to deliver long-winded boring lectures. The public relations officer or whoever is in charge should plan the programme far ahead of time paying careful attention to every detail. The ceremony should begin punctually and conclude in time.

➤COMMISSIONING AND INAUGURATION

Commissioning and inauguration are magic words which fill the hearts of the owners of the hospital with great joy and a sense of accomplishment. It means that they have finally made it, and the hospital is a reality.

Commissioning of the hospital need not synchronize with its formal inauguration. In fact, there are advantages in having the formal inauguration much later than the actual commissioning of the hospital, may be 3 to 6 months later. The hospital is ready to admit patients as soon as the infrastructure is ready and all the things are in a reasonably satisfactory operational state. It will, however, be a terrible mistake to throw open the doors of the hospital when it is in a state of flux. It will do more damage to the reputation of the hospital than any good. The ceremony for commissioning may be a low key affair. However, wide publicity should be given to the services available in the hospital.

The opening ceremony or inauguration of the new hospital is exceedingly important as the event usually makes a lasting impression on those who visit the hospital for the first time. A great number of people will be looking forward to this important event for various reasons. Some people will have a positive attitude and will be interested in the hospital having developed a favourable opinion about it. The attitude of some others will be neutral. They will either have no particular interest in the hospital or have nothing against it. There is a third group of persons which has formed a definite prejudice against the hospital. They will have come with an urge to find fault. The new hospital must be presented to the public at its best, and the programme on the opening day must be such that it creates a favourable impression on all the three groups. A carefully planned and executed programme will result in better understanding and appreciation by the public of the valuable and high quality patient care that the hospital is going to render to the community. It should also engender the wholehearted support and goodwill of the community as a whole which is necessary for the success of the hospital.

The opening ceremony which takes place in the new building requires even more meticulous planning than the ground breaking ceremony. Some of the other elements and arrangements that are involved in its planning are listed below.

- A preview press tour of the new facilities followed by a press conference should be arranged. At this time, a press kit should be distributed.
- A more elaborate programme of the ceremony is necessary.
- An open house for the hospital staff, VIP guests, community leaders and government officials should be arranged.
- News release, press coverage, television news coverage, and video taping of the speeches and the function should be considered.
- The chief guest, speakers and those who will be seated on the dais should be carefully chosen. The chairman (or president) of the hospital governing board, the chief executive officer of the hospital and the chief of medical staff should be among them.
- An attractively designed brochure should be distributed on the occasion. The brochure will include, among other things, a brief write-up of the new building, its salient features and architect's interpretation, photographs of functionaries, messages, landmarks, photographs of some departments, facilities, etc. A programme sheet may also be distributed.

- The public should be shown through the hospital facilities in an organized, systematic fashion, accompanied by intelligent and well informed guides who conduct the visitors from department to department.
- A responsible person should do the follow-up to the invitations. He (she) should make numerous friendly telephone calls on the day prior to the function, reminding VIP guests, officials and the press to be present.
- Security arrangements become necessary if VVIPs are participating in the ceremonies. Passes have to be issued to staff and others, and car passes for the entry of vehicles. Parking arrangements should be adequate.
- If dedication of the new building or *pooja* is to be performed, it should preferably be done in the morning.
- Advertisements should be released to the newspapers announcing the opening ceremony and also the services that will be available to the public.
- As a marketing and public relations strategy, the hospital can announce some benefits to the public as special offers in connection with the opening of the hospital. Some suggestions are given below:
    - Free consultation to all patients who register during the week following the opening of the hospital.
    - Guided tour of the hospital to all visitors for one week. The visitors may be served tea and refreshments after the tour.

    Wide publicity should be given to these programmes.

The inauguration should be held only after the hospital goes on stream and when everything is in place — staff, equipment, facilities, and not when things are in an unsettled state. A guided tour of the hospital facilities, which is one of the highlights of the inauguration, is one thing if it is through an incomplete building, and quite another if it is through the facilities which are fully operational. The latter can be a good marketing strategy. Pictures 1 to 11 (Plates 1–4) show some leading Indian and foreign hospitals.

CHAPTER 2

Organization of the Hospital

CONTENTS

❐ **Organizational Structure**
- **Overview**
- **Governance**
- **Management Structure**

ORGANIZATION OF THE HOSPITAL

For the efficient and effective operation of the hospital or of any institution for that matter, its organization is of paramount importance. An organization is a mechanism — the executive structure of business — which enables people to work most effectively together. It is the framework by means of which the work of an institution is performed. Organization is also the establishment of authority relationships which provide for structural coordination both vertically and horizontally. "Organization structure is like the architectural plan of a building," says William H. Newman, "and the larger and more complicated the building, the more important it is to have a central architectural plan."

In describing the specifics of the planning process and organizational structure of hospitals, we use the general hospital as the primary prototype since it is the dominant health care facility found in any typical community. There are other categories of facilities with major differences because there are factors that affect and make the design, structure, governance and management of hospitals different. There are features specific to the philosophies and objectives of individual hospitals which separate them from other hospitals. Thus we have government, municipal, private or voluntary hospitals which differ from one another to a marked degree in so far as their governance and management structures are concerned. On the other hand, the requirements of facilities and beds and staffing pattern of a teaching hospital are so different from a service hospital that it has to be treated as a class by itself; a specialty hospital like a freestanding psychiatric hospital, or a superspecialty referral hospital has such major differences that it cannot be planned and designed like a hospital providing primary care.

Inclusion of all these different categories is not only beyond the scope of this volume, but would also make it lose much of its relevance and effectiveness through sheer length. Even general hospitals differ widely in their facilities and organization structures depending on local preferences and other special circumstances, and that is how they should be. Each structure should be tailored to the need of the individual organization and the needs of the community it serves. In designing a hospital, the most important consideration should be how well it will serve the primary purpose of providing quality patient care, and how effectively and efficiently will the structure facilitate the operation of the hospital rather than the extent to which it conforms to other patterns.

We shall, therefore, address ourselves to the consideration of one primary model which is the general hospital in the private sector.

This does not mean that the general hospital can be planned in isolation from the broad aspects of total hospital facilities planning described throughout this book. Nor does it mean that the use of this book is limited to those in private general hospitals. Regardless of what category these other facilities belong to, they have the same basic characteristics as the general hospitals owned and operated by private bodies.

ORGANIZATIONAL STRUCTURE

➤ OVERVIEW

At the head of the hospital organization is the governing board, variously called board of governors, board of trustees or board of directors. Regardless of the name by which it is called, the governing board has the same duties, responsibilities and authority everywhere. The governing board is the supreme authority in the hospital, and has the legal authority over and responsibility for the hospital. The board delegates the actual authority of administration to its chief executive officer, also referred to as president, administrator, director or medical superintendent. By whatever designation he is called, the chief of the hospital whom we shall refer to in this book as the chief executive officer (hereafter CEO) is responsible to the governing board for the management and supervision of all hospital operations. The CEO may have several associates to assist him in his administrative duties.

Today's hospital has become so complex that no single person is able to manage all the activities of the hospital because no single individual has all the knowledge and the know-how necessary for managing such a wide range of varied and specialized activities. Therefore, a group of individuals, each possessing some special skill or expertise in a specific area or activity provides the necessary management support to the chief of the hospital. Thus we have associates who are experts in finance, public relations, nursing administration, strategic planning, legal matters, personnel management, purchase and stores, so on and so forth. The CEO, however, has the ultimate legal authority and overall responsibility for making decisions for the organization. Below the associates' level, there are heads of departments who are delegated authority to carry on the work of their departments.

In smaller hospitals, the work of several areas may be combined in one associate, and various departments under a few heads.

The department of the hospital may be grouped under two main headings — those concerned with the professional care of patients, and those concerned with business managements. Service wise, the activities of the hospital may be divided into five groups — medical, nursing, professional or ancillary, business or fiscal and supportive services.

Medical staff may be hospital-based, full-time salaried staff, as is commonly found in a majority of our hospitals, or they may be granted privileges to practise in the areas of their specialization.

The chief of the hospital may be a physician or a non-medical professional administrator. Both patterns abound in our country. The medical chief usually takes the title of director or medical superintendent, and the non-medical chief the title of administrator, director or executive director. In corporate hospitals, a new breed of hospitals for profit, he may be called the president and CEO, chief executive officer, managing director, executive director or general manager.

➤GOVERNANCE

Since the governing board is the supreme authority in the hospital, it is of utmost importance and necessity that its members are selected with great care. Membership of the governing board of a hospital is one of the greatest honours and privileges that may be conferred on any person. To a public spirited man or woman, it offers not only a great challenge, but also an avenue of service to the community. It should be given to those who are willing to devote the necessary time and energy in the work of the hospital, who are competent and qualified to serve on that august body and make a useful contribution. Those who consider board membership as yet another avenue to be used for social prominence or for personal aggrandizement should be scrupulously avoided. It is a wise policy to bar politicians from the hospital boards simply because members of hospital boards should be above political influence. Equally or even more important is the need to eliminate the small town, small-minded "church politicians" who abound in our church-related hospitals. These men and women who are often inspired by petty self-interest are a great menace to our institutions. Some of them "buy" their seats on committees and fight for places in smaller fields because they lack ability and qualifications required for higher offices.

It is to the advantage of the institution to include in the governing board representatives of the learned professions, business, legal and banking professions and even a friendly newspaper editor. Selection of members should be made having regard to their abilities and character so that knowledge, understanding, background, integrity, vision and business acumen may be brought to bear to ensure a dynamic management of the institution. There is no justification in including on the board, a man who has not been successful in his own business. One cannot expect him to successfully guide the destinies of a complex institution like the hospital when he has not been able to manage his own business affairs. Then there are businessmen who, lacking business acumen or flair for details, leave their businesses in the hands of professional managers and do not interfere with their work as long as satisfactory results are produced. These same men, as members of the hospital governing board, feel that they should play an active role in the affairs of the hospital and in their enthusiasm so often overstep the confines of their jurisdiction and do things that undermine the authority of the CEO.

Membership of the board carries with it a consequential responsibility which makes it obligatory for the members to perform their duties conscientiously. They are also prohibited from profiting in any way from their membership of the board or association with the hospital.

How does one appraise the performance of the board both in relation to its deliberative group action and for individual performance? In any governing board, particularly when the board is large and unwieldy, there are always some non-performers who either do not attend the meetings of the board or take only perfunctory interest in the affairs of the hospital. There are also a few overbearing, aggressive and talkative members who dominate the discussions to a degree that the meetings of the board are reduced to the level in which the other members of the group are made to appear as mere onlookers or observers who cannot get a word in edgeways even if they want to. In the end, everyone seems satisfied and talks as if all the members have effectively contributed to the deliberations and decisions of the board. The same thing happens when there is an executive committee that meets and takes decisions in the intervals between the meetings of the board virtually acting as a *de facto* board.

Not infrequently the members of the governing board, especially when they are on board for the first time, feeling that they have had no training and experience in hospital management, allow themselves to be passively manipulated in the hands of a trained and experienced CEO and the medical staff. Conversely — and this is worse — is the other extreme when the governing board arrogates to itself the task of formulating policies and procedures without consulting the CEO and the medical staff, and without the knowledge of and regard for the consequences of such actions on the professional care of patients and the morale of personnel. The board must realize that the management of a hospital is essentially different from that of a commercial organization and requires the guidance of a professional administrator who is trained and experienced in the professional aspects of hospital administration.

There must be some mechanism, even if an informal one, to evaluate the performance of the members of the board, both collectively and individually, and study the differences that exist between the levels of individual and group performances.

One of the important functions of the governing board — perhaps the most important one — is the search for and selection of the CEO of the hospital. The board then delegates to him the responsibility and matching authority to manage the day-to-day operation of the institution.

Good relations between the board and the CEO is a must if the hospital has to function efficiently and merit the support and confidence of the public. The members of the board should not in any way attempt to assume the CEO's functions. Nor should they go snooping around or breathing down his neck. It is generally accepted that the directors of the board should not maintain an office in the hospital, or be there full time. Where this has happened a great deal of harm has been done to the organization concerned. This does not mean that the board should remain inactive. It should guide and help the CEO in the broader phases of operation like the formulation of policies, while giving him full freedom in all administrative matters. In no case should the board relinquish its responsibility to exercise the ultimate control over the operation of the institution.

There must exist a well–defined relationship between the governing board and the CEO. One way of ensuring this if for them to clearly understand their respective roles. Briefly stated, the governing board establishes policies and the CEO executes them. Having granted him the executive authority, the governing board relinquishes the right to deal directly with the staff of the hospital, the line of authority must be through the CEO. Any unwarranted interference of any member of the board in the day-to-day administration will only undermine the authority and effectiveness of the CEO.

Duties and Responsibilities of the Governing Board

The following are some of the duties and responsibilities of the governing board.

- ◆ To formulate and periodically review the mission, philosophy, goals and objectives of the hospital.
- ◆ To determine and establish policies of the hospital in relation to the needs of the community it serves.
- ◆ To raise funds and provide adequate financing through sufficient income and other means, and to enforce business — like management of funds and control of expenditure.
- ◆ To enhance the total assets of the hospital in terms of finance, equipment, personnel and materials.
- ◆ To enforce proper professional standards in the treatment of patients.
- ◆ To fulfil its legal obligations.
- ◆ To exercise its responsibility in the selection and appointment of competent and qualified management and medical staff.

With regard to the selection and appointment of personnel, it is customary for the governing board to delegate the authority to the CEO. In larger hospitals he is generally assisted by a personnel officer and heads of departments in discharging this responsibility. It is, however, not uncommon for the governing board to reserve for itself the right of final approval and discharge of some important personnel like the senior medical staff, administrative officers and heads of departments. But it is a wise policy in such cases to follow the advice of the CEO. More often than not, the CEO may not wish to assume such a weighty responsibility of selecting the senior officers and medical staff all by himself and would want the collective wisdom of the governing board to prevail. However, since the CEO has to work with these people, his recommendations in all these cases must be given due consideration. At no time should the board force on the CEO someone who is appointed not on the basis of merit but because of political connection, favouritism or other extraneous reasons.

Governing Board and Conflict of Interest

It is not unusual to find persons accepting board positions for personal gain or, having accepted these positions initially with good intentions, use them subsequently to further their selfish ends. A conflict of interest may be considered to exist when, for example, the activities of an individual on behalf of the hospital involve securing an improper personal gain or advantage, or when his activities or actions have an adverse effect on the interest of the hospital, or they help a third party to obtain improper gain or advantage. Personal gains may come in various guises. A board member may do business with the hospital on whose board he serves, or he may use his board membership to his advantage in his own business. Examples: a member who is dealing with cement and steel may insist that he supply them for hospital construction, or a member who is the owner of a restaurant may want the hospital to buy meat and provisions from the same purveyors that he deals with, and then receive a discount from these purveyors for his own business; another member may use his board membership to obtain easy loans or credit facilities for his business. At the worst the CEO of the hospital, knowingly or unknowingly, either colludes with the board member concerned or acts in a manner that facilitates the member's activities.

There are other circumstances and activities which may give rise to possible conflict of interest.

* A board member holding a position or having financial interest in concerns from which the hospital buys materials or services, or in concerns which are competing with the hospital in providing services.
* A board member rendering management, professional or consultative services to outside concerns which either do business with the hospital, or are competing with it.
* A board member accepting gifts or hospitality from persons who do business or are seeking to do business with the hospital, or are competing with the hospital, with a view to influencing the members to show favours to their firms.
* Disclosure or use of hospital information, for example, procedures, confidential marketing strategy, etc. for personal gain or advantage of the board member or concerns with which he is connected.

The hospital should protect itself or safeguard its interest by creating a mechanism by which all activities that are likely to or suspected of giving rise to conflict of interest should be reported or investigated. In some cases the investigation may have to be extended to cover board members' immediate family members. In some developed countries there are conflict of interest laws which provide this protection to hospitals.

➤MANAGEMENT STRUCTURE

Management structure is often confused with governance. As a matter of fact, it is the lack of understanding about which functions constitute governance and which constitute management that has led to governing board members so often meddling with the internal management of the hospital. The CEO is the head of the management team. He is the legal representative of the governing board in whom is vested authority for the management of the hospital. Although he may delegate some of his responsibilities and enough authority to his associates and heads of departments to carry out their respective functions properly, the CEO still remains ultimately responsible and accountable to the board for everything that happens in the hospital.

➤Duties, Responsibilities and Functions of the CEO

Without being exhaustive, we delineate here some of the duties, responsibilities and functions of the CEO.

- The CEO submits a plan of organization for the hospital for the approval of the governing board. He also formulates rules and regulations for the proper functioning of the hospital.
- He selects and employs all personnel and fixes their salaries within the approved salary scales and limits of the budget.
- He controls, disciplines and discharges all personnel.
- He prepares and submits an annual budget for approval of the governing board. The budget will show estimated receipts and expenditure, and the anticipated deficit, if any.
- He recommends charges for all hospital services.
- He advises the governing board on the formulation of policies.
- He submits periodic reports to the governing board on the working of the hospital. The annual report will also provide an analysis of the plans for the coming year.
- He directs all activities of the hospital and implements established policies. As executive head of the hospital, he is responsible to the governing board for efficient management of the institution.
- The CEO is the liaisoning officer and channel of communication between the governing board on the one hand and various departments, medical staff and other personnel on the other. He transmits and interprets policies and makes sure that they are followed. Likewise he transmits the ideas and wishes of the staff to the governing board.
- He selects department heads and delegates part of his responsibility to them. The extent of this delegation of authority and responsibility depends largely on the size of the hospital. However, the CEO is ultimately responsible for the management of the hospital while the heads of departments are directly responsible to him.
- The CEO is responsible for employer–employee relations. In smaller hospitals, he may perform all the functions of the human resources department. In larger hospitals, there is generally a full-fledged personnel management department performing the functions of that department under his authority.
- The CEO must exercise sound control over the business management of the hospital. Although he is not expected to have training and expertise in every field of operation, he must nevertheless have sufficient knowledge of all the departments so as to be able to effectively supervise them and ensure that they are efficiently managed.
- The department of purchase should be directly under the control of the CEO. It is here more than in any other department that graft and corruption raise their ugly heads in the form of kickbacks, etc. In smaller hospitals the CEO himself may perform the duties of the purchase officer. However, regardless of the size of the hospital and who does the purchasing, the CEO must be actively involved in purchasing and exercise ultimate control over this department.
- The CEO of the hospital must be a leader in the community. He should identify himself with people and be actively involved in the activities of the community, particularly those which directly or indirectly have a bearing on the work of the hospital. He should be closely connected with public service organizations like the Rotary clubs, women's clubs, etc. whose declared interest is public welfare including the care of the sick.

◆ The duties of the CEO should not be limited to his own hospital, nor even to his own community. He should maintain contact with hospital associations and attend their meetings. He should be a leader in the hospital field and present papers, write articles in professional journals, participate in discussions and pass on the benefit of his knowledge and skill to others. More importantly, he should keep himself abreast of advances in the hospital field. Some progressive hospitals direct, even make it obligatory that the CEO participate in local, regional, state and national hospital affairs when this is to the advantage of the hospital. In addition to this, the CEO must collaborate with various regulatory bodies to ensure current feedback concerning the latest regulations and laws affecting the health care institutions.

Relationship of the CEO with the Governing Board

Although primarily and technically the relationship between the governing board and the CEO is one of employer and employee, it should not be interpreted in the usual sense of these terms. The hospital is a special type of organization where the governing board and the CEO function as partners. Nevertheless, the CEO is an employee whom the board can discharge for just cause.

In the recent times, the relationship between the board and the CEO has become so firm that in many hospitals the CEO is made a voting member on their governing boards. In some boards he acts as a co-chairman, in others he has the title of president under the chairman of the board. This is a common pattern in industry where the CEO is also a member of the board of directors and an equal among equals, and not just a hired employee. With voting privileges, the CEO actively serves on committees including key committees such as planning and nominating committees. He may be either an *ex-officio* or elected member.

The CEO: Where are we Now in Titles?

Everett Johnson and Richard Johnson in their book entitled *Contemporary Hospital Trusteeship* trace the duties and titles of the CEO in the past few decades. In India in the so-called traditional hospitals, largely the church-related and voluntary hospitals, the titles and appellations used for the chief executive officer have not changed much. But in many other hospitals, particularly the new ones, we find titles similar to those in the evolutionary pattern that the Johnsons trace in their book. The following is a brief summary of the transition that has taken place.

◆ In the early stage, the top full-time officer of the hospital was a nurse whose primary responsibility was supervision of nursing care. Other administrative responsibilities were parcelled out among members of the governing board.

◆ As problems and demands on everyone's time increased, a business manager was appointed. As he assumed more and more administrative responsibilities, the board members coordinated the functions of the two positions.

- The third stage saw the appointment of full-time hospital superintendents and administrators. The superintendents and administrators were drawn from the ranks of trustees, physicians and clergymen in church-related hospitals.
- In the next stage, the governing board recognized that the hospital administrator functioned both as the director of hospital activities and as a health leader in the community as well as in the wider health care field. He was then given the title of executive director or executive vice-president or chief executive officer (CEO). This reflected the desire that the hospital should be concerned with the total health care of the community. The board also recognized the need to treat their CEO like his counterpart in industry.
- The position of the CEO entered the final stage with the title of president.

This title "President and Chief Executive Officer" is the common pattern in many hospitals in America and is now coming into use in our corporate hospitals.

"The appropriate title for the chief executive officer... is not a device for satisfying the ego of the person in that position. It is based on the expectations of medical staff, trustees, and the public about the performance level needed by a complex health care institution," say the Johnsons in their book.

The CEO and his Management Team

The CEO cannot be on duty continuously even in a small hospital. Nor can he attend personally to myriads of details pertaining to his office and other departments. More importantly, he is not expected to be an expert and have intimate knowledge of the working of various specialized departments of the hospital. At best, he can be compared to an automobile driver who is not an automobile engineer or mechanic. The driver may not know all about the engine, combustion and functions of other parts. Yet he can be a good driver if he has some knowledge of the various parts of the car including the dashboard, and knows how to change a tyre or replace a fan belt in an emergency. He must have presence of mind and a high degree of ability to coordinate. He must be clear in his mind where he is going and how to get there. This is precisely the job of a hospital CEO.

It is imperative, therefore, that the CEO has assistants to relieve him of his responsibilities and carry out, under his guidance, the duties and functions of major administrative areas as well as act in his place during his absence. There should always be an assistant to the CEO on duty at night. Being responsible to the CEO and as his representative, the night administrator must be given full authority to function. In the earlier times, the night administrator was usually a nursing supervisor whose primary duty was to supervise the work of nurses. Hence her authority in administrative matters was often questioned. If the nursing supervisor is also the night administrator, it is advisable to issue standing orders to the effect that she has full authority in administrative matters during the night.

In small hospitals, the CEO may personally take over the functions of some of the assistants. For example, he may perform the responsibilities relating to personnel management and purchase. He may directly supervise departments like food service under a manager. Likewise, one assistant may combine the work of several assistants.

The assistants have generally the titles appropriate with the chief's. For example, the president's assistants may be vice-presidents, the administrator's assistants may be associate or assistant administrators, the director's assistants may be associate or assistant directors.

Organizational Charts

Figure 2.1–2.5 are organizational charts depicting the management structure commonly found in Indian hospitals. These are, however, for illustrative purposes only and should not be considered as a recommended pattern of organization for any individual hospital. A new hospital has the option of choosing any one of these models, or patterning its individual structure on the basis of these models.

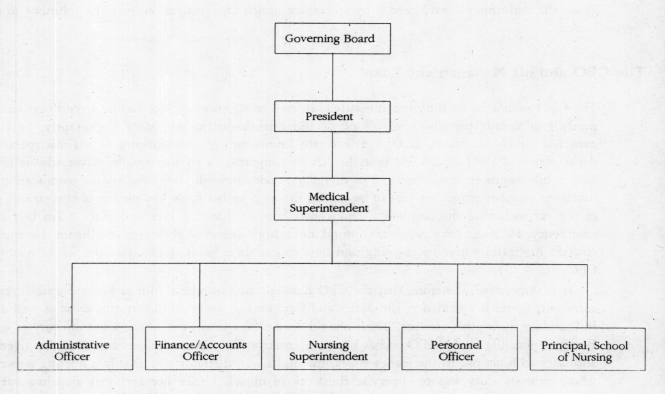

Note: This chart is for illustrative purposes only. It should not be taken as a recommended pattern of organization.

➤ **FIG. 2.1** Organizational Chart of Saint Luke's Hospital, Mangalore

PLATE 5

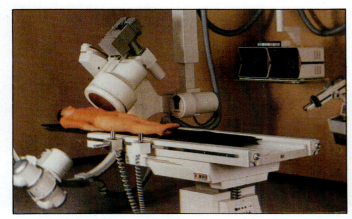

➤ **PICTURE 18** Diagnostic X-ray Systems

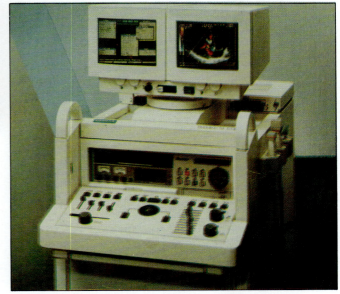

➤ **PICTURE 23** Diagnostic Ultrasound System

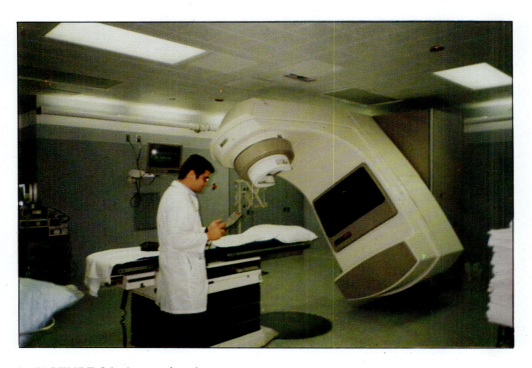

➤ **PICTURE 24** Linear Accelerator

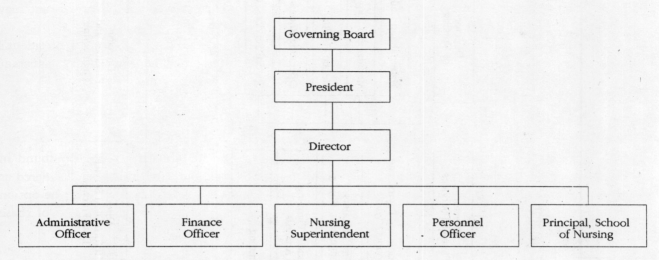

Note: This chart is for illustrative purposes only. It should not
be taken as a recommended pattern of organization.

➤ **FIG. 2.2** Organizational Chart of Francis Xavier Charitable Hospital, Kottayam

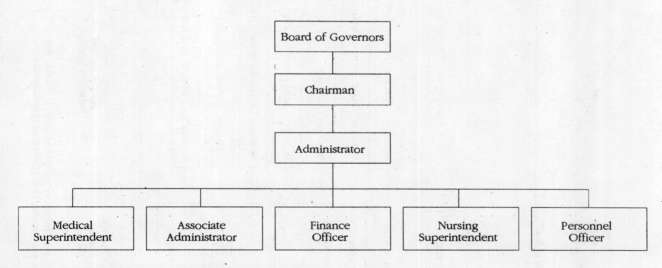

Note: This chart is for illustrative purposes only. It should not
be taken as a recommended pattern of organization.

➤ **FIG. 2.3** Organizational Chart of Ramanashree Hospital, Hyderabad

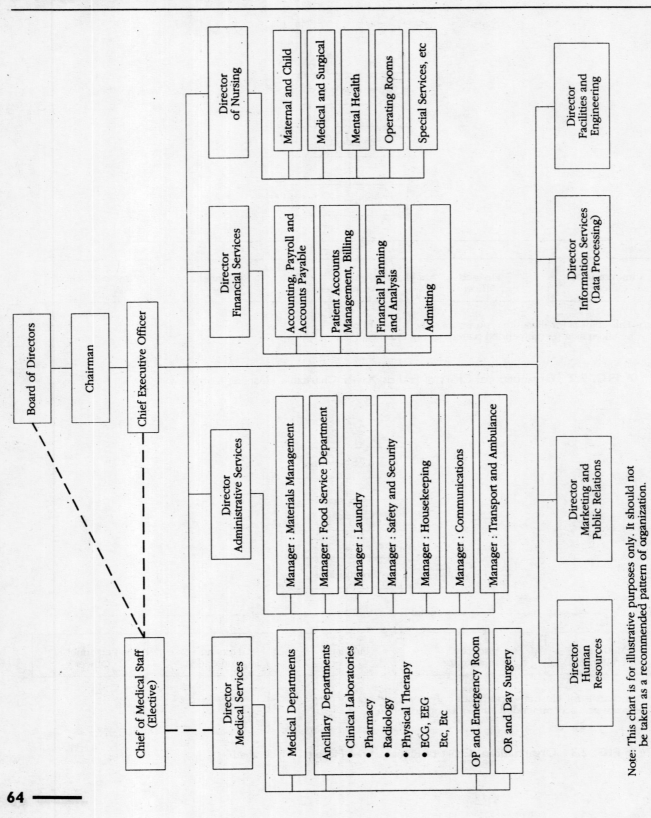

> **FIG. 2.4** Organizational Chart of Cauvery Medical Center, Bangalore

Note: This chart is for illustrative purposes only. It should not be taken as a recommended pattern of organization.

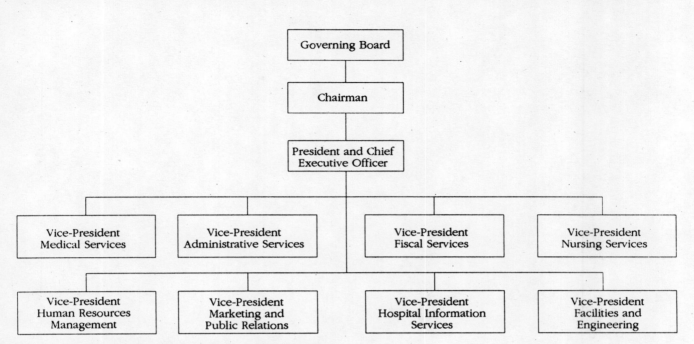

Note: This chart is for illustrative purposes only. It should not
be taken as a recommended pattern of organization.

➤ **FIG. 2.5** Organizational Chart of Sterling Hospital, Chennai

CHAPTER 3

Planning and Designing Administrative Services

CONTENTS

3 CHAPTER

INTRODUCTION

Health care has come a long way since Florence Nightingale tended the wounded soldiers in the Crimean War. At that time, there were no hospitals of the kind we now have. Hospitals of the Christian era, run by religious orders, provided largely religious solace besides food and shelter, but little of medical care. In ancient India, hospitals existed in the sixth century albeit in a rudimentary form. Buddha is credited with having built hospitals for the crippled and the poor, and with having appointed a physician for every ten villages. His son built shelters for the diseased and for pregnant women.

The institutions built by Ashoka around 300 BC were significant as in many ways they were the precursors of modern hospitals. The physicians employed there were adept at every operative procedure. The attendants gave gentle care to the sick, prepared medicines, gave massages and furnished the patients with fresh fruits and vegetables.

The modern hospital is no longer a refuge for the sick, the poor and the homeless. Over the years it has evolved into a complex, specialized and multifaceted social organization rendering preventive, diagnostic, curative and rehabilitative health services to the acutely ill as well as ambulatory patients. The unprecedented advances in the fields of medicine and technology and the proliferation and sophistication of medical equipment which were unheard of a couple of decades ago have revolutionized the management system, requiring professional management, greater specialization among personnel, larger number of employees, and interpositioning of new echelons of supervisory and managerial personnel. In addition, the rapidly changing and all-pervading computer and communication technologies which have rendered obsolete what was available only a few years ago, have resulted in speedier, more sophisticated data handling and processing, and instantaneous and reliable communication systems necessitating special training for personnel.

Consequently, the number and size of administrative facilities required for hospitals have increased. The larger the hospitals, the larger the size of these facilities. In this part of the book we discuss the planning and designing of facilities for administrative services. Because of their related activities, certain administrative facilities which may be called "top management" or "general administration" should be physically adjacent to and grouped with the offices of the CEO.

The administrative block (Fig. 3.1) should be conveniently accessible to authorized visitors and to the hospital's main entrance as well as to vertical and horizontal communication areas. It should be designed to avoid cross traffic from the main lobby. Units generating more traffic such as the human resources department, financial services and, to some extent, the nursing administration

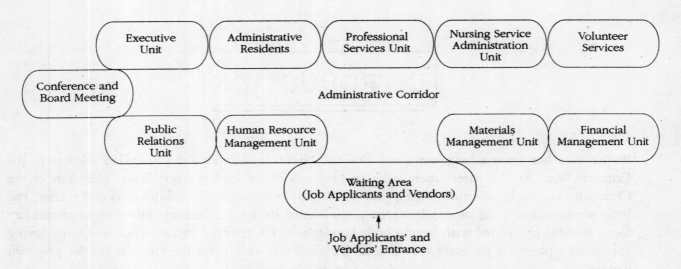

> **FIG. 3.1** Chart Depicting Intra-departmental Relationship in the Administrative Block

should be located nearest the lobby whereas those requiring privacy and generating less traffic such as the executive suite should be at the far end of the main lobby. If possible, the executive suite should have an independent engress to the outside.

Good planning and careful design conserve space. However, space should not be limited or conserved at the expense of efficiency or future expansion. When the operational and functional programmes which we have discussed in Chapter 1 are completed, space needs can be determined as also the budgetary cost estimates.

EXECUTIVE SUITE

➤OVERVIEW

The executive suite is the pivotal point around which the administrative services and activities of the hospital revolve. In direct charge of the hospital and responsible to the governing board is the CEO who acts for it in an executive capacity. In cooperation with his associates, he develops policies, rules and regulations which are approved by the board. He is delegated full authority to implement policies and to conduct the activities of the hospital within the framework of established policies. The CEO delegates authority over the departments to department heads.

➤FUNCTIONS

The functions of the CEO are discussed in detail under Management Structure in Chapter 2 of the book.

➤LOCATION

The executive suite should be located at a focal point on the administrative corridor in the administrative block. As some of the activities in the CEO's office are of a confidential nature, privacy is vitally important. Because of this and limited staff traffic, the unit should be located at the end of the administrative corridor, and yet it should be accessible to the members of the governing board, medical staff, department heads, patients, community leaders, government and public health officials, leaders of civic organizations and others. These relationships typify the nature of the CEO's activities both within and outside the hospital.

➤DESIGN

Since the CEO's office is the place where visiting dignitaries, VIPs and, on occasion, VIP patients are received, the furnishing, decor and flooring of the entire suite should be so chosen as to project and reflect a good image of the hospital. Carpeting of the floor, placing of a sofa set for the VIP visitors and landscaping of the secretarial and other open areas will make the place attractive. Plants provide brightness to the staff and visitors. Similarly, the waiting area should be attractively arranged and furnished. Furnishings in the individual offices should coordinate with the general style in the open area (Pictures 12, 13).

➤SPACE REQUIREMENTS

Depending on the size of the hospital and its activities, space is required for some or all of the following:

- Reception and visitors' waiting area which should be large enough to seat several visitors.
- Secretarial and clerical area.
- CEO's office — large enough to accommodate executive chair and desk with unitized L unit, visitors' chairs, closed management cabinet, bookcases, table for computer, sofa-set, etc.
- Office for director for administrative services (assistant administrator).
- Administrative residents' area (cubicles or desk space).
- Office for director of planning.

➤ **PICTURE 12** Landscaped Open Office

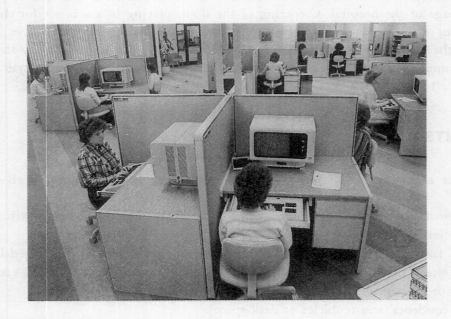

➤ **PICTURE 13** Landscaped Open Office

- Administrative assistant's area.
- Closet and attached toilet within the CEO's office.
- Toilet facilities for personnel of the administrative suite and those attending conferences and board meetings. Proper location of such facilities minimizes the amount of time personnel are away from their offices or out of communication.
- Pantry or kitchenette.
- Supplies storage.
- Place for copier, duplicating machine, TELEX, FAX, etc. with space for supply storage.
- Board room for up to 25 persons, for meetings of the governing board, executive committee, for CEO's weekly meetings with top management people and other similar meetings (Picture 14).

➤ **PICTURE 14** A Board Room

- Conference room or auditorium with toilet facilities, equipment storage room and kitchenette for conferences, heads of departments' meetings, seminars, workshops, medical teaching, nurses' training, continuing medical education programmes, etc. (Fig 3.2). Conference room or the auditorium may be conveniently located outside the administrative block.

Smaller hospitals cannot justify or afford to have a conference room as well as a board room. They may have one room large enough to meet the needs of both.

➤OTHER REQUIREMENTS

In the executive office

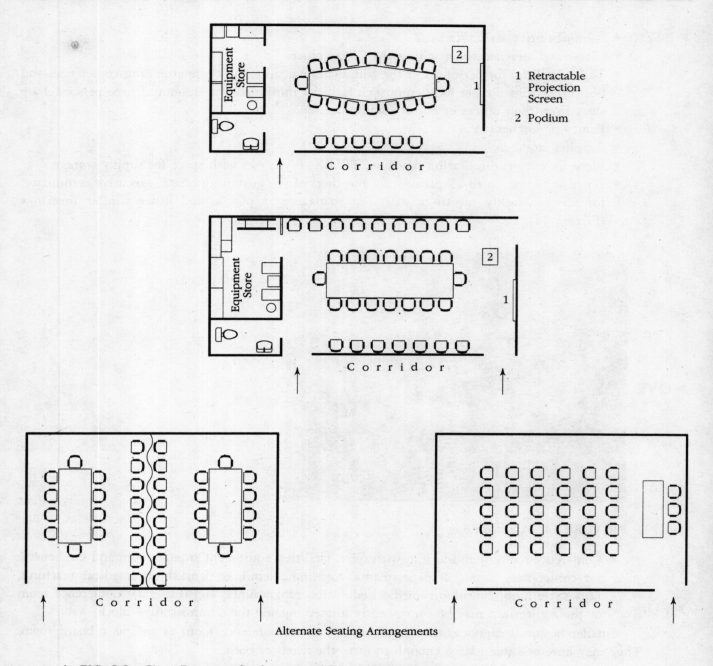

1 Retractable Projection Screen

2 Podium

Alternate Seating Arrangements

➤ **FIG. 3.2** Chart Depicting Conference and Board Meeting Rooms

- ◆ Master phone — multiline instrument with intercept, transfer and intercom capabilities including intercom.
- ◆ A separate private telephone.

➤INTRA–DEPARTMENTAL RELATIONSHIPS

- ◆ The secretary's office should be immediately adjacent to the CEO's office. No one should be able to enter the CEO's office without first stopping at this area. The secretary should be able to supervise the work of other secretarial and clerical staff.
- ◆ If there is an administrative assistant, his office should be within the executive suite as he generally assists the CEO in administrative matters.
- ◆ The offices of the director of administrative services, director of planning, and the cubicles for administrative residents should also be within the executive suite.
- ◆ The reception-control area should be at the main entrance to executive suite. The receptionist will serve all the administrative offices.

PROFESSIONAL SERVICE UNIT

➤OVERVIEW

Under the chief of medical staff or the medical director, the professional service unit directs and coordinates all medical and related activities of the hospital, excluding nursing service, and has overall responsibility for the quality of medical care provided to the patients in the hospital. Hospitals which have facilities stipulated by the respective regulatory bodies offer residency programmes such as internship, junior and senior house officers or house surgeons programme and diplomate of national board programme besides continuing medical education programmes. Where such programmes are offered, a medical education coordinator generally coordinates the work of interns and residents.

➤FUNCTIONS

The following are some of the functions of the professional service unit. The unit
- ◆ directs and coordinates all medical, paramedical and other related activities;
- ◆ establishes standards of medical service;
- ◆ supervises and coordinates the work of the medical and ancillary departments;
- ◆ advises the CEO and through him makes recommendations to the governing board on medical care, medical and administrative problems and on policy matters;

◆ recommends appointments, promotions and discharge of medical staff;

◆ advises clinical staff on difficult medical cases and on a variety of problems;

◆ organizes teaching programmes for interns, trainees and residents;

◆ studies new developments in medical science and practices and applies them in the hospital;

◆ develops or assists in the development of a Formulary for the hospital.

◆ advises the administration on the use and selection of drugs to be stocked in the pharmacy, evaluates new drugs, studies problems or adverse reaction to administration of drugs used in the hospital and helps in the work of the pharmacy.

◆ assists in the preparation of a budget for medical service;

◆ organizes clinical meetings, seminars, etc. and represents the hospital in such meetings, conferences, etc. held outside the hospital; and

◆ prepares the annual report for the governing board on the medical care and on working of the department.

➤LOCATION

It is recommended that the unit is located in the administrative block between the executive suite and the nursing administration unit with access from the main administrative corridor.

➤DESIGN

The flooring, decor and furnishing of the unit should be on the same lines as in the executive suite.

➤ORGANIZATION

The medical staff of a hospital can be organized in two ways. Under one system, found in many of our hospitals, the personnel consists of full-time, hospital-based, salaried medical staff with or without limited consultation practice outside the regular working hours. There is a wide variation in the consultation practice which is sometimes abused. The medical director or chief of medical staff, also a full-time salaried physician, is not generally permitted private practice.

In the other system (Fig. 3.3), medical staff are not salaried employees of the hospital. Except the medical director, and in some hospitals, the pathologist, radiologist, anaesthesiologists and emergency medicine physicians, other doctors are granted privileges to practise in the hospital in their respective specialties. In a well organized system, the medical staff who are granted privileges

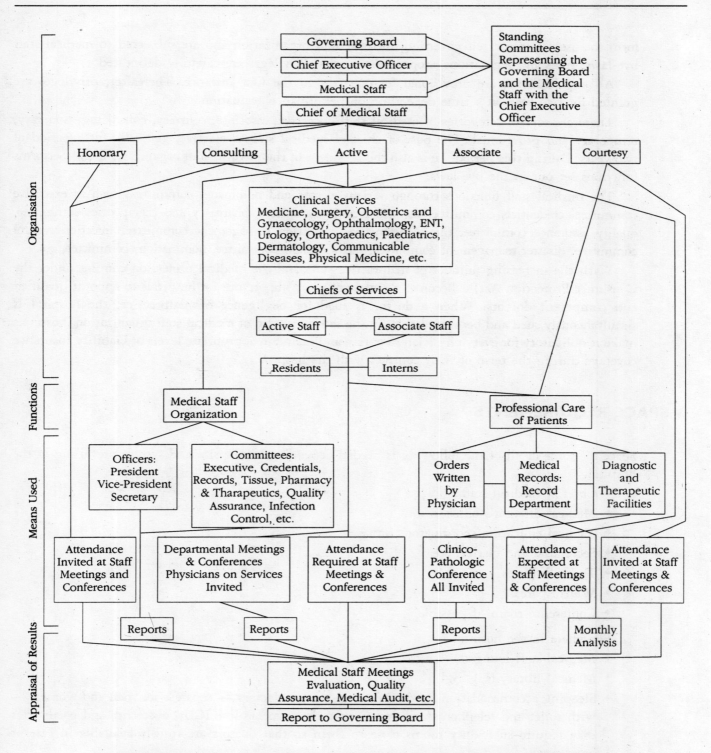

➤ **FIG. 3.3** Chart Depicting Medical Staff Organization in a Hospital where Medical Staff are not Hospital Based, Full Time, Salaried Persons but are Granted Privileges

form themselves into a self-governing medical staff organization and are subjected to medical staff bye-laws, rules and regulations adopted thereunder. Privileges are carefully delineated.

All appointments are provisional for one year in the first instance. Thereafter, privileges are granted for two years at a time depending on performance evaluation.

There are several categories of medical staff — active, associate, courtesy, consulting, honorary, temporary and provisional. The post of chief of medical staff is elective, generally for a period of two years. During this period, he is also the president of the medical staff organization, and performs functions set out in the bye-laws.

The medical staff functions through various elected and nominated committees such as executive committee, credentials committee, medical records committee, pharmacy and therapeutics committee, quality assurance committee, tissue committee, utilization management committee, infection control committee, disaster management committee, medical audit committee, nominations committee, etc.

With the increasing number of malpractice suits and the medical profession coming under the Consumer Protection Act, it becomes a matter of great importance for hospitals to appoint qualified and competent doctors. When a doctor is sued for negligence or malpractice, the hospital is simultaneously sued and becomes liable. In the second system of medical staff organization, hospitals make it obligatory for every physician to obtain and maintain appropriate levels of Liability Insurance coverage during the term of their contract with the hospital.

➤SPACE REQUIREMENTS

Space for some or all of the following is needed depending on the size and system prevailing in the hospital.

- Chief of medical staff's office
- Medical director's office
- Medical education coordinator's office
- Secretarial and clerical area
- Toilet facilities
- Waiting area
- Conference room, if possible.

Elsewhere in the hospital
- Medical staff lounge(s)
- Medical library (Fig. 3.4)
- Sleeping accommodation for duty doctors, on-call doctors — separate for men and women — with toilet and telephone facilities. Certain departments like ICUs, obstetrics and paediatrics may require such duty rooms close to them so that doctors are readily available in case of emergency.

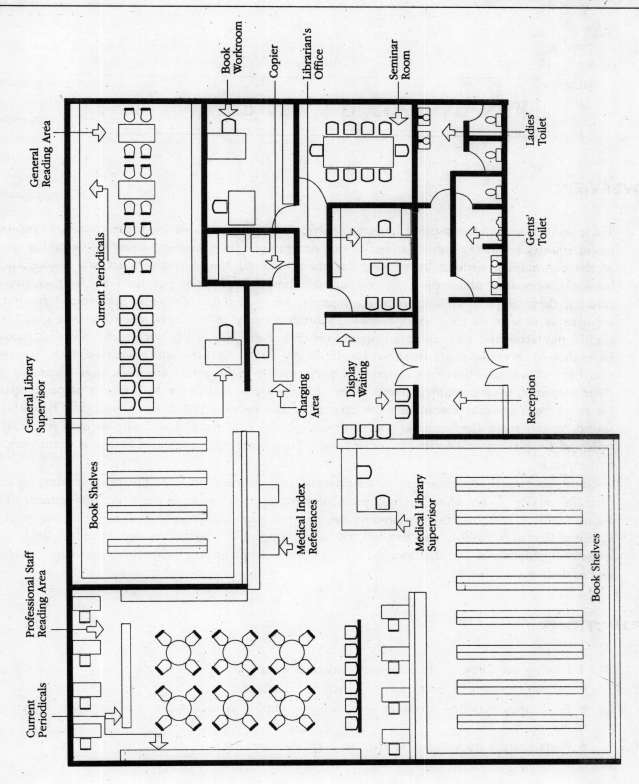

General Reading Area

Current Periodicals

General Library Supervisor

Book Shelves

Professional Staff Reading Area

Current Periodicals

Book Workroom

Copier

Librarian's Office

Seminar Room

Ladies' Toilet

Gents' Toilet

Charging Area

Display Waiting

Reception

Medical Index References

Medical Library Supervisor

Book Shelves

▷ **FIG. 3.4** Layout of a Medical and General Library

FINANCIAL MANAGEMENT UNIT

➤OVERVIEW

There was a time when non-profit hospitals, particularly church-related and other voluntary hospitals in our country set their priorities in health care programmes by responding solely to the health needs of the community without any regard for their financial implications. What was worse, these hospitals, operating under the style of charitable hospitals, did not pay heed to the necessity of running their hospitals in a businesslike manner. Many of them have now been forced to sing a requiem as it were to these compulsions. In order to stay in business, they realized that they should pay attention to economic imperatives. Every decision made within the hospital should be evaluated in terms of its financial implications. Public opinion and priorities of the trustees also have changed. The effectiveness and performance of hospitals are no longer measured by their response to community needs alone but by their ability to maintain a strong, viable financial position that commands the respect of the people. When the hospital is financially sound, it can meet the needs of the community any way. An out of business or a financially unsound hospital does not look after anybody. Hence the importance of financial management in hospitals.

The three main components of the financial management unit — business, accounting and financial service — are often wrongly used interchangeably to refer to the whole department. The business office performs the day-to-day business procedures, primarily dealing with the patient accounts, accounts payable and pay-roll; the accounting office deals with budgeting, auditing and monitoring functions; and the financial service deals with planning, forecasting, reporting, advising and evaluating.

➤FUNCTIONS

The following are some of the important functions of the unit:
- Maintains extensive accounting and statistical records.
- Establishes patients' accounts — a unique number is assigned and a file opened for each patient.
- Posts charges for hospital services to patients' accounts.

- Compiles all charges, and issues bills; also collects cash and unpaid bills.
- Records all financial transactions, including control of cash, recording of purchases, estimate of cost of free service rendered to indigent patients, etc.
- Develops pay-roll records managing and maintaining time cards, absenteeism, leave on loss of pay, payment of salaries, benefits at cessation or termination of service, etc.
- Assists in the preparation of the hospital's budget, estimating the needs of various departments and anticipating patient load and income.
- Prepares costing of departments and services, departmental income and expenditure and comparative statements. The costs of services help in setting fee schedules.
- Handles accounts payable largely on account of purchase of goods, supplies and services. Coordinates with the purchasing department for the purpose of verifying payments for the goods and services received.
- Coordinates legal matters such as statutory reporting of births and deaths, accident claims, free care, etc.
- In some hospitals, the admitting department is under the financial management unit. The unit then supervises and coordinates functions relating to admitting and discharge.
- In some hospitals, the data processing unit is under the financial management unit because originally the data processing was created to automate financial and accounting functions.

➤LOCATION

The unit generates heavy traffic and many transactions occur. Because the department relates directly with the administrative departments and must have a convenient access to the general public and staff of various departments, it must be located on the administrative corridor adjacent to the administrative block from the main lobby. Nevertheless, much of its work requires a level of concentration demanding privacy and freedom from frequent interruptions. One way of providing this is to have a front office for functions like billing, cashiering, etc. and a back office for work demanding concentration and privacy.

➤DESIGN

Computers are being increasingly used in hospitals. Because of this and the need for adaptation of space as warranted by changing accounting practices and changes in personnel requirements, it is recommended that the unit is designed with maximum flexibility with the use of open landscaped areas and minimum use of contained modules except where they are absolutely necessary.

➤ORGANIZATION

The head of the financial management unit is a controller or vice-president or director of financial services who reports directly to the CEO. In small hospitals, the CEO or one of his associates like the director of administrative services or assistant administrator may assume this office and perform the functions pertaining to it. The actual configuration of the department will vary depending on the size of the department and who the top officer is.

Financial operations of hospitals are being increasingly computerized. Computer sophistication may range from automatic posting on mini-computers to complete financial systems. Where an in-house computer network is available, virtually every hospital department interacts with the computer system.

Some of the common activities of the business office in which computer can be used are:

- Open and maintain patient accounts, assign numbers, enter into ledger, etc.
- Post charges to patient accounts using charge codes for specific services. Data can be entered manually. In an in-house network, the system can be programmed to capture and post charges automatically and instantaneously as and when service is rendered.
- Generate bills on discharge of patients in a matter of minutes.
- Perform accounting and auditing functions.

The possibilities are endless. This subject is dealt with in detail under hospital information system.

➤SPACE REQUIREMENTS

Space is required for the following:

- director of financial services;
- senior associates or assistant director(s), business manager, chief accounts officer;
- secretarial and clerical area;
- chief cashier's work area and cashiers' booths. Cashiers must be conveniently located in major areas/departments where cash transactions occur;
- central accounting work area;
- billing section with adjunct cashier's booth. Waiting area for patients and their families waiting to pay bills. Billing area should be adjacent to credit and collection area;
- pay-roll section. Because the work is of a confidential nature, a separate room is recommended;
- credit and collection area;
- internal auditor's room and space for external auditors;
- storage room for supplies, etc.;
- archives storage area for patient records, purchase invoices and other records; and
- a conference room for departmental meetings. Can also be used by auditors.

➤ OTHER REQUIREMENTS

- The chief cashier's office should have a safe for keeping cash until it is sent to the bank.
- There should be a safe or locker for patient valuables given for safe custody during admission if this is not provided in the admitting office.
- A fireproof vault or cabinet for storing valuable records, documents and disks.

➤ PROBLEM SITUATIONS

The following are some of the problem situations which the financial department must tackle:
- high accounts receivables;
- poor cash flow;
- late charges from ancillary and supportive service departments like x-ray, laboratory and dietary especially if the functions are not fully computerized;
- a good percentage of missing bills and lost charges;
- difficulty in obtaining correct information about patients. Many patients give wrong or fictitious addresses. Follow up and recovery of unpaid bills become almost impossible; and
- embezzlement of funds, fraud and theft in various operations of the unit.

➤ INTERNAL CONTROL

The need for a system of internal control in any organization, especially a hospital, is immediately apparent. It is impossible for the CEO to exercise direct and personal supervision over all employees and their activities. He must, therefore, depend on policies and regulations and a built-in system of internal control for the efficient running of the hospital.

Internal control provides a mechanism by which the work of one employee acts as a check on the work of another. For example, the store keeper does not have control over inventory records, employees handling cash do not have access to accounting records, purchasing functions are segregated from accounting, and receiving functions from issuing functions.

It has been estimated that up to one hospital employee in every ten steals habitually. It is incredible in how many different ways people embezzle money and commit fraud. In every organization, some employees are placed in certain strategic positions that makes it easy for them to embezzle money. The accounts clerks may keep two sets of books, write cheques to fictitious suppliers, enter fictitious names in the salary register, give refunds for materials that are not returned, collude with suppliers in obtaining quotations and supplies, tamper with accounts by changing amounts in invoices and cheques after payments have been made — examples can be multiplied.

The only way to check these malpractices is by instituting effective control measures. Any measures taken after the fraud is committed is like locking the stable door after the horse is stolen.

HOSPITAL INFORMATION SYSTEM (HIS)

➤ OVERVIEW

The information system department, also called the electronic data processing department (EDP), is one of the newest and most dynamic departments in hospitals. Originally applied to automate the financial and accounting areas, computer technology has now pervaded almost every activity and has revolutionized the flow of information within the hospital.

The daily flow of information in a hospital is overwhelming. An endless stream of data begins with the outpatient and admitting departments and emanates from every department throughout the hospital. Some of the information is vital to the care and well-being of patients, while other data enhance the efficiency of the hospital itself. The way the hospital responds to the challenges of information resource management, determines the quality of patient care and success or failure of the institution. Crucial decisions must be based upon facts established through management's use of current information.

In the early days of computerization, when fragmentation was the norm, there was no way hospitals could utilize data as a consolidated resource pool. The outpatient department and admitting separately collected specific patient information, and the laboratory stored appropriate data. Many separate systems functioned with little or no sharing among them. Sometimes, subsystems did not agree. We now have a system that puts together all departmental data into a comprehensive database that can be shared on a hospital-wide basis.

Hospitals consist of many diverse groups performing highly specialized functions. It is imperative that these functions be carried out in a well coordinated manner. This gigantic and seemingly impossible task is being performed as a matter of daily routine in many hospitals which have been computerized. The individual systems work as a unified whole that fulfils the needs of both the departments and the hospital.

Today's state-of-the-art computer has a comprehensive clinical and financial database and an advanced database management handling technology. Designed to capture, edit and store information on line and in detail, it delivers maximum responsiveness to on line users. It has capabilities for extensive management reporting for all departments without affecting the simultaneous responsiveness for on line users. It can handle a wide spectrum of hospital's requirements from abstracting of medical records and historical reporting to retroactive processing of a patient's complete financial data to producing a bill.

➤SOME BENEFITS OF THE UNIFIED SYSTEM

* One time data capture is sufficient. This makes repeat entries and duplication of information reduntant, and saves effort, time and expense. It reduces the possibility of human error.
* Since the system is event-driven, it can function independently and notify other departments needing the information automatically.
* The information resource is accurate and up-to-date. One common database makes a powerful information resource available to the entire hospital.
* The existing applications, forming islands of computerization of departments, can be integrated into a unified system.
* The system offers information management for the entire hospital while allowing smooth functioning of departmental systems independently.

➤WHAT TO LOOK FOR IN COMPUTERIZATION

Hospitals will do well to bear the following points in mind while selecting a system.

Integrated System

A system that links all the computerized systems of the hospital so that when complete, there will be an information network that allows on line access to the database which ensures hospital-wide utilization of all features and functions except where security considerations and confidentiality of information prohibit their use. Data should flow smoothly from one system to another without the need for cumbersome interface programs to forcibly fit in unrelated systems.

An On Line, Real time System

In the high-tech health care system, patient records, patient information, etc. should be on line permitting direct, immediate access through terminals, and real time processing in which the computer system records each change and updates all the necessary files, etc. immediately. Example: a computerized on line appointment system or the intensive care computers that monitor patients' heart functions, breathing, etc. are real time.

A Patient-oriented System

Hospital transactions and activities revolve around the patient. The computer system is event-driven which means that each "event" or patient encounter is captured and processed as it occurs. Thus, any time a patient is provided service, the integrated database is updated accordingly.

A Future-oriented System

The computer system should be dynamic and designed to meet the challenges of today's health care needs as well as the requirements of the future. The health care system and the hardware and software technology change so fast that systems become obsolete fast. The system should be capable of being easily upgraded to include future requirements.

Comprehensive But Modular, and Implemented in Phases

Probably no hospital can computerize its entire operations at one stretch, and yet no hospital should create independent islands of computerization that are not integrated into a common database. The design must be modular so that functional modules can be implemented separately as add ons. The system should provide flexibility to first implement those modules most crucial to its environment and to meet the immediate needs, and after mastering them add on other modules. This way the cost can be spread over a planned span of time in a phased manner. The software modularity should be complemented by and integrated with hardware modularity.

Every computerized area representing one module should be complete and no part of it should be left to be handled manually. This will lead to a duplication of efforts.

The System Should be Reliable

Any failure in the system will paralyse the whole hospital operation. The hardware should be capable of 24-hour non-stop functioning. The software should be time tested with a good track record.

System Should be User Friendly

Almost everyone in the hospital including receptionists, clerks, cashiers, nurses, etc. will need to work on the computer. Almost every work is done on it so that this has become a routine operation. The system should be user friendly and simple without needing high computer literacy and in-depth training.

The System Should be Cost Effective

Both the initial cost of installation, etc. and the cost of maintaining the system should be cost effective.

➤APPLICATIONS

Limitations of space do not permit a more detailed discussion of myriads of applications and functions that can be handled through the computer. Without being exhaustive, some of the more important ones are listed here.

Patient management
- Allows a wide variety of interactive enquiries into patient data with the data provided on line and updated in real time.
- Allows monitoring of clinical and financial information flowing into patient database and checks completeness and status of data.
- Provides more efficient registration of outpatients and inpatients. Patient index allows retrieval on line the most current basic patient identification and demographic data.
- Streamlines the admitting process and makes pre-admission bed reservation, provides access to pre-admission data, etc.
- Patients in emergency room, same day surgery, labour room, and ICUs can be registered on line using previous case data which are available on line and admitted directly without going through the admitting department.
- Provides complete, fully integrated, on line real time admission, transfer and discharge processing with immediate and automatic notification of departments involved. Generates and routes all necessary documents.
- Maximizes ability to handle high volume of outpatients. Gives clear picture of outpatient activities and assigns classifications, diagnoses, procedures, and other information.
- Provides daily census and bed occupancy, identifies and maintains on line bed availability and automatically notifies departments concerned.

Medical records
- Authorized personnel can have access to all current and historical data. On line abstracting can be done using screens and conditional editing. All editing is done in real time.
- On line master patient index gives immediate access to essential, episodic patient information. Records can be tracked down and located. Notices can be generated and issued to physicians to return records.
- Medical records reporting gives optimal access to information in the desired format. Reports

can be sorted and sequenced in a variety of ways. They can be generated on a daily, monthly, quarterly, semi-annual and annual basis.

Department level care: service order entry
* This subsystem has comprehensive, on line order processing capabilities. Fully integrated, it saves much time by automatically routing orders to appropriate departments and then automatically capturing charge data for billing.
* It allows to monitor the processing of orders, to see on line if that order has been placed after being approved by the authorized person, received at the concerned department, and if the processing has been completed and charged to the patient concerned. Order status and results of the test can be viewed.
* There is no need for re-entering recurring orders. Enter only once, for example for physical therapy treatment for 20 days. Order requisitions will automatically print the day(s) scheduled for each course of treatment. Single order for any future date can also be placed.
* Orders can be revised. A record of the revised order together with the old order can be retained in the patient's file.
* Allows to view on line any preparatory instructions or other standards of care when the order is placed. For example, when a certain procedure is ordered, preparatory instructions can be displayed to be followed by the nurse or the technician.
* Allows to enquire into entire order history file to see all current and completed orders for any patient.
* Database stores history of all order details used in reporting.

Patient accounting
* Fully automated, reliable inpatient billing/accounts receivable functions.
* Can enter patient information and charges for pre-admission testing and procedures in advance. The information is automatically entered upon admission so that accurate and current billing is effected. Bills can be generated for pre-admitted patients.
* A wide variety of both one-time and recurring charges (for example, bed charges) can be automatically posted upon admission, or at any other time.
* Charges for groups of services, package for a special service (for example, bypass surgery or accommodation package such as daily charges for room, heart monitor, nursing care), physician's visiting fee, medical and surgical supplies, etc.
* On line receivable management, on line cash posting, on line financial patient index enquiry, on line account history, detailed trial balance, bad debt management and write-off, interface with general ledger, bad debt deletion, revenue and statistics reporting and a host of other functions.

General accounting
* General and patient accounting functions are integrated with the general ledger. It gives on line access to the hospital's data and timely reporting.

- Generates a wide range of financial reports, including balance sheet, operating statements and budget projection.

Payroll

- On line enquiry access to payroll and related information and management reports. Payroll automation significantly decreases manual work. In addition to standard payroll functions, other capabilities are automatically calculating benefit accruals with ability to post them to general ledger, integration of payroll and personal data.
- Provides a wide variety of tools from wage and salary administration to management of employee relations and unions besides all standard personnel functions.

Human resources

- Information can be used for managing and utilizing personnel more productively and cost effectively. Total enquiry access to employees' data.
- Helps to plan career development and professional growth of employees, to see skills and proficiency levels of employees, levels of formal education, degrees, study leave programmes, in-service training, etc.
- Information on employees' current and previous jobs with details like salary, experience, etc.
- Details of performance evaluations — dates, performance ratings, deficiencies, development programme arranged, etc.
- Salary expense analysis with automatically updated historical payroll data.

Business office

- Helps to analyse unpaid bills, to detect what is not billed, why it was not billed, and the age of unbilled receivables.
- Define the period of time to hold bills.
- Information on occupancy trends, extent of utilization of hospital facilities, revenue trends, monitor receivables, manage cash, etc.

Collection

- Accurate analysis of aging of receivables.
- Collection letters can be generated automatically when needed. These are fully integrated with accounts receivables.

Nursing service

- Complete, accurate, up-to-date and integrated data, on line, which enhances nursing capabilities for effective patient care.
- Easy access to information, historical retention of data, real time data entry and enquiry and password security to protect patients data from unauthorized access.
- On line data entry helps to collect key patient information for use in patient care activities.

◆ Helps to create and maintain patient care plans for each patient. Standard checklists of patient education plans or self-care instructions can be accessed for preparing patient care plans.
◆ Provides summary of patient at a glance. Derives and combines information from assessment, care plans, etc.
◆ Assesses a patient's need for nursing care based on degree of acuity.
◆ Creates personnel schedules with variable shifts, rosters, employees and nurse station reports, etc.

Results reporting

◆ On line enquiry will allow instant access to results of tests, investigations and procedures that have been captured and stored only moments before in other automated systems such as laboratory, x-ray, etc. Test results are automatically routed to concerned locations. Results may be viewed on line, or printed for inclusion in the patient's chart. Results can be revised as needed.

Materials management

◆ Processes departmental requisitions interactively, determines purchase requirements, identifies the best vendor, generates purchase order, and monitors the status of the order. Can marry receipts and purchase orders, update on-hand stock, check status of stock, adjust inventory levels, etc.
◆ Producing purchase orders is a complex task requiring a comprehensive data base of all the stock and non-stock items, a quick cross-reference to a vendor file and a purchase contract file. A department look-up capability is required to check department codes, and an open order capability with a receipts function to close out files.
◆ Inventory stock status maintained in real time.

Pharmacy

◆ Automated master files for drugs, medication administration record and patient's medication profile.
◆ Allows orders to be entered by predefined common orders, order sets, drug number, etc. High volume drug orders can be established as predetermined common orders. By using a code, appropriate drug, method of distribution, dose, directions and times for administration can be selected.
◆ Maintains comprehensive medication profiles giving an accurate to-the-last transaction, on line summary of the patient information, prints labels, gives previous quantity of drugs supplied, date delivered, etc. for refill orders. This ensures that correct drugs are dispensed and in correct quantity.
◆ Inventory is automatically updated as transactions occur. On line displays show current quantity on hand, reorder level, safety level and quantity on order for each item.

➤SOME ADVANCED COMPUTER APPLICATIONS

In the Area of Diagnostic Imaging

Exceptional developments have taken place in the application of computer and data communication technology to diagnostic imaging leading to new types of medical information systems.

The total digital imaging system incorporates the latest technology including optical fibre and laser disks. Its purpose is to achieve and display a large volume of medical information centred on diagnostic images. When required, the data can be retrieved and immediately displayed on high resolution screens; image manipulation and comparison can be carried out.

Bedside Computer System

The bedside computer system (see Picture 54) for nursing documentation is a revolutionary new concept in nursing care. It allows nurses to record assessments of patients and nursing intervention at the bedside where care of patients takes place. Data is entered on a bedside terminal using specially designed screens that prompt appropriate entries with instantaneous response time. This information is then automatically communicated to a computer located at the nurses' station where it is arranged in a format, summarized and printed on a laser printer for insertion into the patient's chart. In addition, nurses and physicians can enquire into the bedside or nurse station cathode ray tube terminal (CRT) at any time to review information about any patient. The primary benefits to nursing are improved productivity and enhanced documentation.

The use of this advanced bedside computer eliminates the need for manual data transcription, calculation and graphing by automating the entry of patient data into the chart. This saves the nurse's time. Since data input is automatically encoded with the date, time and initials of the nursing professional, the patient's chart shows accurate information as to "who did what to whom at what time."

A host of information hitherto not possible to obtain in this manner can now be captured by the system. Some of them are — vitals (temperature, pulse, respiration, blood pressure and central venous pressure (CVP), intake/output (I/O) fluid levels, diet, hygiene, dressings, weight, etc.

➤DESIGN

The information system department should be designed to meet the functional needs of various sections of that department and other departments of the hospital. If there is a centralized department, the design should include glass walls with visibility into the department. This will promote public and employee interest. This will also help to present computerization of the hospital to the general public in a good light as a modern management tool for better care of patients.

➤ORGANIZATION

The head of the hospital information system department is a person who is qualified and experienced in computer system. Graduate and postgraduate computer diploma/degree holders are available. Depending on the set-up and the extent of computerization and its sophistication, the department may have some or all of the following staff in addition to the head of the department.

- ◆ Computer operators who load and unload input and output devices.
- ◆ Console operators who monitor or direct activities of the system.
- ◆ Supervisor who should be informed of all problems in the computer room.
- ◆ Administrative staff who are responsible for daily operators and for interfacing with administrative personnel in other departments.
- ◆ User coordinators who are responsible for training the operators or staff and also act as liaison between technical staff and user department staff.
- ◆ Data entry personnel or key punch operators who enter data largely from departments which are not computerized or where data is not entered on line.
- ◆ Programmers who code instructions of programs run on the computers.
- ◆ Analysts who analyse methods and design systems.
- ◆ System programmers who install and maintain the master program.

➤GUIDE TO PURCHASING SOFTWARE AND HARDWARE

We are living in an era of great changes in which our environment is being restructured as an information-based society. Paradoxical as it may seem, managements are clamouring for more and more data only to become bogged down with an information overload. There is an unprecedented demand for skilled personnel who have the capability to retrieve, organize and analyze massive banks of information which is increasing by leaps and bounds by every passing day. We all recognize the realities of this information age but few of us find ourselves really equipped to respond to its needs.

The processing power of the sophisticated computer of our times is truly awesome. It has entered our work place in a big way affecting every aspect of our environment.

Managers are generally faced with two difficult questions with regard to the purchase of software and hardware: (a) Should I buy or should I wait? (b) What are the selection criteria for purchase? The market is swamped with hardware and software products. With the electronic technology continuing to advance at an accelerated pace, new products are introduced making the earlier products obsolete. This applies to medical equipment as well, but strangely an administrator would not think twice before purchasing a medical equipment as he does about the computer. When it comes to the computer, he procrastinates. A commitment is necessary because the longer he procrastinates the more difficult it is for him to make up his mind. To make the selection process easier, experts advise that the buyer learns the computer jargon. A well-informed buyer is the best buyer who will not be gullible and be an easy victim for a smart computer salesman. Only when he acquires computer literacy and masters some basic terminologies does he find himself in a good position to explore the various options before him.

The most important thing that should be done prior to evaluating a software is to define the hospital's needs. This is a time consuming process which is not easy for a novice in the field. He should consider, among other things, the volume and capacity requirements, functions and features that are important to the hospital's needs, output and report specifications and a price range that comes within the hospital's budget. It is better to avoid newly introduced products until they have established a reputation.

Acquire Software First

The most common mistake the computer purchasers make is to buy hardware first and then look for software programmes to run on it. This is like putting the cart before the horse or catching the wrong train. You may probably have a set of excellent equipment but the purpose will not be served.

There are two kinds of software that guide and control the computers:

- One is system software which controls the basic computer devices and provides the necessary structure for memory and file formation. This is generally included in the package price when one buys the hardware.
- The second is application software which is specifically designed to help one to perform one's tasks. This includes electronic spread sheets, data base management, word processing, graphics and special function packages. The application software should be the buyer's first concern.

Software can be canned packages or custom-made. Packaged software is mass produced and is relatively inexpensive. But, in reality, it may turn out to be more expensive because the programme may need to be modified to suit the institution's needs. Modifications are much more costly, difficult and time-consuming than most of us realize. With the lack of full-time and skilled programmers in hospitals, there is a good chance that the original packaged programme will not work after the attempted modifications and tinkering.

The custom-made software is tailored to meet specific needs and data processing requirements of an institution. It generally takes months to develop and is expensive. There are excellent canned packages offered by reputed firms which customize them to a particular hospital's requirements.

Selecting Software Selecting reliable software for hospital applications that complements existing procedures (and hardware) is not easy. The more critical the procedures that will be automated, the more important it is to follow a well thought out selection process.

Hospitals will do well to use the following criteria as a starting point:

- Define the needs for software functions. Some of the points that should receive attention are — size of the user group, frequency of repetitive tasks to be automated, volume of database required, functions and features important to the hospital's needs, specific goals, etc. It should also be remembered that software is written and indexed by industry — for finance, banking, manufacturing, business, accounting, health care, etc. What hospitals need is the last one.

- Assess sophistication level required. Software applications may range from very basic to extremely advanced ones. Packages may be available, but in most cases they need to be customized to suit the needs of the users. A large, integrated system will call for larger expenditure.

- Keep the future requirements and trends in mind. The user may soon be saddled with an obsolete system as the requirements may change and more sophistication may be needed.

- Evaluate the software documentation — operation, installation and instruction manuals. The operating manual should define instructions step-by-step in clearly written and understandable language. The vendor may also supplement instructions with tutorial diskettes, cassettes and training workshops.

- Have a hands on demonstration. Do not make your decision on the basis of the demonstration model that comes with the software package. Demonstration models always work to the customer's satisfaction. Try a section of the package and see how well it works and how "user friendly" it is. The things you should look for are help screens, option menus and on line tutorials on how to run the program.

- Enquire about the antecedents of the vendor and his strengths and weaknesses. Find out the firm's background and product development trend. Check if it provides after sales service, support and training for staff. Ask if changes can be made to the software, if necessary, to meet the institution's present and future requirements.

- Determine whether or not the software is proven in the field. Check when the software was first offered and how many programs have been installed.

- Compare the costs and terms with other similar software programs.

- Check availability of training, installation, other services, modifications and custom-made programming.

- Make provision for changes in application and product developments. Then select the software that will work satisfactorily at present and be expandable to meet the needs of the future.

If hospitals match the features of the software program with the needs of their institutions, they will derive two major benefits in the long run, namely, time and money.

Some Common Mistakes Hospital Administrators Make When Selecting Personal Computer Software Frequently administrators who purchase personal computers complain about their limitations for hospital applications and leave them unused. This is because they fail to realize the importance of selecting the proper software for the intended tasks. Purchasers should avoid some of the mistakes mentioned below while selecting personal computer software.

- Many purchasers think that the programs that are complex and hard to use are the best programs and have the most capability. Not so. Most powerful programs can surprisingly be the easiest ones to use. While selecting the software, go in for one which is easy to use. It may be indicative of the depth of thought that has gone into its preparation.

- Purchasing the canned package off the shelf which needs to be modified. This has many disadvantages as we have seen earlier.

- Using general purpose programs for highly specific applications. For example, trying to accomplish special tasks such as purchasing with word processing or spread sheet software.

- Using floppy disks with software designed for hard disk systems. Keeping the massive data generated in the hospital on multiple floppy disks is not a satisfactory answer. Programs operate faster on a hard disk with higher capacity. Hard disk units are necessary in many areas of the hospital.

- Automating manual procedures. An inexperienced manager thinks of tasks that are better done manually as the ones he is going to automate. This means starting on the wrong foot. Good computer programs are designed for specific tasks which are done better on a computer such as some of the applications we have seen earlier.

- Not defining objectives clearly or defining them improperly before selecting the software. There is no justification for investing money on software if the objectives are not clear.

- Purchasing the costliest software assuming that it would be the best. Since software is basically the same, the managers should consider selecting the least expensive package. They should realize that often there is no direct relationship between complexity, capability and price.

Selection of Hardware

We have made no effort to include a detailed discussion on the selection of hardware as we feel this is best left to experts who should advise hospitals in this regard. However, we offer a few general tips that may be borne in mind.

- Check compatibility which has plagued computers since their inception. Popular software packages do not generally operate except on a few microcomputer models. So your options are limited. Start your hardware evaluation process by testing the software package you have selected, and see if it meets the hospital's needs relating to memory size, storage capacity and compatibility with peripheral equipment requirements.

- Consider the hospital's future requirements. Can the equipment be upgraded or added on to as work requirements and technologies change? Not all computer manufacturers offer such an option. Remember, in many models, the specifications you get are what you get when you buy — nothing more. You cannot later purchase additional memory, disk drives or operating systems.

- Whatever model you buy, maintenance should receive a priority consideration. Purchase your computer from a reputable dealer who can offer reliable local service.

- Make a careful assessment of your peripheral equipment requirements. Consider carefully the kind of printer, number of disk drives, storage capacity requirements, furniture needs, etc. If you do not, you may end up incurring some extra cost.

◆ Printer cost varies widely depending on the quality of the printer and the additional features it has. It is here where you should not be tight-fisted, or else you will regret later. Depending on the functions for which the printer is to be used, consider these questions before selecting a printer.

◆ Is it compatible with your software and your computer?

◆ How is the quality of printing (try actually printing on it), the noise it makes and the speed at which it prints?.

   ◆ Can it print bold face, underlining, italics, graphics, equations, special characters and columns?

   ◆ Can it use form-fed paper or only a sheet feeder for a single sheet?

   ◆ Can it print on paper wider than 8 1/2"?

   ◆ Does it have "pause" feature?

   ◆ Does it have ability to work on the computer while the printer is printing a file?

◆ There are two kinds of computer breakdowns — electronic and mechanical. The latter is more common, more difficult to diagnose and more expensive. Check available options regarding extended warranty on disk drives, printers and other moving parts.

◆ Do not entertain in your mind the popular misconception that computer operation is easy. Remember, computer is a tool, and like any other tool, calls for training and a certain amount of skill to operate. In order to put the computer to good use and to get the best return on the investment, every manager should devote sufficient time and effort for training in computer operation.

➤ TRENDS AND ADVANCES

To understand the trends and advances in the fields of computer hardware and software, let us take a look at what was available three or four decades ago and what we have now. To most of us who are lay persons in this field, the changes that have taken place within a span of 30 years or so appear to be mind boggling.

Computational capability or computing power of the computer hardware has increased enormously in a way that is difficult to comprehend. At the same time it has greatly reduced its physical size. The computational power that required a room full of equipment is now available in a self-contained cabinet, and that which required several disk size modules may now be contained in a portable hand-held calculator. The result is increased overall performance, lower power requirements, and higher reliability. Larger systems took advantages of these developments to increase their capability to handle bigger problems at a faster speed.

Earlier, large computer systems all over the country and the world handled their own computing tasks independently without any link with other systems. Now a single central computer with its all-encompassing capability can receive inputs through communication links from outlying locations and send back outputs to these locations through the same communication links. In advanced countries like America, hospitals and research centres throughout the country are linked in this way making it possible to share data.

The increased capability in smaller packages (microprocessors and microcomputers) and decreased cost have resulted in computational ability being applied to a large number of sophisticated uses such as seen in microwave ovens, portable calculators and electronic toys in addition to personal and home computers.

In the area of software, unfortunately, the software written for one application had to be rewritten to apply to another application although the applications were similar in function because programs were not "transportable". As we said earlier, modifying a program is difficult. Apart from the people who write it, few others understand it. These problems led to efforts toward software development of high level languages for structured programs and programming for ease of maintenance and, more importantly, programming that would allow sub–programs to be used again in new applications without rewriting them. There was a mushrooming of applications and programs and so were libraries of software. Original developers shared and even sold these programs and applications. More universal software became the objective. Simultaneously, an era of pre-programmed software was ushered in. Programs that were already programmed were written by manufacturers to make it easy to use their machines for many applications. Alternate programs could simply be "plugged in" to alter the base machine to handle problems in many areas of operation.

Advances in software have not kept pace with those in hardware and electronics which have moved far ahead technically. There is reason to believe that there will be significant advances and innovations providing greater computing capability in a small space in the years to come. In order to capitalize on these advances, computer scientists and programmers will learn how to use the hardware more effectively and accomplish better use of systems through software. It is also a matter of concern that the software cost in a system application is quite high. This should be brought down. The tasks before computer scientists are the effective use of hardware and software, reuse of software and the control of software development costs.

No hospital administrator who has anything to do with computers can afford to be oblivious of or indifferent to the rapidly changing world of computers. In real life, whether at home or at work, many facets of our daily lives are touched somewhere somehow by computers.

➤OTHER CONSIDERATIONS

Cost of Computerization

The cost of computerization is often underestimated or miscalculated. There is heavy investment especially if a networking type of system is chosen. Besides, the initial years of operation will increase total hospital expenditure. There is a prevailing misconception that computerization will result in the reduction of staff. This is not true. The seeming reduction in some areas is offset by the addition of personnel in computerized areas and the centralized computer department if there is one. There may be apparent savings as a result of computerization in respect of missing bills or lost charges, but they are matched by the high cost of operation and maintenance of computers. The question that is often asked is whether computerization at that cost is justified. The answer is "yes" if one considers increased accuracy, efficiency and speed in patient care and convenience to patients and staff, not to mention the elimination of a vast amount of manual record keeping and written communication.

➤ PROBLEM SITUATIONS

Confidentiality of Information Regarding Patients

Maintaining confidentiality of patient care information is a primary responsibility of the hospital. In the integrated information system (networking system), where all end users share the common database, there should be three levels of security.
   ◆ For operators and terminals: An operator's password which can be changed regularly, and a terminal level entry control.
   ◆ For patients' records: Each record should contain a field specifying either total access or limited access to that patient's data.
   ◆ For nursing stations: Security handled at the intelligent terminal level and limiting procedures to that nursing station only.

Computer Down Time

There will be occasions when the computer will be down. All departments using computers, particularly in essential services, should have a contingency procedure for such situations. When the computer is available, pending data must be entered.

ELECTRONIC DATA PROCESSING (EDP) GLOSSARY

BASIC: (Beginner's All–Purpose Symbolic Instruction Code) — A high-level language developed to provide an easy to use and learn interactive language for time-sharing or dedicated computer systems.

Bit: The smallest possible piece of information. A specification of one out of two possible alternatives. Bits are written as 1 for "yes" and 0 for "no".

Bug: An error in the program or in the system.

Byte: A sequence of adjacent binary digits operated upon as a unit — usually 8 bits.

Cathode Ray Tube Terminal (CRT): Common instrument for transmitting data to a computer, retrieving data from the computer and displaying them in a visual manner on a TV type screen so that anyone can read them. Cathode Ray Tube Terminals have typewriter keyboards for communication.

Central Processing Unit (CPU): That part of the computer system which contains the main storage, arithmetic unit, and special register groups. It performs arithmetic operations, controls instruction processing and provides timing signals.

COBOL: Acronym for Common Business Oriented Language, a computer language widely used in business operations.

Coding: Using symbols and abbreviations to give instructions to computer. Synonymous with writing the program.

COM: Computer Output Microfilm System This is the process of translating computer-generated information into a miniature image on film.

Computer Interactive Processing (Mode, Conversation): This is an operation of data processing systems which allows the user to carry on a conversation with the system at an input-output terminal. Since a prompt response is obtained from the system as each unit of input is entered, a sequence of runs can take place between the user and the system typical of a conversation.

Computer Primary or Main Storage (Main Memory): A device in a computer in which the binary bit representations of a program instructions and data are stored. The memory device is closely linked with the CPU so that individual program instructions and data elements may be obtained from memory very rapidly.

Computer Program: A set of instructions stored in the computer which directs it to perform a specific process.

Computer User Manual: A procedure manual for computer applications usually prepared in the computer system design process as part of the documentation of the system.

Computerized Word Processing System: A system, based on microprocessor technology, in which stenographers type dictation directly into computer terminals. Errors can be easily corrected by inserting items or lines when appropriate. This information, once it is proofread, is then printed out on an impact printer.

Confidentiality: Status accorded to data or information which is sensitive for some reason and therefore must be protected against theft or improper use and disseminated only to individuals or organizations authorized to have it.

Data: Another name for information.

Data Collection Network (DCN): A computer coordinated system that collects, stores and disseminates information from a computer centre.

Data Entry: inputting information into a computer for processing.

Data Security: The policies and procedures established by an organization to protect its information from unauthorized or accidental modification, destruction, and disclosure.

Data File: A major unit of information that is stored. Examples of data files include accounts receivable, payroll master file, and general ledger.

Debug: Identify and correct errors in a computer system or program.

Disk Pack: A device that contains a set of magnetized disks.

Editing: The process of deciding what data to accept, examining them for accuracy, and rejecting those that do not meet predetermined parameters.

FORTRAN: Acronym for Formula Translation. A computer language widely used in scientific and engineering applications.

General Purpose Computer: A computer that is designed to handle a wide variety of problems.

Hardware: The physical machinery and equipment that comprise a computer system.

Hospital Computer System: A hospital's electronic data processing and communications system which provides on line processing with interactive responses for patient data within the hospital and its outpatient department, including ancillary services such as clinical, laboratory, x-ray, pharmacy, etc.

Input: Data entered into a computer system for processing.

Input/Output Devices (I/O): Computer hardware by which data is entered into a digital system or by which data are recorded for immediate or future use.

Magnetic Tape: A tape that has been coded with a magnetized material. It is used to record information in the form of polarized spots.

Memory: A device on which data can be stored for retrieval at a later time.

Microcomputer: The name usually given to a small computer which uses a microprocessor for its CPU.

Minicomputer: A small computer system with limited resources and most often used for specific applications.

On line: A device that currently is an operating part of the computer system. A terminal is on line if it is logged into the system. An idle service is on line if it may be activated by the computer.

On line Information: Computerized patient records stored on a storage device which permits direct, immediate access through terminals.

Operation Manual: A manual that gives detailed instructions to a computer operator on how to perform computer related tasks.

Output: An information signal going out of a system or part of a system.

Peripheral Equipment: Equipment that is not under the direct control of the computer, such as printer, card reader, or cathode ray tube.

Personal Computer: A microcomputer system used for home and small business applications.

Primary Storage: Storage in the main storage area of the computer itself.

Printout: Printed data document from a computer operation. An on line patient admission system in a hospital may create a printout of the admission information to be used as an identification and summary sheet in the hospital patient record.

Priority Interrupt: A method of providing some commands to have precedence over others.

Programming: The advance preparation of instructions for use by the computer.

Random Access: A storage device by which access time in retrieval is made to be independent of the location of data or sequence of input.

Real Time: Processing of data instantaneously as they are received, enabling the user to have immediate control.

Real Time Processing: A form of interactive processing in which the computer system records each change and updates all the necessary files, etc. immediately. In health information, a computerized on line appointment system could be real time; intensive care computers that monitor the patients' heart function, breathing, etc. are real time systems.

Software: A set of computer programs, procedures, and possibly associated documentation concerned with the operation of a data processing system, for example, compilers, library routines, manuals, circuit diagrams.

Sub-system: An identifiable portion of a main system.

Terminal: Devices for input (usually a keyboard) and output (usually a printer or CRT screen) operated by a person. These are some distance from the computer and are often connected to the computer by telephone lines.

NURSING SERVICE ADMINISTRATION UNIT

➤OVERVIEW

The nursing service which normally constitutes the largest single group of hospital personnel and is the mainstay of the organization, is responsible for providing comprehensive and continuing nursing care to all patients in collaboration with other health care personnel. Good nursing care is the result of coordinated administrative and clinical planning. To be effective, the nursing service should be self-governing to the extent that it exercises responsibility for establishing and maintaining standards and for supervising the work of professional and supportive nursing service personnel. The department is also responsible for teaching programmes for nursing and auxiliary personnel. Because it is difficult for most lay people to judge the hospital by way of practice of medicine, the reputation of the hospital is often based on the nursing care it provides. Heading the department is the director of nursing, also called nursing superintendent, who reports directly to the chief executive officer.

➤FUNCTIONS

The nursing administration is responsible for
- ◆ establishment of objectives for the department of nursing and organizational structure to achieve these objectives;
- ◆ formulation of nursing service policies and procedures, and for keeping them up to date;
- ◆ putting into effect and interpreting the administrative policies established by the governing board;
- ◆ maintenance of stable staffing pattern;
- ◆ selecting and assigning nursing personnel;
- ◆ planning and directing orientation and in-service training programmes for professional and non-professional nursing staff;
- ◆ constantly evaluating and improving nursing care of patients and establishing nursing standards;
- ◆ maintenance of proper nursing records for clinical and administrative purposes;
- ◆ assisting in the preparation of and administering the budget for the department;
- ◆ coordination of the activities of various nursing units;
- ◆ promotion and maintaining effective and harmonious relationships among nursing personnel, and between the nursing service department and medical staff, patients and public; and
- ◆ participating in community health and health education programmes.

➤LOCATION

The nursing administration unit generates moderate to heavy traffic. It should be centrally located in the administrative block convenient to the executive suite from where it can also improve coordination of nursing services on the floors. With a view to decentralizing nursing administration and to improving patient care, communication and administration staff relationship, many hospitals place supervisory nursing staff in the patient care areas.

➤DESIGN

There is an acute and perennial scarcity of nurses in our hospitals and their turnover is generally high. Since the nursing unit directly recruits the nursing staff and since that is usually the first impression of the hospital that the prospective nurses receive, the nursing administration office should be carefully designed with special attention to the waiting area, furniture, decor, etc.

A separate section in this book is devoted to planning and designing of nursing units on the floor.

➤ ORGANIZATION

How a hospital's nursing service is organized is determined by many factors — the size and kind of hospital, education and skill of the nursing staff, type of medical staff organization, relationship between the nursing staff and other departments and the extent to which personnel in various departments come in contact with the nursing department.

Although positions in the nursing service hierarchy vary from hospital to hospital, one can usually see the following nursing positions in most hospitals. Listed in descending order of responsibility are director of nursing (or nursing superintendent), assistant director of nursing (assistant nursing superintendent), supervisor (departmental sister), charge nurse (ward sister, head nurse), senior staff nurse and staff nurse. The director of nursing may have one or more assistant directors and several supervisors to cover all areas of the hospital and shifts since the nursing service works and is staffed round the clock. There may be specialist supervisory staff or departmental sisters for paediatrics, maternity, psychiatry, operating rooms, CCU, ICUs, etc. The head nurse or ward sister who is next in the hierarchy is in charge of the ward or unit and is responsible for nursing in her respective unit or ward. There is generally a night supervisor and, in larger hospitals, a supervisor or assistant director in-charge of in-service education.

The head nurse is a key member of the nursing staff. Within the organized nursing care unit, she is responsible for the administration and coordination of patient care and other activities including preparation of nursing care plans, instruction of nurses, and supervision of personnel in the unit. She is responsible for the ward 24 hours of the day in the sense that personnel on evening and night shifts report to her and that she assigns duties to them. The head nurse must exercise good judgement, adhere to the policies, rules and regulations of the institution, and make sound decisions that result in good care of patients.

The night supervisor supervises and coordinates the activities of the nursing personnel during the night so that continuing care is maintained round the clock. She visits nursing units to oversee nursing care, ascertain the condition of patients, and give advice to nurses regarding treatment, medication and on any problem they may have. When she goes off duty, she informs supervisors of the subsequent shift of the patients' condition, and other matters of importance.

Education, training, experience and other qualifications required for various nursing service positions vary from institution to institution, but one can see some similarity in them. For example, the director of nursing may require an M Sc degree in nursing and 8 to 10 years of experience 5 to 6 of which at least in progressively responsible management positions. Similarly, the assistant director should have preferably an M Sc degree and 6 to 8 years of experience, the supervisor and head nurse be at least graduates with adequate experience, and the staff nurses in charge of special units must have, in addition, special training in their specialties, like operating room techniques, psychiatric nursing, CCU and ICU nursing care, CSSD, etc. The registered staff nurse, who may be a graduate or certificate nurse, has received a 3-year training in a recognized college or school of nursing and is registered under the Nursing Council.

There are other categories of staff in the nursing service.

* Auxiliary Nurse Midwife (ANM) is a staff member who has received a 2-year training in some aspects of nursing and has a certificate awarded by the Nursing Council. Some hospitals use them to relieve the staff nurses of some of the routine work.
* Unit manager is a professional manager with a college degree who supervises the administrative functions of the nursing unit.
* Nurse aide/orderly is a non–professional staff member who assists the nursing staff in various activities of the ward.
* Ward secretary and ward clerk are staff members who assist in routine clerical and paper work.

➤SPACE REQUIREMENTS

In the nursing service administration unit, provision should be made for the following:
* Nursing director's office
* Assistant director(s)' and supervisors' offices
* Reception control area with waiting area for visitors
* Secretarial and clerical work area
* Conference room
* Toilet facilities
* Storage room for active files, inactive files and office supplies.

➤OTHER CONSIDERATIONS

Staffing

Finding and retaining nursing staff in adequate numbers is a serious problem for many a hospital. To overcome this problem many hospitals establish their own schools of nursing which assuredly supply nurses regularly year after year. Many of the nurses have to execute a bond to serve the hospital for one or two years.

Salary is the most important reason for the high rate of turnover of nurses. By and large nurses are not paid adequately, and what they are paid is meagre when compared to prevailing salaries in the Gulf countries where many of them are attracted to find much greener pastures. Indian hospitals can in no way match Gulf salaries.

Salaries apart, there are other ways by which hospitals can increase their rate of nurse retention. The most important is by increasing their professional satisfaction. Some suggestions:

◆ The initial interview is important. Be honest. Tell the applicant what you can offer and what you cannot. False promises create disappointment and the nurse may quit after a little while.

◆ Plan the induction and orientation programme carefully. Make the transition smooth for nurses fresh from school by making the programme comprehensive and by providing intensive management skills and practical training. Remember, even an experienced nurse will be new to your set-up, surroundings, procedures and personnel.

◆ Evaluate the performance of new nurses and interview them at various times after employment. Discuss areas of weakness, concern or dissatisfaction. Show a positive attitude. The purpose of performance appraisal is to help the nurse to improve her job performance. At the same time, if she has done well, give recognition.

◆ Maintain good communication links through in-house newsletters, department meetings and unit level meetings. Involve all nurses in these activities. The director of nursing and her senior staff should be directly involved in these meetings.

◆ Organize educational programmes. Many nurses have a strong desire to learn and upgrade themselves.

◆ Provide promotional avenues for deserving nurses. Reward leadership qualities and clinical excellence. Nobody likes to stagnate in one position. Remember, for nurses more than for others, opportunities for horizontal mobility are as important as the usual vertical mobility.

◆ Follow a well-defined and strictly observed transfer policy which is fair. Use it for the purpose of professional growth and development.

◆ Make sure that your salaries and benefits are competitive.

Staffing Pattern

The complement of nurses required for a hospital, generally referred to in terms of nurse-bed ratio is a much debated and misunderstood subject. The Indian Nursing Council's stipulations are so high that not many hospitals adhere to them. Be that as it may, it is good to keep those stipulations in mind so that managements do not keep their hospitals grossly understaffed. Table 3.1 shows the nurse-bed ratio as stipulated by the Nursing Council.

For a 150-bed hospital:

◆ Nursing Superintendent — 1 (for minimum of 150 beds)
◆ Deputy Nursing Superintendent — 1
◆ Assistant Nursing Superintendent — 2
(for every additional 50 beds one more assistant nursing superintendent).

TABLE 3.1

Nursing Staff for Wards and Special Units
(Excluding Outpatient Department)

	Staff Nurse	Sister (each shift)	Departmental Sister/Assistant Nursing Superintendent
Medical Ward	1:3	1:25	1 for 3–4 wards
Surgical Ward	1:3	1:25	1 for 3–4 wards
Orthopaedic Ward	1:3	1:25	1 for 3–4 wards
Paediatric Ward	1:3	1:25	1 for 3–4 wards
Gynaecology Ward	1:3	1:25	1 for 3–4 wards
Maternity Ward (including newborns)	1:3	1:25	1 for 3–4 wards
Intensive Care Unit	1:1 (24 hours)	1	
Coronary Care Unit	1:1 (24 hours)	1	
Nephrology	1:1 (24 hours)	1	1 Departmental Sister/ Assistant Nursing Superintendent for 3–4 units clubbed together
Neurology and Neurosurgery	1:1 (24 hours)	1	
Special Wards — Eye, ENT, etc.	1:1 (24 hours)	1	
Operation Theatre	3 for 24 hours per table	1	1 Departmental Sister/Assistant Nursing Superintendent for 4–5 operating rooms
Casualty and Emergency Unit	2–3 Staff Nurses depending on the number of beds	1	1 Departmental Sister/Assistant Nursing Superintendent for Emergency, Casualty, etc.

Note

The Nursing Council recommends 30% leave reserve posts because a nurse is entitled to 30 days of earned leave, 10 to 12 days of casual leave and a certain number of days of sick leave and maternity leave, and in some hospitals, 24 days off duty in a year in addition to one or one and a half days of weekly off. It is observed that on any working day 25% of the nursing staff is off duty on casual leave, earned leave, etc. A nurse, therefore, works for about 240 days in a year whereas the hospital requires nursing service on 365 days of the year and 24 hours of each day which means that 30% more nurses are required on the rolls.

➤PROBLEM SITUATIONS

Fluctuating Census and Mix of Patients

Fluctuations in census and mix of patients needing various degrees and kinds of nursing care often pose problems to hospitals. To overcome these problems, hospitals in advanced countries adopt a system of classifying patients by grading them according to the amount of nursing time and skill they require. The patients are classified on the basis of acuity of illness and the care required such as minimal, partial, moderate and intensive, and, on the patient's part, his (her) capability to meet his (her) physical needs to ambulate, bathe and feed himself (herself). The classification is also used to determine the category of personnel — registered nurse, ANM, or nurse aide — who should provide the required care.

- ◆ Category I — A patient who requires only minimal amount of nursing care — an average of 2.8 hours of nursing per 24 hours.
- ◆ Category II — A patient who requires an average amount of nursing care — an average of 4.3 hours of nursing per 24 hours.
- ◆ Category III — A patient who requires above average nursing care — an average of 5.8 hours of nursing per 24 hours.
- ◆ Category IV — A patient who requires maximum nursing care — a average of 8.6 hours of nursing per 24 hours.

Examples of what the patient can do himself (herself) and what is required to be done for him(her) are listed under each category.

➤NURSING SERVICE AND COMPUTERS

Computers are being increasingly used in hospitals and even in our country they are making steady inroads into patient management and patient care areas. The cost of information handling by the nurse in terms of the time spent by her on this activity is high, and more computerization means more time spent on it. This has become a major problem preventing her from caring for the patient. Increased nursing skill is accompanied by increased information handling. Nurses' notes are the single most time consuming written communication.

The change from a manual record system to a computer-based record system requires good planning and thorough orientation and training for nursing personnel. Those who may have come from different backgrounds — little work experience, rural upbringing, no prior computer contact — may be reluctant or negatively inclined toward computer applications. They need to be reassured and given thorough training.

Among the functions that nurses may be required to perform are — admitting the patient into the computer system, daily entering of clinical information, entering doctors' orders, changing previously entered data, discharging the patient, etc. They would also have to learn how to access and manipulate information.

When computerization is considered, some basic questions which are pertinent to the nursing service should be addressed. Some of them are:

- ♦ What would be the feasibility and cost of including clinical information in a computer–assisted system?
- ♦ What orientation and training programmes would be needed to implement the system?
- ♦ Who would input the data?
- ♦ If a non-nurse inputs clinical data, should it be checked or validated by the nurse?
- ♦ What information should be available and to whom? What about confidentiality of patient information?
- ♦ What information should be temporarily stored, and what information permanently stored on line or by other means?
- ♦ What is the value of information as far as patient care is concerned, and for decision making?
- ♦ What information needs to be retained, for how long, and for what purpose?
- ♦ What information would be retained as nurse input?
- ♦ What steps should be taken:
  - ♦ to facilitate nurses' acceptance of computerization and data processing?
  - ♦ to minimize or counter nurses' dissatisfaction or stress arising out of role conflict?

HUMAN RESOURCES DEPARTMENT

➤ OVERVIEW

A sizeable percentage of a hospital's operational expenses is used up by the payroll. Besides, employers have come to recognize that employees are the organization's most valuable asset, but the most difficult one to handle as they can make or mar the institution. Consequently, in the recent times, the department of human resources has received a greater degree of attention than ever before by health care management. One significant offshoot of this growing recognition is the trend to designate what was once called the department of personnel management as the human resources department concomitantly with a higher status and a larger role assigned to it. This is reflected in many institutions elevating the chief of personnel to the rank of director or vice-president of human resources or as associate administrator. The fact that in a large and well-organized modern hospital there may be as many as 300 different jobs is indicative of the complexity and importance of the department's functions unlike they are in any other organization.

The position of the human resources department in the organization is one of staff relationship with the management. It also functions as a service department for the employees. In the former role, it assists the various department heads in their responsibilities as far as their personnel are concerned. It assists the hospital's management with a variety of activities such as employing new staff, determining their pay, evaluating performance, interpreting policies, initiating disciplinary procedures, handling grievances and employee relations, and negotiating with labour unions.

➤FUNCTIONS

By and large the human resources department has four major areas of responsibilities: employment, compensation, benefit administration, and labour relations. A myriad of specialized functions branch from these four major areas. Some of them are:

- Employment: the department screens, interviews, administers tests where necessary, and refers potential candidates to hiring managers or heads of departments. It conducts reference checks and sets up pre-employment physical for those considered for employment.
- Inducts and orients new employees: New employees receive orientation to the hospital, their departments and their jobs.
- Wages and Salary Administration: This consists of the development of individual job descriptions. Positions are evaluated using a pre-determined group of factors such as education, skill and experience needed, responsibilities of the job, working conditions, etc. Salary scales are fixed using a point value. The department is responsible for establishing and maintaining an equitable and competitive salary programme that will help in attracting and retaining competent employees.
- Benefit Administration: In addition to compensation, hospital employees receive fringe benefits in the form of hospitalization benefits, contributions to provident fund, gratuity and retirement benefits, subsidized housing, travel allowance, etc. which have become a major expense for many hospitals. The department administers these benefits.
- Conducts or organizes training and in-service programmes.
- As management's representative, serves as the chief negotiator for the hospital in union disputes or assists in the preparation of papers and materials for professional external negotiator or labour lawyer.
- Keeps complete personnel files on all employees in addition to master employee file, employee history record, master position control file, master personnel policy file, termination information file, job evaluation file, follow-up reminder file, etc.
- Maintains organizational charts and staffing patterns.
- Develops procedures to control absenteeism, turnover and sick leave.
- Position control administration: Assigns code numbers for all authorized positions and verifies vacancies before they are filled.

- Prepares manpower planning, assists in the preparation of departmental budget and in the establishment of employees' budgetary control.
- Develops personnel polices for approval by the management or governing board.
- Is responsible for employee and labour relations in the hospital, watches union activities or drive for unionization.
- Establishes health and safety programmes. Oversees employees' health programmes.

➤LOCATION

The human resources department generates relatively light to moderate traffic within the hospital with respect to the hospital personnel. However, there is heavy traffic from job applicants. A street level entrance to the department is recommended so that the department is easily accessible to the job applicants. Such an arrangement will also eliminate the stream of these applicants interfering with the regular hospital traffic.

Because of the close working relationship between the director of human resources department and the CEO under whose direction the former works, the department should be in close proximity to the executive suite. It should also be centred in the administrative block with convenient access to payroll records of the financial service unit but adjacent to a waiting area directly accessible from the outside to accommodate job applicants.

➤DESIGN

As the human resources department is usually the first impression of the hospital that the prospective employees receive, it must be carefully planned and designed. The offices, reception and waiting area, their decor and furnishings should be pleasing. They should project a positive image of the hospital and its philosophy.

➤ORGANIZATION

The organizational structure of the human resources department varies depending on many factors, largely the size and sophistication of the hospital and the department. In small hospitals, the director may have one or two clerical assistants and may perform just the traditional functions of the department; or, in addition to being responsible for employee matters, he may be given other responsibilities like public relations. In others, the functions of this department may be combined with other administrative responsibilities or handled by the CEO himself.

The director of human resources reports directly to the hospital CEO. In large hospitals, he may have an assistant director, and managers placed in charge of major areas like wage and salary administration, employment, employee relations and training. These positions in turn have specialists under them such as job analysts, interviewers, benefit analysts and instructors. There are also personnel assistants, secretaries, clerks, receptionists, etc.

The organization of the department revolves around its most important function of finding and retaining qualified and competent personnel to staff the various sections of the hospital. However, in most hospitals, the nursing services department is responsible for recruiting its own staff.

The term recruitment in personnel management refers to the process of finding potential applicants for employment. It is the process of locating and attracting applicants for various positions. This is the primary task of the human resources department. It does not select. "Selection" means choosing the person best qualified for a specific job. It is only when an adequate number of well-qualified applicants are available that a good candidate can be selected. The department sets the stage for selection which is done by the heads of departments or selection committees.

➤SPACE REQUIREMENTS

Space is required for
- ◆ Director's office
- ◆ Assistant director's office
- ◆ Employment manager's office
- ◆ Employee relations manager's office
- ◆ Training manager's office
- ◆ Space for analysts, interviewers, instructors, etc.
- ◆ Secretarial and clerical area
- ◆ Waiting area with reception-control and tables or counter tops for applicants
- ◆ Testing area for written tests, typing tests, etc. and for making ID cards
- ◆ Multipurpose room for orientation programmes, lectures, demonstrations, training classes, etc. Other departments like nursing services, finance department, volunteer service, public relations and purchase department (for demonstration of new products) can use it. There should be a door to this room from the corridor.
- ◆ Interviewing area which provides one or more offices or cubicles for the interviewers
- ◆ Storage room for audio-visual equipment, extra furniture, equipment, portable black board, speaker's podium, projection stand, etc.
- ◆ Storage room for supplies, stationery, etc.
- ◆ Adequate space for file cabinets for personnel files, office files, etc. Most of these files are of a confidential nature.

➤OTHER CONSIDERATIONS

Director of Personnel

Most of our personnel managers come from an industrial background with a strong industry orientation. They come with such a fixation, rather a preoccupation with the affairs of the union that they fail to pay adequate attention to other functions of human resources development. Too often they are out of tune with the work, philosophy and mission of the hospital, and there are not many who feel comfortable working in hospitals with lesser pay and fewer perquisites. The CEO and the governing board must remember this when they select their director of human resources.

Computerization

Computers are being increasingly used in the human resources department. A new hospital must take advantage of this facility from the beginning. A variety of employee-related information can be maintained on the computer — personal data of employees, attendance, absenteeism, tardiness, payroll information — the possibilities are endless. The recent addition to the list is a computerized time punching machine using a magnetic ID card which in addition to registering the time of arrival and departure, provides complete and advanced time attendance information such as attendance, absentee report and statistics, leave statement, overtime calculations, data for payroll, and many more functions instantaneously. This will eliminate the cumbersome exercise of manual data entry for pay–roll applications.

Accessibility of Information

Except for the personnel director and the administration, information about individual employees is restricted and is available to managers only on a strict need-based basis. Managers can have information only on their own employees, that too statistical and not unlimited personal information.

PUBLIC RELATIONS DEPARTMENT

➤OVERVIEW

Public relations is a relatively new service in hospitals. Nevertheless, its rapid growth in recent years and the fact that many hospitals now have public relations programmes and public relations officers on their staff indicate that it has earned an important and permanent place in hospitals.

The importance of and the need for public relations in hospitals can be appreciated when one considers some of the problems today's hospitals have to contend with — high cost of medical care and the growing public criticism of hospitals, problem of delivering quality care at affordable cost, need for efficient and professional management and increasing involvement of governmental agencies and consumer protection forums in patient care and internal management of hospitals. Different strata of the public — community, employees, medical staff, patients, visitors, etc. — form their opinions about the hospital according to the source of their information. These opinions can be influenced by a good public relations programme.

A good public relations programme is essential both inside and outside the hospital — inside the hospital to maintain dedicated staff who will provide warm and personal service to patients, and outside, to communicate the activities of the hospital to the people and to interpret people's perception of the hospital and its policies to the hospital's management.

Health care is changing at an alarming pace; with it, the role of public relations in hospitals is changing too. There has been an explosion in technology that has revolutionized the practice of public relations in terms of its acceptance and sophistication. Many public relations directors who remained insulated in their self-contained hospitals performing the traditional public relations activities such as the publication of house journals and functioning as hospital's media agents, have suddenly found themselves ill-equipped and unprepared to face the challenges of the new concepts of marketing, positioning and advertising that have been introduced in public relations. In some hospitals the role of public relations has been superseded by the role of marketing communications; in others, it has been relegated to the background by marketing. Whether public relations practitioners have adapted to the challenges facing their profession or not, nobody denies that all communicators need basic understanding of public relations, its tools and marketing strategy if they are to succeed in today's competitive world. In India, things have changed a great deal in the so-called corporate hospitals, and in others changes are in the offing.

To be effective, public relations must depend on the governing board's appreciation of the special public relations needs of the hospital as well as on the support of all departments. By and large, this latter support is readily forthcoming because public relations cuts across organizational lines, and managers are more than willing to do in the interest of the hospital what they normally would not do for other departments.

In recent times, there has been a pronounced trend toward the CEO of the hospital to assume an expanded public relations role. Consequently, CEOs are seen spending more time on public relations functions. These activities include public speaking, dealing directly with the news media, lobbying in favour of or against issues that affect health care and hospitals, dealing with consumer protection forums, promoting industrial or occupational medicine by meeting corporate bosses, and cultivating community and top opinion leaders. The public relations director, of course, arranges these meetings and prepares promotional materials for the CEO's use.

The major responsibilities of the public relations department are interpreting, advising, marketing and communicating. To carry out these responsibilities effectively, the public relations director should be fully informed of everything that goes on in the hospital. In other words, he should be in the inner circle so that he has direct access to information as to what goes on there. He should be a member of the top management team and should attend meetings of the governing board. This requires that the director is a person of the highest personal integrity, stature, judgement and personal discretion.

Public opinions about a hospital are formed first and foremost on the personal experience of patients and their families, then on the opinions of the staff and their families, former patients and their families, persons who have first-hand information about the hospital, visitors, and only then on information derived from other sources. This primacy of the individual as a source for opinions has a profound influence on the hospital's public relations. It underscores the importance of a sound internal relations programme as the basis for all external public relations.

It is rightly said that employees are the first line of public relations. It is the responsibility of the management that all employees are made aware and are constantly reminded that the hospital is judged by their actions, appearance and the quality of service they render. Outside the hospital, they are the representatives of the hospital and by their actions and words, they can make or break the hospital. It is important to understand, therefore, that good hospital public relations begins with the employees, and that it is a product of positive employee relations. Progressive personnel policies which lay emphasis on employee–oriented working conditions, an attractive environment, adequate pay scales, caring supervision and an enlightened management result in high morale, productivity, internal harmony and motivated, loyal and contented employees who take pride in their organization.

From another angle, it is necessary to ensure that all employees know their hospital. A well-informed employee is the best public relations representative of the hospital. Providing a good orientation programme and keeping the employees continuously informed of all aspects of the hospital are the combined responsibility of the human resources and public relations departments.

►FUNCTIONS

It is the responsibility of the public relations department

- ◆ to interpret to the management the different viewpoints and attitudes of various strata of the public which it has identified, and recommend actions to effectively solve the problems arising from attitudinal changes on issues relating to the hospital;
- ◆ to study the actions and activities of the management and the impact they may have on the public's perception of the hospital, appraise the management of any harmful effect and recommend suitable action to avoid it;
- ◆ to gather and analyse data on an on-going basis to improve communications between the hospital and the public;

- to assist the institution to secure support and guidance from the public in the development of the hospital and, to that end, help in conducting market surveys;

- to develop communication materials such as internal and external newsletters, publications, audio-visuals, media releases consisting of materials on hospital services, health topics and human interest stories to promote the hospital's goals and objectives, and direct them to reach specific target audiences;

- to establish channels of communications between the hospital's public and the management team;

- to help in raising funds for the support of the hospital;

- to develop and maintain good relations with the media and communicate with the press as authorized by the hospital's CEO, always keeping in mind the sound principle that public relations practitioners should be open and honest with the press;

- to organize hospital's speakers' bureau consisting of the CEO, senior members of the staff including medical staff, and governing board members to speak to civic groups, clubs and organizations in an effort to promote the hospital; and

- to participate in community affairs that have a bearing on the well-being of the hospital, be member of service organizations such as Rotary, Lions or other groups, and become respected community leaders.

➤LOCATION

The office of the public relations director should be located in close proximity to the office of the CEO. Proximity to the CEO's office is necessary because generally everything that is of significance emanating from the public relations office requires the approval of the CEO. It also helps the public relations director to have personal consultations with the CEO and accomplish things speedily. In addition, many of the activities of the department such as planning events and programmes, news releases, etc. must be done in cooperation with the CEO and sometimes other senior administrative officers. The department is subject to a minimum amount of personnel traffic.

➤ORGANIZATION

The public relations department is supervised by a director of public relations or a public relations officer (PRO) as he is generally called. He reports directly to the CEO. In smaller hospitals, the functions of this department may be combined with other administrative responsibilities or handled by the CEO himself. The size of the department depends on the size of the hospital and such other factors as sophistication and extent of departmental activities.

➤ SPACE REQUIREMENTS

Space and facilities are required for:
- ◆ Office of the director of public relations
- ◆ Secretarial and clerical area
- ◆ Office of the copywriter who writes the copy of advertising and promotional materials, brochures, and audio-visual materials.
- ◆ Workshop or work area for preparing layouts and montages of photographs, films and display materials. Space is also required for storage of files, photos, brochures and clippings.

➤ OTHER CONSIDERATIONS

Hospital and News Media Relations

News media relations is one of the most important roles of the hospital's public relations. Hospital news media relations is built on credibility, professionalism and mutual trust. This is not accomplished in a day. The hospital and the media must respect and appreciate each other's role. News media has the responsibility of disseminating news of public interest to its readers, viewers and listeners. In this it expects the hospital to serve as a news source — news of VIP patients, for example. On the other hand, the hospital has the primary moral and legal responsibility to protect the right of the patient to privacy in addition to his right to proper care. Balancing these two commitments, hospitals generally use such standard terms as good, fair, serious and critical to describe a patient's condition.

Every hospital must develop and enforce an official policy for dealing with the press and the kinds of information that may be released. The policy should specifically state who speaks for the hospital, whether the CEO or the public relations officer, or any other person, and in their absence as during the night when generally accident victims are admitted, who may be called to make a statement to the press.

Good news media relations is built on mutual understanding and personal contact with the media established over a period of time. It is good to follow certain sound principles in dealing with the press. The important ones are:
- ◆ Be honest — one lie or misleading statement might destroy the credibility of the department and the hospital.
- ◆ Be accurate — figures, statements, etc. should be absolutely correct.
- ◆ Be concise and professional — chances of stories, etc. being accepted are good when the material given to the press is prepared in good journalistic style and format or professionally prepared.
- ◆ Be appreciative — A thank you note or telephone call to the editor who published the hospital's story or write-up will be greatly appreciated and will be rewarding.

Disaster Preparedness Planning

When faced with an external or internal disaster, the hospital's ability to respond is directly proportional to its preparedness planning. Unfortunately, the hospital's preparedness to deal with disaster is too often inadequate. Public relations has a big role in dealing with disasters. The time to think as to how to respond to a disaster is before it strikes.

Every progressive hospital must establish a disaster plan for the hospital stating who does what. There should be simulated drills to train the staff in this exercise. The public relations must prepare its own disaster plan and develop a checklist which spells out what the public relations director and his staff should do during a disaster. This topic is discussed again in detail in a later chapter.

Promotional Tools

The following are some of the promotional tools that public relations can use to promote the hospital.

- ◆ Brochure
- ◆ News releases
- ◆ Internal newsletter
- ◆ External newsletter
- ◆ Public service announcements
- ◆ Promotional materials by direct mail
- ◆ Displays on buses, etc.
- ◆ Health care and educational seminars, workshops, conferences, health fairs, exhibitions
- ◆ Videotape capabilities, programmes
- ◆ Press conferences
- ◆ Bulletin boards, information racks
- ◆ Hoardings
- ◆ Hospital speakers' bureau
- ◆ Feature stories
- ◆ Insertions in newspapers
- ◆ Open houses
- ◆ Posters
- ◆ Hospital's public address system
- ◆ Hospital day, founders' day celebrations
- ◆ Annual reports
- ◆ Signs and graphics
- ◆ Hand-outs and flyers (small advertising leaflets that ate widely distributed).

MARKETING AND GUEST RELATIONS

➤INTRODUCTION

When this book was still in the draft form, some well-meaning friends of the authors who browsed through the book expressed disappointment at what they thought was a serious omission of not including in it such an important topic as marketing which has assumed great importance in the operation of modern hospitals. The omission was intentional at that time, not because we thought it fit to relegate marketing to a secondary position but because we believe that there are other programmes which are equally if not more effective and much less expensive than marketing that even the charitable hospitals and other not-for-profit organizations whose sensibilities are generally hurt by marketing can put into practice. An effective public relations programme coupled with a well-planned guest relations programme is one such. By and by, we veered round to the idea that some discussion on marketing might be useful. So in deference to the wishes of those friends and for the benefit of those who want to introduce marketing programmes in their hospitals, we have subsequently included a brief disscussion on marketing. We have included a section on marketing and guest relations which most of our hospitals can practise with great benefit and without much cost.

MARKETING

Marketing has a useful role in the operation of hospitals of all kinds. Marketing is no longer the dirty word that it used to be. As a matter of fact, hospital marketing is slowly but surely coming of age even in our country and is being woven into the fabric of hospital's planning and public relations programmes. Much of the negative attitude towards hospital marketing stems from the salesman's image it conjures up in the minds of people, particularly among the physicians. Marketing has also the image of being primarily for profit-making which is the life-line of business and commerce. At the same time, the concept of marketing in the hospital set-up is a grossly misunderstood idea. Marketing is not selling as most people are inclined to believe, and certainly it does not involve creating a demand that is not needed. It is seeing what the public perceives as its needs, having the hospital identify the services it is capable of delivering and then developing those services. In other words, when a hospital embarks on a marketing programme, it seeks to identify the needs of present and prospective patients, and tries to meet those needs effectively. It then seeks to provide information about the special services it has to offer and promotes them to attract people. Advertising is one of the methods that contributes to this task.

Today's health care market has become consumer driven. Patients are better informed and know more about health and medical services. They ask for high quality care and, having become cost conscious, demand quality care at a reasonable price. They personally shop for and choose their hospitals and take responsibility for their judgement and decisions. This was done by others for them earlier.

On their part, hospitals are developing and instituting new consumer-oriented programmes that need to be promoted to the public. Quite often, people are ignorant of what special services are available in hospitals. They need to be informed of this. A more educated consumer wants to be informed about medical procedures, wants to know what alternatives he has and wants to be involved in the medical care process.

It should be remembered, however that all the lofty ideas about marketing in hospitals that we have neatly packaged and presented here are not so simple and not so laudable in real practice as we have made out. In the Western countries where almost every hospital has a full-fledged marketing department, there is growing concern that the industry is going too far and that what the hospitals are really concerned is to offer what sells rather than what is needed, and that business strategy, not quality care, is the overriding consideration in marketing. There are hospitals in which advertising is carried out not differently than what is done in trade and commerce; it has often degenerated to selling without truthfully revealing the relevant facts.

Earlier, health professionals scorned the very idea of marketing. In countries like America if a licensed member of certain professions — medicine is one of them — engaged himself in any explicit marketing, solicitation or advertising in pursuit of his clients, he ran the risk of losing his license. As for these professionals, they were proud of their value–based professional practice and excellence and felt that they were above "selling". They were simply available to those who needed their services. Things have changed,. In some cases, the pendulum has moved to ridiculous limits in advertising. As someone humorously said, one day we might see an advertisement like this : "This week's special — kidney transplant — only $3000." Between the two extremes of this kind of advertisement and the Puritanical administrator who said, "I would rather use prayer than marketing," there are various shades of marketing.

In the final analysis when a hospital practises "healthy" marketing, the patient should be the beneficiary. the campaign should result in market research that analyses the community and its health needs. Based on this, the hospitals should rearrange their services to meet the needs of the people. Only then will they be in a position to take their message to the public and tell them who they are and what they offer. Marketing also lays increased emphasis on health care education and fitness programmes. An informed consumer will be a better health care buyer. The majority of people who read educational and promotional materials may not be in immediate need of hospital care. The context of overall well-being is the best time to drive home the hospital's message to the people, in that they have time to check out hospitals before they actually need to check into one.

There is a misconception in the minds of people that marketing is advertising. This is not true. Advertising is only one aspect of marketing. It can also be expressed as an added dimension of public relations. Public relations people, they say, gave birth to marketing in the health care industry. Although an important aspect of marketing, advertising does not bring patients to hospitals if other necessary ingredients are absent.

There are two kinds of advertising — image building advertising and product advertising. Opinions as to which of these two is more productive vary greatly. Image building advertising is used for what in public relations they refer to as positioning the hospital. It identifies and reinforces the market position of the hospital by projecting a good image of the organization. It fosters patients' trust in the hospital so that when they really need hospitalization, they readily go to the hospital of their choice.

Product advertising, as the name indicates, is promoting the hospital's services by extolling its merits. Hospitals need both kinds of advertising, image building advertising particularly when a new hospital is opening its doors to the general public or when a new programme or service is being introduced. First of all they must create a place for themselves, carve out a niche as it were, and then show to the people what services they offer.

Public relations and advertising are both essential to successful marketing. Public relations builds up goodwill and an image to influence opinions; advertising generates a message which will lead consumers to choose a hospital or its services.

Advertising should have a specific message which should be exciting and appealing to the people's emotions. Repetition of the message in multiple media ensures better chances of its being remembered and understood. Just to say "we are a great hospital" or "we are health care people who care about you", or "our clients are our best references" or talking of the organization's hoary past will not have any impact on the target audience unless the services offered have integrity. If the services offered are just ordinary, patients will really spread the word as to what kind of service they received.

Hospitals should remember that their public relations departments are not skilled and equipped to handle advertising jobs. That should be best left to a competent advertising agency which has the expertise to translate ideas into messages and to select proper media. The public relations department would provide the agency the necessary data, information and insight into the hospital's philosophy, services and products. The agency, with its rich experience and the resources at its command, translates them into a powerful message which will influence the purchasing behaviour of people. The services must, however, be of high quality for the advertising effort to make a powerful impact on the target audience. Two factors which will yield good results are quality products or services and marketing communications.

Marketing efforts will not succeed unless there is a commitment on the part of the top management to change existing programmes, services and, if necessary, its philosophy to be in consonance with the needs of the people it serves.

GUEST RELATIONS

Traditionally, hospitals were judged largely by the quality of medical and professional services they provided, and so their marketing strategy was focused on that aspect. Today that is not enough; people judge hospitals also by the quality of "people" service they get, that is, personalized service rendered by courteous, caring and friendly people. To patients, service is tantamount to quality. Informed patients now shop around for hospitals where they can receive such quality care and personalized service. So more and more hospitals institute guest relations programmes in their hospitals aimed at "humanizing" the hospital stay for their patients. Hospitals believe that these programmes influence patients' decisions as to where to go for their treatment, and help to position their hospitals as an institution of the people's choice.

The concept of guest relations, borrowed from the practice of guest relations in first class hotels, is based on the premise that every patient is a guest of the hospital and should be treated no differently from how he is treated in a star hotel or in our homes for that matter. With the growing interest in guest relations across the country and elsewhere, many hospitals are paying close attention to the hotel industry's strategies in pampering their customers. Hospitals, unlike hotels, can be intimidating to both the patients and their family members. Often, patients are afraid and act unnaturally, and their relatives and friends are worried and distraught. The first-time patients do not know where to go and in the unfamiliar surrounding, many of them are confused. So a special kind of care is required in hospitals. For this, members of the staff who are already technically competent to provide quality care to the patients, should be additionally trained in aggressive guest relations. They must be taught the art of conversation, how to communicate with and read body language, how to handle guest and employee encounters, how to deal with problem guests even under provocation, how to handle stress, and lastly how to please the guests. The underlying principle of guest relations is the ability of the staff to greet guests and to carry on pleasant conversations with a smile showing sincere concern for the patients through proper service and adequate attention, and a supportive management to back the staff in their work.

For the programme to be successful, the guest relations culture should pervade the whole hospital, and there should be a commitment to the programme on the part of all staff — from the top management to the last of the employees. The following are some of its components.

- *Person to person training*: Every one of the staff is introduced to the guest relations programme and trained in all its aspects.
- *Removing organizational obstructions to good guest relations*: There are many things in a hospital set-up which act as blocks to good patient relations, such as the interminable and tedious

waiting the patients have to do at several points, complex procedures, slow registration and admitting procedures with unnecessary and cumbersome forms to fill, slow and inefficient staff, non-availability of medical staff, lack of signage system particularly in large hospitals where patients lose their way — to name just a few.

The staff will identify these problems and prepare a plan of action to remedy them.

◆ *Special programmes and services*: Under this, special programmes and services are developed to give additional services not provided through the ongoing programmes. Examples are the centralized patient service centre and the volunteer movement.

◆ *Evaluation and feedback*: Patient care and patient satisfaction are evaluated regularly and services are improved in the light of these evaluation reports. Suggestion boxes and questionnaires to patients eliciting their opinions about the services rendered can be used as a part of this campaign. (See Patient Opinion Poll Questionnaire at the end of this section.)

◆ *Communication and promotion*: Promote hospital services and the concept of guest relations among the external and internal public through internal and external newsletters, marketing communications and by means of various other promotional tools which we have discussed under public relations. All these publications should show that the guest relations programme is an adjunct to high quality patient care.

◆ In addition to these components of the guest relations programme, the staff should be told that the patient is the most important person for the survival of the hospital. They should be constantly reminded that the hospital is judged by the kind of service they render and that their livelihood depends on it. The patients do us a favour when they come in. We are not doing them any favour by waiting on them and pleasing them. Lastly, patients come to us with their needs and their wants: it is our job to meet them.

◆ In some hospitals performance evaluation includes a review of employees' behaviour towards the patients, how they greet the patients both in person and on the phone, how they handle patients' complaints, and so on. Merit pay and promotion are based on employees' guest relations skills.

Body Language

Body language consisting of facial expressions, hand and shoulder gestures and body postures has a big role in the guest relations programme. It markedly influences the hospital–patient relationship. These expressions can be positive, friendly or negative. Positive body language like a smile or friendly gesture can build up an immediate rapport with the patient who will place his trust in the person with whom he is dealing. At the same time, body language enhances the intended meaning of a message, or it can distract from or totally distort the intended meaning. Consider the following scenario: A patient approaches an employee and seeks information about a certain procedure in family planning and is told "I think you should find that out from the concerned department." With one set of facial expressions, tone and gestures, this reply could mean, "I do not know the procedure. It would be better if you went to that department. Someone there would be glad to explain it to you."

PLATE 7

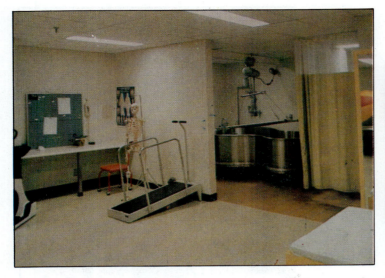

➤ **PICTURE 38**   Physical Therapy Department: A part of the gymnasium with ramp with handrails for exercise. At the right hand side is a Hubbard Tank

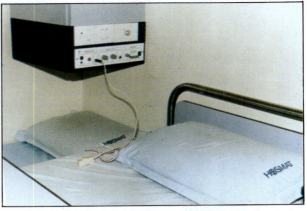

➤ **PICTURE 42**   Nurse Call System with detachable bedside cord

➤ **PICTURE 43**   Nurse Call System Bedside Integrated Panel

➤ **PICTURE 44**   Nursing Station with Colour-Coded Directional Sign

PLATE 8

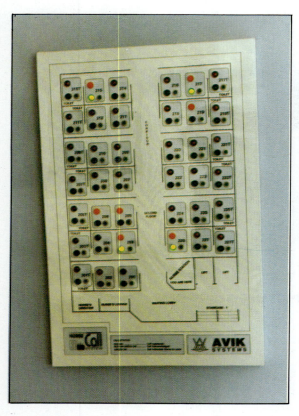

➢ **PICTURE 45** Nurse Call System: Nurse Station main panel

➢ **PICTURE 50** Paediatric Play Room area

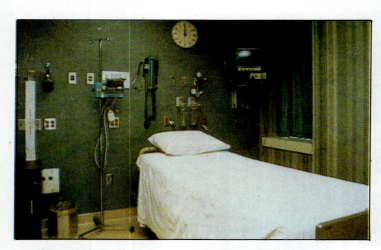

➢ **PICTURE 51** View of an Intensive Care Unit

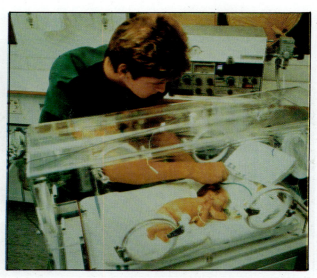

➢ **PICTURE 55** A sophisticated Ventilator that can be used for any application is respiratory care of premature infants. Used in the operating rooms and intensive care units

With a different set of facial expressions, tone and gestures, for example, a negative attitude with a frown on the face, this could mean, "Why are you bothering me with this matter? How would I know? I am not a family planning expert". On the other hand, a deadpan (expressionless) look and some meaningless gesture could leave the patient bewildered not knowing what the message is.

Body Postures Body postures are also important. Shoulders should be square and ahead of the guest. Imagine a typist–receptionist who does not face the guest but sits in some other direction, and then turns her head to the guest and speaks to him; worse still, answers the guest's questions even without turning her head and continues typing. What kind of a signal will this lack of concern or casual attitude send to a sick person? The correct body postures for the receptionist is to face the guest with her body squarely and totally. Employees should be told and trained in these simple but important rules which ensure effective guest relations. Other tips are: Be self–confident. Do not be nervous, be relaxed. When you are nervous, shoulders become stiff and your body becomes stiff. When you relax, your guest also relaxes.

Use of Names Use a persons's name as soon as you have heard or read it, and continue using it during your conversation with him. You will find that this will make a marked difference in building up cordial relations with patients. Conversely, forgetting the names of people whom you know — a repeat patient, for example — can be a source of great embarrassment to you, even a liability. A person's name is important to him; it is all that distinguishes him from other people. Nobody likes to be addressed as a nameless person. On the other hand, addressing him by his name gives a personal touch to your conversation — your tone can make it even more personal — and creates a warm and friendly feeling.

Patient Contact and First Impression Patient contact and the first impression are crucial. Remember the following rules:
- Make eye contact with the patient and maintain it throughout your encounter with the patient.
- Greet the patient in a friendly manner.
- Ask an open–ended question such as "How may I help you, sir?" or "How may I be of help to you?"
- Use the person's name as soon as he introduces himself or his name is written, and use it often after that.
- Introduce yourself to the patient and tell him your title.
- Sense the person's feelings and show empathy.
- Do not criticize him or find fault with him.
- Be courteous and friendly in speaking.
- Use correct body language, facial expressions and gestures that will demonstrate your friendly nature.
- Remember to keep the patient's case and contents of his medical records relating to his illness and care confidential. Do not discuss a patient and his illness in public places such as elevators and corridors.

◆ Tell the patient what care and services to expect from the hospital. Encourage him to ask questions about his illness and the hospital services available.

Telephone Courtesy

Good telephone etiquette enhances the effectiveness and value of public and guest relations. Know and include it in the guest relations programme for all staff. The telephone operator or whoever receives the calls can be the number one public relations officer of the hospital. Since she is the first contact that outside callers have with the hospital, the manner in which she handles the calls is important.

The hospital is full of emergencies and life-threatening situations; so every call may be a patient's life-line on which his life depends. The telephone is not only a public and guest relations instrument, it can also be a life-saving device. Needless to say that the operator should be highly professional, sensitive, and resourceful as well as prompt, courteous, kind and respectful. Some rules:

◆ Make a prompt connection; answer the telephone within three rings.

◆ Speak with a pleasant voice — a voice with a smile.

◆ A prompt connection and a pleasant voice will have an important psychological effect on the caller. They will make the caller feel immediately that he is dealing with an efficient and dynamic organization.

◆ Do not give the impression that you are bored or tired.

◆ Greet the caller as appropriate to the time of the day and give the name of the hospital — "Good morning, Cauvery Medical Center."

◆ Be courteous, speak gently, distinctly, and slowly. Your mood and attitude should be cheerful.

◆ Listen attentively to avoid the necessity of having to ask the caller to repeat.

◆ Do not use time-consuming and worn-out words such as "Hello", "Yes", "Okay", "Yeah", "Hold on", "See ya", etc. Identify yourself or the department where you are working.

◆ Screen the calls for your boss or officers with the phrase, "May I know who's calling?"

◆ Use the caller's name if you know it.

◆ Do not eat, drink, smoke or chew gum, tobacco or betel leaves while you are talking on the phone.

◆ When you have completed the conversation, replace the phone on the receiver gently.

◆ If the caller says, "Thank You," answer "You're welcome," not "Yes" or "All right."

◆ Know your hospital. Most of the time 'the calls you receive are in the nature of enquiry — about the hospital services, doctors, patient care, etc. or the calls may be for others. As a guest relations officer of the hospital, you should be familiar with and know everything about the hospital, its activities and personnel. It will make bad public relation if you plead ignorance about these things or start checking them, keeping the caller waiting.

Telephone etiquette is for everyone in the hospital, from the chief of the hospital to the last of the employees. Plenty of books are available which deal with all aspects of telephone manners as applicable to all categories of staff such as telephone operators, secretaries, department employees, officers and the boss.

PATIENT OPINION POLL QUESTIONNAIRE

This is a sample for illustrative purpose only.

ADMISSION

Were you admitted to the hospital via:

Regular admitting	☐	Emergency room	☐
Transfer from another hospital	☐	Other	☐

How much time did the admitting process take from the time of entry into the hospital until admission to your room?

About 30 minutes	☐	60 minutes	☐
1–2 hours	☐	2–4 hours	☐

Did you need assistance to find the admitting office?

Yes	☐	No	☐

Was assistance given?

Yes	☐	No	☐

At the time of admission, were you and/or your family asked for information which would help to plan your care?

Yes	☐	No	☐

Comments:

_____

_____

COMFORT

Was your room clean and comfortable?

Yes ☐ No ☐

Were your room and toilet cleaned daily to your satisfaction?

Yes ☐ No ☐

Was the atmosphere restful and relaxing?

Yes ☐ No ☐

Was all of your equipment (bedrails, nurse call system, TV, bed controls, taps, fixtures, etc.) in good working order?

Yes ☐ No ☐

Was your nurse call answered within a reasonable time?

Yes ☐ No ☐

Were you disturbed by noise during your hospital stay?

Yes ☐ No ☐

If yes, what was the source of noise or disturbance?

Roommate ☐ Guests/Visitors ☐
Hospital Personnel ☐ TV ☐
Equipment ☐

During what time of the day

6–9 am ☐ 9 am–noon ☐ 1–3 pm ☐
3–6 pm ☐ 6–9 pm ☐ 10 pm–6 am ☐

Comments:

_____

_____

FOOD SERVICE

Were you on a special diet?

Yes ☐ No ☐

If yes, please check the appropriate one:

Low sodium	☐	Low fat/low cholesterol	☐
Calorie control	☐	Cardiac diet	☐
High fibre	☐	Low residue	☐
Bland	☐	Other _____	

If you were on a special diet, did you receive an explanation on what to expect?

Yes ☐ No ☐

Was your food from the hospital kitchen or from home?

Hospital kitchen ☐ Home ☐

If from the hospital kitchen, was it tasty and attractively served?

Yes ☐ No ☐

Was it served hot and in time?

Yes ☐ No ☐

Were food trays delivered and removed within a reasonable time period?

Yes ☐ No ☐

Comments:

_____

_____

BUSINESS OFFICE

Did you need to see the finance officer?

Yes ☐ No ☐

Was the billing procedure explained to you at the time of admission?

Yes ☐ No ☐

Did you ask for an estimate of your hospital expenses?

Yes ☐ No ☐

Did you receive it?

Yes ☐ No ☐

Comments:

_____

_____

CARE

When you arrived at your assigned hospital unit:
Were you taken to your room immediately?

Yes ☐ No ☐

Did you have to wait near the nurses' station prior to admission to your room?

Yes ☐ No ☐

When you were taken to your room, did you find your room and bed clean?

Yes ☐ No ☐

Were you shown how to use the nurse call?

Yes ☐ No ☐

Were you shown how to use the telephone?

Yes ☐ No ☐

Were you taught how to use the TV?

Yes ☐ No ☐

Were you taught how to use the emergency call button and the pull cord in the bathroom?

Yes ☐ No ☐

Did hospital personnel smile and introduce themselves upon entering your room?

Doctor	Yes	☐	No	☐
Nurses	Yes	☐	No	☐
Residents	Yes	☐	No	☐
Housekeepers	Yes	☐	No	☐
X-ray Technicians	Yes	☐	No	☐
Laboratory Technicians	Yes	☐	No	☐

Did the following hospital personnel explain the treatment or procedure that they would perform on or for you?

Doctor	Yes	☐	No	☐
Nurses	Yes	☐	No	☐
X-ray Technicians	Yes	☐	No	☐
Laboratory Technicians	Yes	☐	No	☐
Respiratory Therapists	Yes	☐	No	☐
Physical/Occupational Therapists	Yes	☐	No	☐

Did the nursing staff inform you of your daily progress?

Yes ☐   No ☐

Do you feel the nurses caring for you showed interest in your personal needs?

Yes ☐   No ☐

Do you feel that the nursing staff have made your hospital stay as pleasant as possible?

Yes ☐   No ☐

Have the nurses been willing to listen to you and your concern?

Yes ☐   No ☐

Overall, how would you rate your nursing care?

Excellent   _____
Above average   _____
Average   _____
Below average   _____

Do you feel satisfied that enough attention was given to your illness?

Yes ☐   No ☐

Have instructions been given to you so that you know what is expected of you after you were discharged?

Yes ☐   No ☐

Were you treated considerately by those individuals who drew a sample of your blood?

Yes ☐   No ☐

If you received physical/occupational therapy services, did the therapist show a caring, considerate attitude?

Yes ☐   No ☐

Did hospital employees solicit illegal gratification or tips?

Yes ☐   No ☐

Were you constrained to pay?

Yes ☐   No ☐

Comments:

_____

_____

DISCHARGE

Was your discharge a pleasant experience?

Yes ☐   No ☐

Were there hassles?

Yes ☐   No ☐

Did you have to wait for a long time after the doctor said you could go home?

Yes ☐   No ☐

If yes, how long?

Under 60 minutes ☐   1–2 hours ☐
2–3 hours ☐   More than 3 hours ☐

If it is more than 3 hours, was it the hospital's fault such as delay in billing, discharge papers, etc.?

Yes ☐   No ☐

If no, was it yours like you could not settle your bills, relatives did not come?

Yes ☐   No ☐

Comments:

_____

_____

OPTIONAL

We monitor complaints and compliments and try to rectify the complaints. If it would help us to know your name, please complete the following:

Name _____

Address _____

_____

_____

Ward and Room Number _____

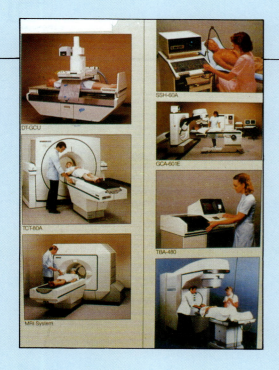

CHAPTER 4

Planning and Designing
Medical Services

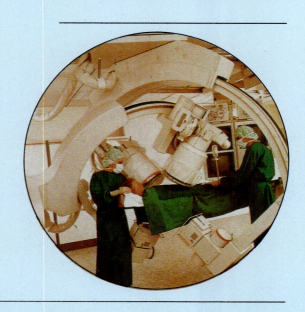

CONTENTS

- ☐ **Outpatient Services**
- ☐ **Emergency Services**
- ☐ **Clinical Laboratories**
- ☐ **Radiologic Services**
- ☐ **Diagnostic Radiology**
- ☐ **Radiation Therapy Department**
- ☐ **Nuclear Medicine**
- ☐ **Surgical Department**
- ☐ **Labour and Delivery Suites**
- ☐ **Physical Medicine and Rehabilitation**
- ☐ **Physical Therapy**
- ☐ **Occupational Therapy**
- ☐ **Speech and Hearing Therapy**
- ☐ **Pulmonary Medicine**

OUTPATIENT SERVICES

➤OVERVIEW

Outpatient care was once on the side-lines, and having been originally designed with a limited scope, it offered only basic, minor services. In a significant move all over the health care world, outpatient care has changed as a major service encompassing a wide range of treatment, diagnostic tests and minor surgeries, some of which required hospitalization earlier.

Hospitals must pay as much attention to proper planning, designing, organization and functioning of the outpatient department as for any other department. Some facets of the outpatient department are maintained separately from the inpatient services. Nevertheless, the two should be integrated physically, functionally, and from the clinical and administrative points of view. This is because in most cases the patient is studied and given treatment in the outpatient department up to the time he is hospitalized. He is then admitted and cared for as an inpatient until he is referred back to the outpatient department where his treatment is continued. Besides, as an outpatient he has his diagnostic tests and procedures in the ancillary and adjunct services which are integral parts of the hospital. The advantage of treating the patient in the outpatient department is that it eliminates the need for or reduces the length of hospitalization and consequently the cost to the patient.

Caring for the sick is the primary function of the outpatient department; thereby it renders an essential community service. In addition, it functions as a centre for imparting education to professional staff as well as to patients. It provides medical students, house physicians and other professional staff such as nurses and technicians with valuable and diversifred clinical experiences. Nowhere else in the hospital does the health educator have an opportunity to drive home facts about health to a captive but receptive audience as in the waiting rooms and clinics of the outpatient department.

The emergency room is an integral unit of the total outpatient department. The next chapter is devoted to a fuller discussion of this unit. The expression "ambulatory care" is sometimes synonymously but wrongly used for outpatient department. Ambulatory means able to walk, and applies to both outpatients and inpatients. While some inpatients are ambulatory, not all outpatients are.

➤LOCATION

The outpatient department should be conveniently located adjacent or in close proximity to vital adjunct services such as registration and medical records, admitting, emergency and social services. It should also be easily accessible and rapid services should be available from the laboratories, radiology, pharmacy and physical therapy departments since practically all of the diagnostic and therapeutic departments are used by the patients during every visit. Attention should be paid to circulation which should result in the smooth flow of the various traffic lines traversing the department (see Fig. 4.1).

In larger hospitals, if the laboratory is located at a distance or on another floor, it is recommended that a laboratory substation is located in the outpatient clinics area. The outpatient department should be on the ground level preferably with a separate entrance and adequate parking facilities.

➤DESIGN

Special attention should be given to the design of the following areas of the outpatient department.

- ◆ Adequate parking facilities. Entrance located at ground level and designed to handle wheelchairs and stretchers.
- ◆ Storage area for wheelchairs and stretchers, neatly alcoved and out of stream of traffic but conveniently accessible.
- ◆ Lobby and lounge to provide seating accommodation for the largest foreseeable number of persons. A large number of persons crowding in one single area creates an impression of cold impersonality and inefficiency. To improve the atmosphere, patients should be dispersed to subsidiary or sub–waiting areas adjacent to the clinics.
- ◆ Elevators easily accessible to the lobby are specially important for cardiac and obstetric patients who require immediate care in their respective areas.
- ◆ Minor operating room(s). If the hospital has an outpatient surgery programme, a room should be equipped for minor surgeries not requiring general anaesthesia, and for general diagnostic procedures not requiring overnight's stay. A recovery room to hold the patients for observation after surgery should be adjacent to the minor surgery room. It should be visible from the nurses' station for nurses' supervision and observation. Other facilities that may be provided are toilets, a place where patients could change from their street clothes to hospital gowns where necessary, space where they could keep their valuables, a patients' preparation area, relatives' waiting area, etc., and, of course, scrub and change rooms for doctors and nurses.
- ◆ Arrangement for keeping a major portion of the outpatient department locked after outpatient hours for security reasons.
- ◆ Doors between emergency services and the outpatient department for certain adjunct services like medical records, etc. to be available to the staff and emergency patients during the evening and night shifts.
- ◆ Main lobby and main waiting area and sub-waiting areas in the immediate vicinity of various clinics. (Fig. 4.2).
- ◆ Doctors' offices (Fig. 4.2 and Fig. 4.3).

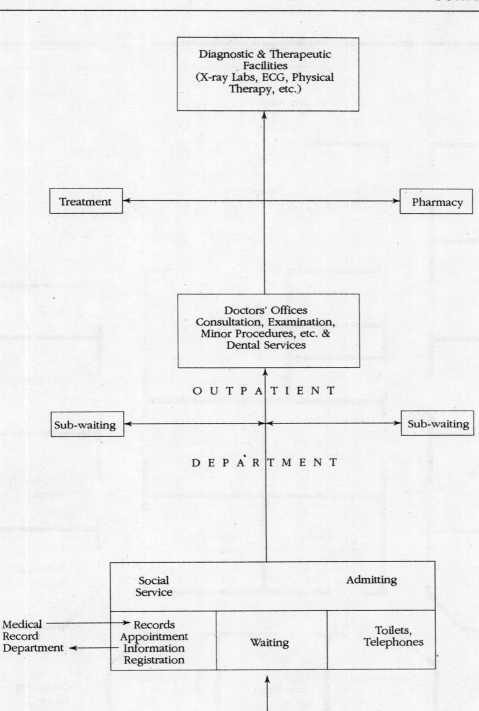

> **FIG. 4.1**   Flow Chart of Outpatient Department

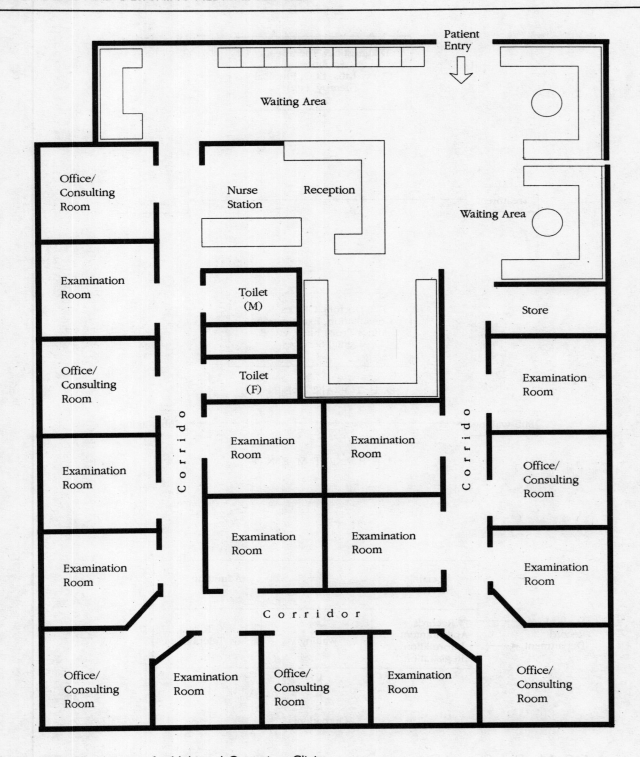

> **FIG. 4.2** Layout of a Multi-pod Outpatient Clinic

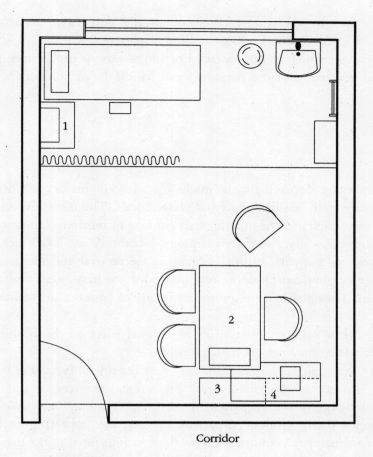

Corridor

1. Work Table
2. Physicians Desk
3. Wall Mounted Cabinet
4. X-Ray Viewer

➤ **FIG. 4.3** A Typical One-room Outpatient Consultation-cum-Examination Facility

- Registration cubicles designed to afford privacy and maintain patients' self-esteem.
- Directional signs freely used throughout the department.
- Public areas with provision for public telephones, patients' toilets, water coolers or drinking fountains, enquiry desk or counter, cashier's booth, coffee shop or snack bar, gift shop, florist's shop, meditation room, and if possible, a separate play area for children.
- Information, medical records and cashier's office should open into the waiting room but designed to offer some degree of privacy.
- Admitting and social service offices open directly off the main waiting area. Require sub-waiting areas adjacent to them as well as privacy.

◆ Efficient patient flow. The key to operational efficiency in the outpatient department is efficient patient flow. (Fig. 4.1) All systems affect the flow of patients.

If the units are arranged in a manner which facilitates coordination of various procedures, patients need not waste time in prolonged waiting, and the personnel can handle large numbers with ease and efficiency.

➤ ORGANIZATION

Typically, the staff in the outpatient department is made up of four major organizational components — medical staff, nursing staff, ancillary staff and clerical staff. The most important and central to the organization is the medical staff. The nursing staff consists of registered nurses, ANMs (MPWs) and nursing or hospital aides. Ancillary staff are radiology, laboratory and ECG technicians. The clerical staff carry out registration, patients' billing, cashiering, secretarial and medical record functions. In teaching hospitals, senior physicians holding teaching positions have house staff (interns and residents) working under them. They assist in the professional care of patients in the outpatient clinics.

Many hospitals use volunteers in a variety of activities in various parts of the hospital. The outpatient department is one area where they can be used profitably.

The management structure of the outpatient department may vary considerably among hospitals. In larger hospitals, the medical director or the director for professional services or the medical superintendent may be directly in charge of the outpatient department. He may have assistants in various units such as the emergency room, clinics, outpatient surgery, etc. reporting to him. In smaller hospitals there may be a coordinator to coordinate the work of various units of the department reporting to the medical director or medical superintendent.

➤ COMMON ELEMENTS OF OUTPATIENT
FACILITIES AND SPACE REQUIREMENTS

The following facilities are required in the outpatient department.

Public Areas and Administration

◆ Wheelchair and stretcher storage alcove
◆ Reception and information desk
◆ Registration counter and cubicles for staff to interview new patients on a one-to-one basis
◆ Lobby and waiting lounge
◆ Public toilet facilities

- Public telephone(s)
- Water coolers or drinking fountains
- Space/office(s) for private interviews with social workers
- Multipurpose room(s) for conferences, meetings, health education purposes, etc.
- Employees' facilities including lockable drawers and cabinets for valuables
- General storage for supplies, equipment, etc.
- Coffee shop/snack bar
- Gift shop — best located off the lobby in a conspicuous location with adequate display space
- Meditation room/retiring facility — for meditation or for use by anxious or bereaved relatives
- Other facilities — doorman's station, magazine stands, display racks, room with STD/ISD facilities, etc.
- Bank extension counter where out-station patients can deposit their cash while hospitalized — desirable but not essential. Can be offsite.
- Security post at strategic location

Clinical Facilities

- General purpose examination rooms — minimum floor area 80 sq. ft. excluding vestibules, toilets, closets, etc. Wash basin and a counter top for writing. At least 2' 8" clearance at each side and at the foot end of the examination table
- Special purpose examination rooms — for specialty clinics such as eye (dark room required), ear, nose, throat — facilities as required for special procedures and equipment. Wash basin, counter/work top, etc. as for general purpose examination rooms.
- Treatment room(s) for minor procedures and cast work
- Observation room(s) for isolation or disturbed patients — conveniently situated near the nurses' station to permit close observation
- Nurses' station with work counter, communication system, space for charting, supplies, refrigerator, locked storage for drugs, narcotics, etc.
- Clean storage for storing clean and sterile supplies, cabinets and shelves
- Containers for collection, storage and disposal of soiled materials.
- Sterilizing facilities — may be located off site, and flash sterilizers in some clinics
- Wheelchair storage space, out of the direct line of traffic.

►PROBLEM SITUATIONS

One of the most frequently heard criticisms of the hospital is the prolonged and seemingly interminable waiting a patient is subjected to at various stages of his visit to the outpatient department — in front of registration, at the doctor's office, at the cashier's, laboratory, pharmacy, and for and in between appointments. This is annoying to patients and bad public relations for the hospital. Large numbers of patients attending clinics are not the only reason that makes people wait. Incompetent and inadequate staff, cumbersome and time consuming procedures and forms, poor design that has not taken into account the circulation and work flow in the department are some of the causes that contribute to this malady. There is no denying the fact that things can be better. Hospitals should study the problem. Expeditious handling of a workable and efficient system and records completion and transportation contribute to the efficiency of the staff and satisfaction of the patients.

EMERGENCY SERVICES

➤ OVERVIEW

In planning and designing emergency services, hospital planners must be aware of the fact that patients often seek emergency services for situations other than acute medical conditions. While the hospital might define an emergency as a life and death situation involving an injury or acute illness threatening life, limb or sight, a situation which warrants immediate treatment, the patient's perception of what constitutes an emergency is often less well defined. For example, to him his inability to contact his family physician at night to treat his constipation of two days may be an emergency. Consequently, the emergency department is required to render a comprehensive range of services right from the elementary first aid and general outpatient services to sophisticated management of surgical and medical emergencies and full-scale trauma care.

There is a growing public awareness that the hospital is the most appropriate place where round the clock care is available for unexpected illness or injury. General practitioners have increasingly come to accept the hospital's emergency department as the safest place to refer trauma and other life-threatening cases which they cannot tackle. There is also a perceptible rise in the number of accident cases which naturally come to the emergency department post-haste. Consequently, the emergency department is in great demand.

Hospital planners must take cognizance of these changes in the concept, functions and demand on the emergency department and incorporate flexibility in the design of the department to handle a wide range of cases in the most economical and efficient manner in terms of space, equipment, personnel and supplies. If this is not done, the department may turn out to be a liability to the hospital, because even in the best of circumstances, maintaining a 24–hour emergency service with its high fixed cost and periods of low utilization can be costly. On the other hand, a well designed and efficiently managed emergency department is an important source of revenue to the hospital. Emergency patients use ancillary services of the hospital to a considerable extent and this brings in revenue.

➤LOCATION

The emergency department should be located on the ground floor with easy access for patients and ambulances. There should be a separate entrance to the department which is away from the main hospital and the outpatient entrance; it should be well marked with proper lighting and signs, and should be easily visible and accessible from the street. Since the emergency department becomes the main entrance for the hospital during the night, it must relate to the public and vehicular transportation.

The department should be close to admission, medical records and cashier's booth. Where possible, admitting functions, cashiering, registration of new patients and creation of their medical records should be done in the department. A good percentage of emergency patients — studies show over 40 per cent — require x-ray. The portable x-ray unit is not satisfactory. So close proximity to the radiology unit is essential to facilitate the movement of accident cases and to save every minute which is precious. The laboratory services including the blood bank should also be accessible to emergency since a sizable number of patients need this service. In the location of the emergency department, proximity to elevators is also important in order to proceed to surgery without loss of time.

➤DESIGN

The entrance to the emergency should be sheltered to protect ambulance patients from the weather while unloading. There should be paved access to permit discharge of patients comfortably from ambulances and cars. Adequate parking space for ambulances and cars of patients and medical staff should be provided. The ambulance entrance should be large enough to admit one or more ambulances negotiating with stretchers. If there is a raised platform for ambulance discharge, ramps should be provided for wheelchair and pedestrian access.

With injured patients, accident victims and their distraught relatives around, emotions frequently run high in the emergency department. So traffic control within the department is critical. The design should facilitate good public relations. Unsatisfactory arrangements, congestions, delays and inefficient operation due to faulty design will present a poor image of the hospital in the eyes of the emergency patients and their relatives. The design should also facilitate quick access to the patients by staff and supplies.

➤ ORGANIZATION

As mentioned earlier, levels of emergency care range from elementary first aid to sophisticated surgical procedures. Not all hospitals can maintain an emergency department which will offer full-scale trauma services. However, emergency first aid should be available at every hospital. It is important to remember that having a full-scale emergency trauma facility without proper equipment and competent staff round-the-clock may be more dangerous than having no such facility at all and that lives may be lost because, in the absence of specialists, staff and equipment, no reasonable care can be provided. Either the casual staff will mismanage cases or by the time they direct them to other hospitals, it may be too late.

The following are the essentials of well-organized services for the care of fractures and other traumas in emergency and accident department as listed by the American College of Surgeons through its Committee on Trauma. We have somewhat adapted it to suit our purpose.

- An efficient, promptly responding, well equipped ambulance service with competent personnel in charge.
- A well equipped emergency operating room with supplies always ready for use.
- A small recovery ward.
- Efficient personnel including at least a competent physician, nurse, and attendant on round-the-clock duty or on call.
- Supervision of treatment of fractures and other injuries by well qualified and competent surgeons in their respective fields.
- Adequate diagnostic and therapeutic facilities under competent medical staff.
- A well documented medical record for every patient which includes immediate record of all injuries, physical findings, treatment, etc.

Although the emergency is a part of the total outpatient services, the current organizational trend is for the unit to be a separate and distinct department under the direction of a full time physician. The reporting relationship is as in any other medical department of the hospital. The trend is for full time emergency room physicians even in hospitals where medical staff are not hospital-based salaried physicians, but are granted privileges to practise. It is most important, however, that the staff of the emergency department are highly skilled and competent persons. In addition, they must be endowed with patience and understanding, and be able to work calmly and with self-control even under provocation. They should be given orientation in dealing with the public. More importantly, they should know how to communicate with people who are bereaved and those who are under stress, fear and anxiety.

One important responsibility of the department is to formulate, document, and periodically review and update policies and procedures of the emergency department. These must be compiled in a manual and made available to all staff members of the department.

Policies and procedures should cover, among other things, the following subjects:

- medico-legal cases such as road accidents, assaults, attempted suicide, poisoning, industrial accidents, deaths resulting from criminal acts, etc.;
- police procedures and reporting;
- notifiable deaths;
- patient brought dead or in dying condition;
- disposal of bodies, autopsy, morgue procedures;
- accident and emergency room register; and
- medical record and release form procedures.

➤ PHYSICAL FACILITIES AND SPACE REQUIREMENTS

Facilities in the emergency department can be considered broadly under two categories: (a) administrative and public areas and (b) clinical facilities. In the clinical facilities four functional areas can be identified. These are

- trauma care area where the severely injured surgical cases are handled;
- medical examining area;
- splintage and casting area for orthopaedic cases: and
- observation beds for patients who need to be kept under observation for neurological and other medical reasons. The patient may later be transferred to inpatient area.

The following facilities are required:

Administrative and Public Areas

- Reception-control — for observation and control of access to the treatment area, public waiting area, and pedestrian and ambulance entrance area. Should be equipped with communication including intercommunication system. One of the activities included in this area is the triage function (discussed elsewhere) during internal or external disaster;
- Space for stretchers and wheelchairs adjacent to the entrance but out of stream of traffic. Stretchers sometimes serve as examination tables in an emergency and should, therefore, be sturdy and provided with wheel-locks;
- Public waiting area with toilet facilities, water coolers or drinking fountains, public telephones, STD and ISD facilities and vending machines, if possible. The waiting area should be separated from the working or treatment area;
- Space/room for security staff, police, ambulance driver and attendant;
- Office for the night administrator/night supervisor — can be off site but not too far away; and
- Coffee/snack bar in the vicinity.

Clinical Facilities

- Trauma room(s) for emergency trauma procedure or emergency surgery, with resuscitation and life support equipment and drugs, medical gas outlets (oxygen, vacuum and compressed air) if central gas is provided, examination/procedure table, examination lights, x-ray film illuminators, cabinets, supply shelves, and if the room is used for orthopaedic and cast work, closed storage space for splints and other orthopaedic supplies, a plaster sink, traction hooks, etc.

- Examination/treatment room(s) with examination tables, examination lights, work counters, cabinets, wash basins, x-ray film illuminators, medication storage facilities and medical gas outlets; (Picture 15).

- Scrub stations conveniently located close to each trauma and orthopaedic room;

- Additional adjustable space for triage, treatment, observation, etc. in the event of disaster handling;

- Staff work area and charting space with counters, cabinets, medication storage facilities, Dictating facilities, etc.;

- Storage space for equipment such as portable x-ray and "crash carts" (cardio-pulmonary resuscitation emergency carts) which should be easily accessible;

- Separate soiled and clean utility rooms;

- Toilets for patients;

- Janitor's closet;

- Rooms for duty/on-call doctors, separate for men and women, with sleeping accommodation, shower and toilet facilities; and

- Locked cabinets, etc. for staff's personal effects.

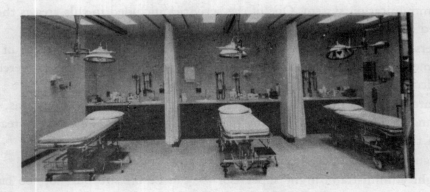

> **PICTURE 15**   A section of Emergency Department showing curtained treatment
> cubicles and resuscitation and life support equipment, medical gas
> outlets, etc. mounted on the wall behind

➤ OTHER CONSIDERATIONS

Legal Concerns

Emergency department staff, particularly the medical and nursing staff, including residents and interns, should be conversant with the legal aspects of emergency services. There are a number of reporting laws aimed at detection and appropriate action by police and other governmental agencies in cases involving accident, suicide, assault, public health and safety. It is mandatory for doctors to comply with these laws and also testify in courts when required to do so. The emergency procedure manual should cover policies and procedures relating to such cases.

Physicians and nurses must also be conversant with consumer protection forums and the Consumer Protection Act, and liability suits arising out of malpractice or negligence, and be aware of ways of protecting themselves and the hospital legally. Good medical and administrative practice including maintenance of well documented medical records are a good legal protection.

Disaster Preparedness

Every hospital must have a well established and well rehearsed contingency plan to cover any internal or external disaster when several emergency cases arrive at the hospital simultaneously, for example, victims of a bus accident, major fire or some other disaster. The emergency department has an active role to play in such a situation in coordinating with the hospital's disaster committee. The emergency staff (and others who participate in disaster plan procedures) are trained to recognize the nature and relative severity of a patient's condition. In the triage area, patients are rapidly sorted and sent to appropriate treatment areas, for example, hyper-acute (life-threatening)cases to emergency room, serious casualties to surgery area, ambulatory care (non-life-threatening) cases to outpatient department, waiting room or observation area, the emotionally disturbed cases to the chapel or meditation room, and the dead on arrival to the morgue. A triage sorting system establishes priorities for treatment of critical patients. The priorities are based on the degree to which the patient's life is threatened.

Code Blue Procedure

Another area where emergency staff is active in handling cardio-pulmonary resuscitation (CPR). "Code Blue" is a term used in hospitals to announce an emergency of serious nature such as cardiac arrest. There is a pre-established procedure and a pre-appointed team which promptly responds to such emergencies. A cardiac arrest or similar emergencies may take place anywhere in the hospital. When the emergency staff is busy coping with a disaster, personnel outside the emergency department may be instructed to respond to Code Blue call so that timely patient care is provided in such situations.

$$\boxed{\text{CLINICAL LABORATORIES}}$$

➤OVERVIEW

The primary function of the clinical laboratories is to perform laboratory tests which will provide information to clinicians in arriving at correct diagnoses and in the treatment and prevention of diseases. The practice of modern medicine requires more and more laboratory examinations.

Some of the laboratory tests are specialized tests. Others are routine ones such as urine analysis and blood cell counts. In addition to the usual tests, obstetrical patients may require Rh factor and blood grouping. For surgical patients, they may include examination of organs and tissues removed during operations. The laboratory plays an important role in the hospital's infection control and surveillance programme.

With the advances in technology, today's medical care may be said to have entered an era of laboratory medicine. A couple of decades ago, laboratory determinations were done manually covering only basic diagnostic tests needed for patients' assessment. Most laboratories are now equipped with, to a varying degree, sophisticated automated instruments such as automated analysers which have increased productivity. Tests are performed in a matter of minutes and with the highest degree of accuracy and reproducibility. The workload of the laboratories and the number of tests they perform have increased enormously.

Today, in larger hospitals, major sections of the laboratory are specialized to the extent that they engage specialized staff and perform all the tests relating to their respective disciplines within the sections. The number of staff has also increased greatly. However, in smaller hospitals, some of the areas may be combined and staff may be trained to work in more than one section. Some may be rotated through biochemistry, haematology, urine analysis and blood bank, and others through microbiology and histology. It is not unusual to find smaller hospitals preferring not to establish and staff some sections like histology and cytology, but sending specimens to other larger institutions.

Most larger laboratories provide 24-hour service. In smaller hospitals, the staff may be on call during the night shift.

➤FUNCTIONS

With their specialized sections, the laboratory makes a complex department. Its functions can best be depicted by describing the functions of its major sections.

The haematology section performs tests and procedures which pertain to the examination of blood and blood forming organs of the body.

The blood bank is involved in the blood donor programmes, typing, cross matching, processing, etc.

The biochemistry section performs quantitative and qualitative analyses of body fluids, secretions and substances found in tissues.

Other clinical pathology sections perform a variety of studies that include urine analysis, semen analysis, etc.

The histopathology section prepares and examines tissues in order to provide data on the cause and progress of diseases. It does microscopic examination of tissue pathology.

The bacteriology section identifies micro-organisms found in body fluids, skin scrapings and surgical specimens. It tries to identify, grow and perform sensitivity tests of organisms to different antibiotics.

The cytology section examines body cells for diagnosis of malignancy and other diseased conditions, effects of hormones, and so on.

The clinical laboratories perform other functions such as teaching and research programmes. Some hospitals have a school of medical technology.

Autopsy is the responsibility of the pathology section of the laboratory.

➤LOCATION

The laboratory should be conveniently located on the ground floor to serve the outpatient, emergency and admitting departments. It should also be close to or easily accessible to surgery, intensive care, radiology and obstetrics.

In large hospitals which have a large number of outpatients or when the main laboratory is not within walking distance, there may be a laboratory sub-station in the outpatient department.

The autopsy area should be a little removed from the emergency and inpatient areas. Access to the autopsy area as well as to the morgue should not be from patient areas and the removal of dead bodies to the outside should be through non-public corridors as this should be protected from patients' and visitors' view for psychological reasons.

➤DESIGN

While designing the laboratory, the use of modules is recommended both for work stations and for piping layout for essential utility services. For work stations, the modules may be 10 feet by 20 feet with work benches 12 feet long and 30 inches high. Equipment by and large requires space at the back for air circulation. This should be borne in mind when determining the depth of work benches, particularly if there is a wall at the back. Modules should be open within sub-divisions and closed between them. Open areas include haematology, urine analysis, biochemistry and autoanalysers as well as clerical and

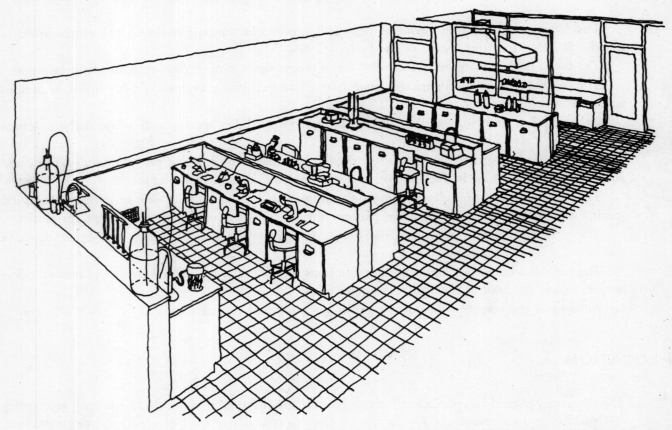

> **FIG. 4.4** Perspective View of a Laboratory in Small and Medium Sized Hospitals Showing Work
> Benches/Work Stations and Piping Layout for Utility Services

secretarial work. Closed areas include bacteriology, serology, pathology–histopathology, sterilization, glass washing, blood bank, and offices. (Figs 4.4, 4.5, Picture 16).

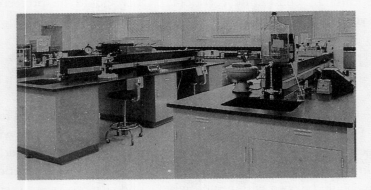

> **PICTURE 16**   Inside View of a Laboratory in a Small or Medium Sized Hospital

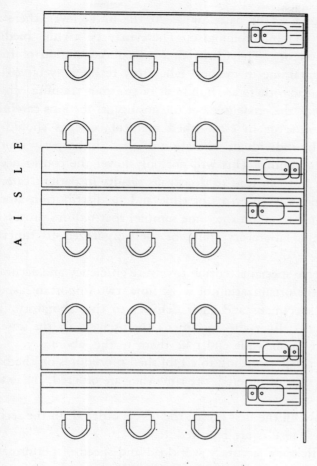

Layout Plan of Work Station

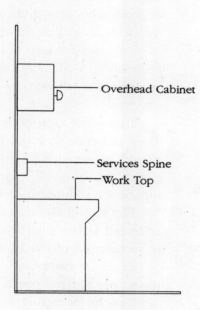

Side Elevation of Work Station

➢ **FIG. 4.5** A Typical Laboratory Work Station

The laboratory needs cold, hot, distilled and deionized water in certain sections. It is, however, recommended that distilled water is piped to all sections of the laboratories where water is used.

➤ORGANIZATION

A pathologist is generally in overall charge of the laboratory. Larger hospitals may have one or more assistant pathologists, and chiefs over the divisions of microbiology and biochemistry. The microbiologist may be an MD in microbiology (medical doctor) or an M Sc in microbiology trained under the faculty of medicine. The chief biochemist may be an M Sc in clinical biochemistry trained under the faculty of medicine. In some hospitals, microbiology is organized as an independent department under a medical doctor not reporting to the pathologist. At the lower level, there are medical technicians/technologists, biochemists and microbiologists. There may be a chief medical technologist reporting to the chief of the laboratory. The Christian Medical Association of India (CMAI) has an excellent two-year training programme in medical laboratory technology offered in their affiliated member hospitals. A word of caution: there are hundreds of spurious training centres all over the country. Administrators should check the credentials of the applicants for jobs carefully. Candidates trained under the CMAI programme or in established medical colleges should be employed. Too often, laboratory work is done by individuals who are untrained, improperly trained or those who were given superficial on-the-job training. This will not only lower the confidence of the medical staff in the reports made by them but can lead to disastrous results in patient care.

Requests for tests reach the laboratory randomly in terms of time and in distribution among types of tests. Many of the orders are for multiple tests (on the same sample) that require processing at several work stations. This means that the laboratory must be well organized to function effectively.

Two major areas of the laboratory must receive special attention to ensure efficiency and accuracy. They are work flow and information flow. In the organization of work flow, two important aspects should be noted. One is the procurement of specimens and their delivery to the laboratory. The outpatients generally go to the laboratory and deliver the specimens themselves. In the case of inpatients, technicians collect blood samples at the bedside and take them to the laboratory.

The second aspect is that the tests are performed according to established procedures on the basis of several factors such as the number and types of tests ordered, the time they are ordered, the extent of automation of the laboratory, etc.

There are two phases in the information flow. In the first phase, the test is ordered. In the second phase, the ordered test is logged in and the test results are reported.

Two major factors which determine the efficiency, accuracy workload and speed of performance of the laboratory are the extent of automation in the production cycle, and the extent of computerization of the laboratory and its various activities.

The type of sophistication of automated equipment varies from hospital to hospital. Biochemistry and haematology are the two areas which are largely automated. Many hospitals have now autoanalysers in chemistry which handle high volume procedures, and equipment that does low cost profile testing which consists of a predetermined set of a dozen or more tests from a patient's single specimen. Some sophisticated equipment can do multiple tests in various combinations.

The laboratory processing system may be manual, partially computerized or fully computerized. In the manual system, everything is done manually — orders which are in writing are delivered by messengers from various points of the hospital, logged in the register and passed on to the technician. Tests are performed, results are recorded in the register and in a report form, and are transmitted to the departments that requested for the tests or sent to be filed in the patient's medical records.

A partially computerized system uses minicomputers with data terminals located in the laboratory and other parts of the hospital. They can be used for a wide range of applications — order entry and reporting, charge capturing, etc. The system eliminates paper work and missing charges. Clerical time is saved and response to emergency requests becomes faster. The computer system greatly improves the efficiency of the processing of information by the laboratory.

The fully computerized system is a dedicated system which performs most of the clerical work. The system organizes the work flow, provides work assignments, reports data and permits instant on line access to data. The advantages are increased production at lower unit cost (excluding the cost of computerization), availability of comprehensive data, reports and statistics. The disadvantages of a fully computerized system is the cost which may exceed the benefits. Hospitals must consider to what extent they should computerize the laboratory for the increased accuracy, efficiency, speed and convenience they get in return.

Laboratory Procedures

The following are the usual procedures for requesting and handling laboratory examinations:
- Request for examinations should be in writing.
- The physician, nurse or laboratory technician may be responsible for obtaining specimens.
- Instituting time schedule for accepting certain types of specimens will facilitate the operations of the laboratory.
- Emergency requests are normally accepted at all times, and shall have priority over all other requests. The hospital should, however, ensure that this facility is not misused. Misuse of laboratory's facilities occur when physicians order tests as emergency even though they are not emergency tests. Sometimes tests are ordered as emergency when they should have been requested on the previous day. There is a danger of laboratory staff losing faith in such emergency requests.
- Specimens should be sent to the laboratory in properly labelled containers.
- Nurses should be familiar with the proper time for taking specimens (fasting, non-fasting, etc.), minimum volume necessary and the proper container. A list of commonly requested investigations with all these details should be posted and be readily available at the nurses' station.
- All requests should be in well prepared request forms which should be uniform in size.
- Different colours may be used for different types of forms for easy identification. Then doctors and nurses may also easily locate them.
- In the inpatient area, the nurse is responsible for making out all forms according to established standing orders. If the specimen is taken by the laboratory technician, the nurse will hand over the form to him (her).
- No specimen or request should be left in the laboratory. It should be given to the laboratory staff who will receive and enter it in a daily record ledger. A numbering system should be followed to identify specimens.
- Results of the tests are entered in both copies of the request form; the original is then returned

to the nursing unit to be attached to the patients' records. The results are also entered in the daily record or register.

- The laboratory reports should be given only to authorized medical personnel or sent to be attached to the patient's records. They are never given to patients except when requested by doctors who are not members of the hospital staff and the requests are brought by the patients who may then personally receive the reports.

➤FACILITIES AND SPACE REQUIREMENTS

Space required for laboratory services will depend on the size and type of the hospital and the extent of sophistication of the work of the laboratory in which it functions. As the volume of work increases, more space will be needed. It is recommended that as the hospital grows in size, departmentalization of the laboratory should begin. At 200 beds, bacteriology and serology may be combined into a separate unit and rest of the laboratory into another. When the bed capacity increases further, it will be necessary to provide a separate unit for each of the major divisions.

It has been found by experience that, as with the number of laboratory tests, space requirements of the laboratory tend to double every eight to ten years. Hospital planners should provide for future expansion of every section of the laboratory.

Space is required for the following:

- Work counters with space for equipment, microscopes, incubators, centrifuge, under the counter and overhead cabinets. Work stations should be equipped with vacuum, gas, electrical services, sinks and water.
- Counter sinks for hand washing and for disposal of non-toxic fluids.
- Specimen collection area for blood, urine and faeces. For blood collection area, work counter, space for patients' seating and wash basin for hand washing. For urine and faeces collection area, toilets (men and women separate) with wash basin, and counter top to place the specimens. Hatch windows may be provided through which the specimens can be passed through.
- Storage facilities for reagents, standards, supplies and stained specimen microscopic slides. Refrigerators may be necessary.
- Appropriate storage for chemicals and flammable liquids.
- Facilities and equipment for terminal sterilization if there is no incinerator on site.
- Blood bank for refrigerated storage of blood — discussed separately.
- Administrative areas including offices for pathologist(s), secretarial and clerical work area, space for filing and records.
- Staff facilities — some of these may be outside the department.
- Sterilizing area.
- Culture media preparation area.
- Glass washing area — dirty area which should be separated and closed.

- Storage for surgical specimens.
- Space for autopsy and morgue (details below).

Blood Bank

It is said that there is no greater therapeutic tool than the administration of whole blood when it is needed, and perhaps no more lethal weapon at our disposal than administering contaminated blood or improperly given blood.

Every hospital should have a committee, of which the pathologist is a member, to establish written procedures for the proper use of blood and blood derivatives, including identification and compatibility testing of blood, criteria for use, and review of all transfusion reactions occurring in the hospital. Storage facilities under adequate control and supervision are necessary. An alarm system should be instituted to notify personnel of the loss of electric power and faulty temperature.

With a view to modernizing the blood banking system in the country, the Government of India recently introduced amendments to the Drugs and Cosmetics Rules, 1972. Under these amendments, existing blood banks and those which intend to apply for a license to operate a blood bank are required to fulfil the conditions set out in the amendments. The salient features of the conditions are:

- Seven rooms within a space of 100 sq. mtrs.;
- Two laboratories, one for blood group serology and another for screening the blood for Hbs Ag, HIV antibodies and Syphilis. The two laboratories and the blood collection room are to be air-conditioned;
- Two refrigerators maintaining temperature between 4 to 6°C with recording thermometer and alarm device, one for the blood collection room and another for the laboratory;
- Personnel — a medical officer trained in blood banking for six months, a registered nurse and two trained technicians (MLTs);
- For AIDS test, the hospital can have its own testing facilities or can avail the facilities of the laboratories of the Central Government.

The rules specify procedures and other requirements relating to:

- licensing
- list of equipment and supplies needed for the blood bank
- refreshment services
- laboratory equipment
- reagents
- general supplies
- personnel
- testing the whole human blood
- expiry date
- records, labels and labelling.

Hospitals are advised to write to the State Director of Drugs Control for more information regarding this.

Autopsy Room and Morgue

Autopsy has greatly contributed to the advancement of scientific knowledge and to the conquest of disease. Medical staff should be interested in the advancement of knowledge through autopsy.

It is mandatory that the police is notified immediately regarding all deaths from actual or suspicious violent causes, including sudden deaths without obvious causes, while under anaesthesia, from abortion, poisoning and deaths from contagious diseases. All cases of unclaimed boides should also be notified. In all these cases, the bodies are sent to the nearby government hospital for autopsy. The authorities would want to know not only the immediate cause of death, but all the circumstances surrounding it as it is necessary to document them for court action. For this, the hospital must maintain accurate and complete medical records in each case. The hospital has a duty to cooperate with the police in such cases. The hospital has also the responsibility of issuing a death certificate at the time of death.

When the hospital performs a post-mortem examination, it generally faces difficulty is securing the permission of the deceased person's next of kin or guardian before the actual post-mortem is performed. It is advisable to obtain a written permission for legal protection. This can be in a standard form. Tact and diplomacy must be exercised in seeking permission. If the hospital has gained the respect and confidence of the relatives, this permission may be more easily forthcoming.

All deaths in the hospital should be reviewed by the medical staff or a committee of the medical staff.

The following facilities are required.

- Autopsy table made either of porcelain or stainless steel with a sink at the footend and a shadowless lamp fitted over the table.
- Work counter with sink and hand washing facilities, wall mounted x-ray viewing box, writing table, instrument cabinet and sterilizers.
- Storage room for equipment, supplies and specimens.
- Doctor's office for writing reports, discussions, etc.
- Visitors' waiting room.
- Changing room(s), toilet, shower and lockers.
- Janitor's receptacle or service sink and housekeeping facilities.
- Refrigerated facilities for storing bodies.

Ready-made multi-tiered cold rooms for storing bodies are available in the market. These may be easily installed.

If the body is to be transported to far away places, it may be embalmed. Frequently relatives like to bathe, dress up the body and perform last rites or pre-burial funeral service before the body is taken for burial or cremation. There should be space for these activities around the morgue.

RADIOLOGIC SERVICES

The main function of the radiologic services is to assist the clinicians in the diagnosis and treatment of diseases through the use of radiography, fluoroscopy, radioisotopes and high voltage acceleration.

In large hospitals, the radiologic services may be organized as three separate departments, namely, diagnostic radiology, therapeutic radiology and nuclear medicine. In some hospitals, diagnostic radiology and therapeutic radiology are two independent departments with nuclear medicine tagged on to diagnostic radiology. By and large in typical medium and small-sized general hospitals, only the services of diagnostic radiology may be available.

DIAGNOSTIC RADIOLOGY

➤ OVERVIEW

Good medical and surgical care depends, to a great extent, on the availability of prompt, thorough and skilful diagnostic services. Among the many modern diagnostic techniques that are so necessary for the effective treatment of the patients, x-ray examinations are of vital importance. A well planned diagnostic radiology department ensures an efficient flow of service (Fig. 4.6), prompt scheduling and minimum of movement and distance to the patients and staff. In an average general hospital, almost 90 per cent of the work in the x-ray department consists of radiography and fluoroscopy.

The radiographic procedure basically consists of the exposure of the x-ray film, development of the exposed film and interpretation of the x-ray. While the first two steps in this process are carried out by technicians, the interpretation of x-ray films is always done by a radiologist.

The following are some of the typical radiological examinations [Picture 17, (Plate 5: Picture 18), 19]. Each of them requires different procedures.

◆ Routine x-ray, normally performed by an x-ray technician, involves x-ray procedure (film exposure) of spine, neck, chest and extremities (legs, arms, hands and feet).

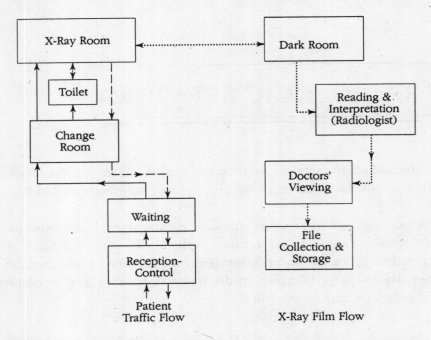

> **FIG. 4.6** Activities Flow Chart in the Diagnostic X-ray Department

> **PICTURE 17** A Versatile X-ray Diagnostic Motor–operated Unit which permits a wide range of applications including all kinds of gastrointestinal (GI) procedures, having state-of-the-art technology with digital electronics

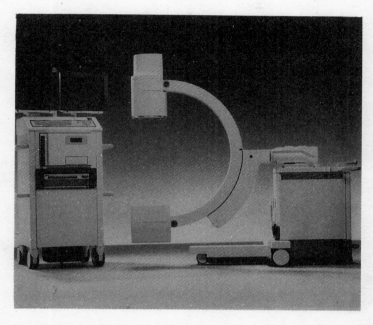

➢ **PICTURE 19** A Mobile C-arm Dual Field X-ray Image Intensifier for surgery, traumatology, orthopaedics, endoscopy and intensive care

- Fluoroscopic procedure is performed to study the digestive tract. Performed by a radiologist, a fluoroscopic procedure includes an injection, an enema or swallowing of a radio opaque medium, film exposure, and examination of the digestive tract by the radiologist. A routine exposure of the abdomen, called a scout film, is taken before fluoroscopic procedure. Another or a series of routine exposures of the abdomen (similar to a scout) is taken after the fluoroscopic examination. Common fluoroscopic examinations are gastrointestinal series (upper and lower GI series), barium enema, barium meal, gall bladder, and intravenous pyelogram (IVP).

- Special procedures, also performed by the radiologist, are complex fluoroscopies. These require special preparation and are generally scheduled in advance. In some cases, these procedures may require minor surgical procedures. Some of the special procedures are cardiac catheterization, neuroradiology, etc.

- Mammography is a simple, safe and reliable radiographic examination of the internal structure of the breast (Picture 20). The low dose x-ray study using the most modern equipment can detect breast cancer years before the patient or the doctor can feel it. This early detection can allow removal of the tumour before it spreads. Over the years, mammography has added new techniques and highly sensitive film to greatly reduce the amount of radiation needed.

In advanced hi-tech hospitals, the diagnostic radiology department performs other procedures that are part of or related to the x-ray department. Highly sophisticated equipment is used in the

—— 157

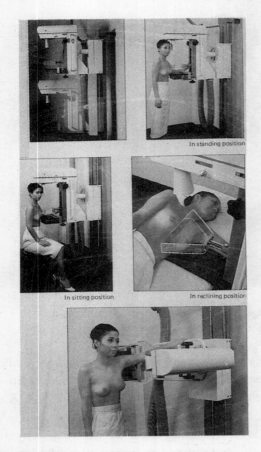

➢ **PICTURE 20** Mammography System

performance of these procedures. Some examples of the various kinds of equipment are diagnostic ultrasound, computerized axial tomography (CT or CAT scan) and magnetic resonance imaging (MRI).

Diagnostic ultrasound (Plate 5: Picture 23) is an imaging technology which is becoming increasingly popular because it does not require potentially harmful radiation.

Computerized axial tomography (Picture 21) is an x-ray technique that uses a special scanner and computer to produce cross-sectional images of head or parts of the body. Unlike standard x-rays which take a picture of the entire part of the body or head, CAT scan has the ability to image it one "slice" at a time. In x-rays, dense tissues, like bone can block the view of parts lying behind them. In CAT scan, the various slices clearly show both the bone and the underlying soft tissues in the body and the skull and the underlying brain tissues in the head. Looking at these images in sequence, the radiologist can create a three-dimensional picture of the part being examined.

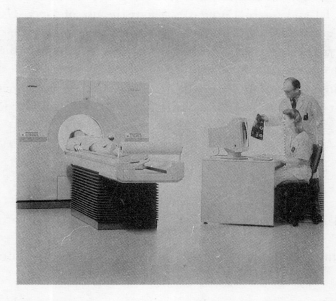

➢ **PICTURE 21**   A Computerized Axial Tomography Unit. The Patient positioned for scanning and each anatomical structure is viewed from different perspectives and displayed on the screen

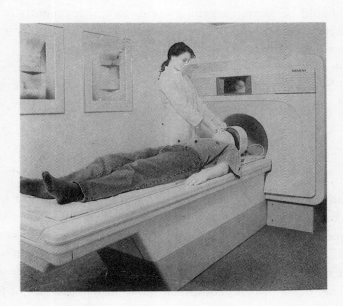

➢ **PICTURE 22**   Magnetic Resonance Imaging (MRI) System: A Patient is positioned and readied for scanning, these sophisticated machines are known for their superb image quality, high throughput capabilities and advanced clinical applications

The MRI (Picture 22), a marvellous gift of science and technology to medicine, allows the radiologist to see soft tissues (muscles, fat and internal organs) without the use of x-ray. Using two natural, safe forces — magnetic fields and radio waves — this unique imaging technique can look through hard bones to examine underlying soft tissues. The MRI can help the physician diagnose a variety of conditions faster and safer than ever before. It has no known side effects.

Ultrasound scanning, CAT scanning and MR imaging are relatively new technologies which use advanced, sophisticated equipment which is expensive. Not many hospitals can afford to have them.

➤FUNCTIONS

In addition to taking, developing and interpreting x-rays, and performing a variety of procedures described in the foregoing sections, the radiology department has the responsibility of engaging in research and participating in educational programmes for residents, interns, nurses and technicians.

➤LOCATION

The diagnostic radiology department should be located on the ground floor, conveniently accessible to inpatients, outpatients and emergency patients. It is also desirable to locate the department close to the elevators and near other diagnostic and treatment facilities.

Functional needs of the department are best served by locating the x-ray rooms at the end of a wing. In this way, the activities within the department will not be disturbed by traffic to other parts of the hospital. See Figs 4.7 and 4.8.

➤DESIGN

Due to the complex and highly specialized nature of radiological services, equipment and facilities, it is imperative that a radiologist and other experts are called and consulted in the earliest stages of planning of a new hospital. These experts will be familiar with the mandatory regulations of the Atomic Energy Regulatory Board (AERB) and the Division of Radiological Protection, Bhabha Atomic Research Centre (BARC) in designing the department.

The programme, the number of x-ray machines and equipment, and anticipated use of facilities will largely govern the size of the department, its staff, location and space. They must be planned in relation to other services (Fig 4.7, 4.8).

Flexibility in design is of vital importance, and it is necessary to provide for an increase in the

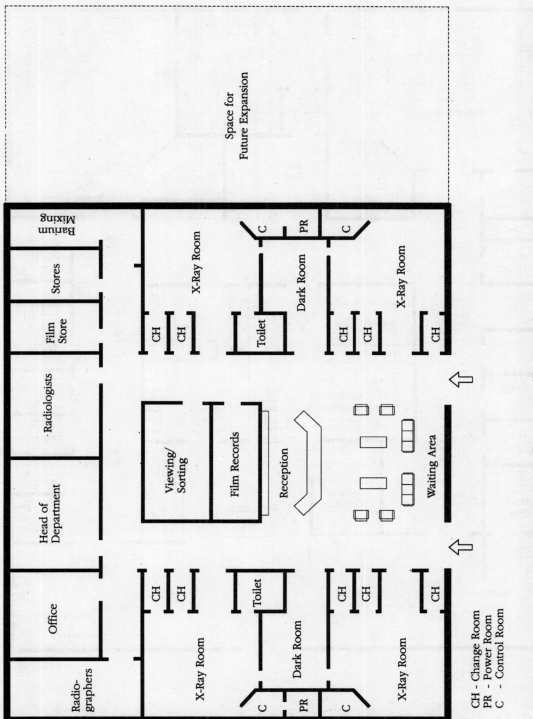

▷ **FIG. 4.7** Layout of a Diagnostic Radiology Department: Illustration 1

CH - Change Room
PR - Power Room
C - Control Room

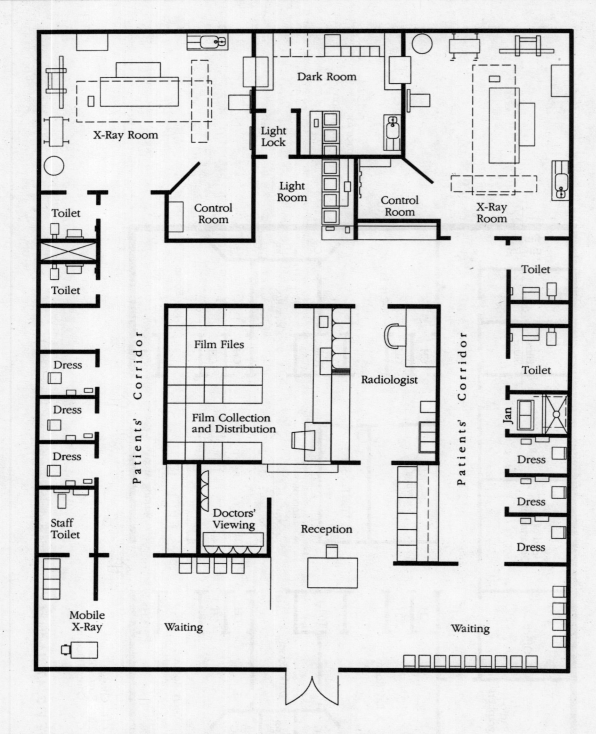

> **FIG. 4.8** Layout of a Diagnostic Radiology Department: Illustration 2

volume of work load and for expansion of the x-ray services. Studies have shown that as in the case of the laboratory, space requirements of the radiology department have tended to almost double every 10 years. Many hospitals having allotted inadequate space to the department, later find that expansion is practically impossible. Lack of adequate space results in needless waste of effort and time and in inefficient scheduling and operation of the department.

The diagnostic area should be built for efficient work and traffic flow (see Fig. 4.7 and 4.8). These rooms should be designed around a central core. The optimum size of the x-ray rooms is about 14 × 18 feet. The ceiling height varies depending on the type of x-ray machine. However, a minimum of 9 ft. 6 in. is recommended.

A control-monitoring generator room should be located between two diagnostic or fluoroscopic rooms. A holding area where stretcher patients are observed until films are checked is strongly recommended for radiography rooms. Each fluoroscopic room should have a toilet nearby but outside the room. One diagnostic room close to the entrance should be designated as a chest x-ray room.

There should be adequate number of dressing rooms for the patients so that equipment and staff can function without delay. In addition to clothes hooks, chairs, etc, lockers may be provided for patients' valuables.

An improperly designed and operated darkroom probably spoils more films than any other activity. A light-excluding entrance to and proper ventilation of the darkroom are necessary. It is desirable to have a film transfer cabinet between the radiographic and darkrooms unless there is a daylight automatic processor.

The reception control area should be located at the entrance to the department adjacent to cashier and billing and the waiting room. Toilets, lockers and the dressing area should be part of the waiting area. A wheelchair and stretcher storage alcove should be within the reception control area.

➤ORGANIZATION

The diagnostic radiology department is under the direction and supervision of a qualified radiologist. He may be a full-time or part-time salaried staff of the hospital. In small hospitals where a competent full-time radiologist's services are not available, he may visit the hospital two or three times weekly and go over the case histories with the physician in charge. He may receive a fee for services rendered. Fundamental relationships of a radiologist are with the medical staff, patient care units and outpatient department.

The department may have a chief radiographer and several radiographers or x-ray technicians. Radiographers and x-ray technicians are generally trained in two-year hospital-based training programmes. Medical colleges and the Christian Medical Association of India offer excellent radiographers' training programmes, the latter in its affiliated member hospitals.

Medical colleges offer training programmes for darkroom technicians. But in most cases, darkroom technicians and x-ray aides are usually trained on the job. X-ray aides provide support for the technicians in mixing chemicals and positioning patients. They may also escort or transport patients to and from patient rooms.

Other staff in the department are x-ray clerks and x-ray transcriber who, as the names indicate, perform clerical and filing work and transcription of reports respectively.

The radiology department provides 24-hour service. In smaller hospitals, the late night shift (11.00 p.m. to 7.00 a.m.) is generally staffed by on call personnel.

The activities flow of the department is as described below:

+ Reception and registration, billing and cashiering, scheduling.
+ Radiographic examination — x-rays are taken.
+ Development of film, manually or through automatic processor.
+ Reading and interpretation of film and dictation of report.
+ X-ray report transcription and distribution to all concerned.
+ Indexing and filing.
+ Patient escort or transportation.

Policies and Procedures

The department must have a written policies and procedures manual. The following polices, among others, should receive attention.

+ The radiologist should be responsible for all examinations and treatment as well as for x-ray interpretation and consultation.
+ There should be adequate number of trained and registered radiographers. At least one should be on call at all times.
+ Under no circumstances should the radiographers or x-ray technicians attempt to provide interpretation and diagnosis.
+ All reports should be signed by the radiologist.
+ Films and reports of examinations and treatment should be made available to the referring physicians.
+ Films are the property of the hospital.
+ Reports should form a part of the patient's medical records.
+ All films should be preserved for at least five to seven years or for the period covered by any statutory regulations.
+ All films should be legibly and permanently marked using markers or other appropriate devices.
+ Routine complete physical examination should be made periodically on all personnel of the department with blood counts/checks for those who are exposed to radiation.
+ The standards and regulations of the Atomic Energy Regulatory Board (AERB) and the

Division of Radiological Protection, Bhabha Atomic Research Centre (BARC), should be adhered to.

Physical Facilities and Space Requirements

Physical facilities and space requirements are as follows:
- Reception-control and registration area.
- Patients' waiting room.
- Secretary and file clerk's area for assembling, sorting and filing of films, reports, etc. and for transcription of reports.
- Doctors' viewing room near the office of the radiologist so that he can be available for consultation. The room is also near the film files. Privacy is needed so that doctors' comments are not overheard by patients. There should be adequate number of x-ray film illuminators.
- Radiologist's office, conveniently located near the x-ray rooms, should have provision for viewing and charting of films.

General Facilities
- Dressing rooms
- Patients' toilets
- Staff toilets and lockers — can be off site.

Diagnostic X-Ray Rooms
- Facilities are needed for routine diagnostic x-ray and fluoroscopic procedures. Rooms should accommodate equipment, patients, stretchers, staff and staff work area. Shielded control areas should have provision for viewing the entire table and patients and for two-way audio-communication during film exposure;
- Sink equipped with goose-neck spout and draining board for hand washing and rinsing utensils and barium equipment;
- Storage cabinets; and
- Writing counter.

Film Processing and Distribution Areas
- The darkroom should be located between two x-ray rooms with utility sink and draining board. If a through-wall processing unit tank is provided, it permits the radiologist or physicians to read wet films in the light room area without interrupting darkroom procedures.
- Film processing area: Processing begins at the developing tank in the tank room and continues in the light room.

◆ Collection and distribution area: After sorting and assembling, films are passed on to the radiologist for interpretation, and then temporarily filed for viewing by the doctors.

Barium Preparation Area

Provision for a two-compartment sink in a counter close to x-ray rooms. Not needed if premixed commercial preparation is used.

Storage Facilities

◆ Space for storage of active films in the collection and distribution area. Adequate space for five to seven years' storage.
◆ Inactive film storage area. May be off site.
◆ Storage space for unexposed films. Should not be warmer than the air of adjacent occupied area.
◆ General storage for bulk supplies.
◆ Daily linen supplies storage.
◆ Drugs/pharmaceuticals storage and storage for sterile supplies.
◆ Storage for crash cart.
◆ Storage for gowns.
◆ Janitor's storage.

Other Considerations

Radiation Protection

Protection against ionizing radiation in excess of tolerable limits is mandatory. Although modern x-ray equipment may have built-in safeguards to some extent, protection against wandering rays is necessary. Lead screens with lead-glass windows must be provided for the protection of the radiologist and technicians. The barrier design should be checked by qualified experts before the construction plans are approved and the completed installation checked with radiation detecting devices before the facility is put to use. Primary barriers should be provided on all surfaces of the x-ray rooms which are exposed and secondary barriers on surfaces of other rooms. Adequate protection to the darkroom is also necessary. Hospitals are advised to study the mandatory regulations of Bhalha Atomic Research Centre at the planning stage. They should also be familiar with the requirements in relation to staff once the facility becomes operational.

Theft in Radiology Department

Theft of x-ray films — the only easily marketable commodity in the department — is rare. A more frequent problem is the fraudulent use of x-ray equipment and films largely by technicians, sometimes in collusion with the radiologist. In small and medium sized hospitals where there may be only one technician on duty, particularly during evening and night shifts, and during weekends, it is easy for unscrupulous technicians to take x-rays and give away the films without charge. Internal control measures should be instituted.

RADIATION THERAPY DEPARTMENT

➤ OVERVIEW

Radiation therapy services are clinical specialties devoted to the treatment of patients with cancer and other tissue growths. The treatment is basically a tissue-destroying procedure. Cancer is treated by ionizing radiation either alone or in combination with other forms of therapy.

The department varies in size depending on the types of services and programmes offered and the sophistication of the equipment it has. Radiation therapy department in a well established modern hospital may have two treatment machines: a cobalt unit and a linear accelerator. Cobalt unit is used for deep radiation therapy, and the linear accelerator (Plate 5: Picture 24) uses recent technology to treat a wide range of oncology cases. Some hospitals may also have a superficial x-ray machine which is used to treat skin cancers. Currently, however, this is being done by electron energy from a linear accelerator. A simulator [(Plate 6: Picture 26), 27, 28]

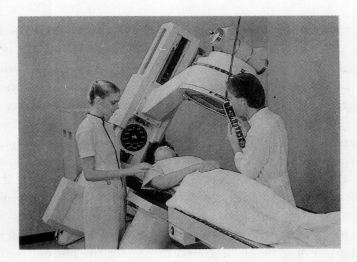

➤ **PICTURE 25** Radiotherapy and Radiotherapy Planning System which integrates x-ray, CT, radiotherapy planning and radiotherapy systems into a unitary concept

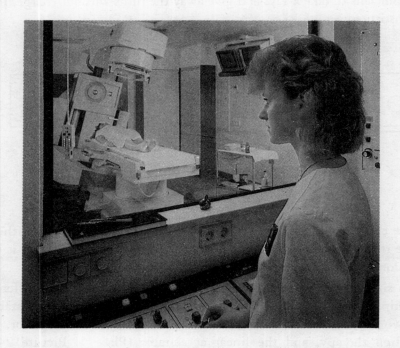

> **PICTURE 27**   Therapy Simulator: A C-arm Imaging System which achieves
> superior simulation accuracy for use in therapy treatment
> planning and tumour localization

is another important piece of equipment which is used to set the teletherapy fields to the location of the tumour. The simulator is also used to plan the treatment with accuracy so that the cancerous area receives the maximum radiation dose while the vital structures such as spinal cord, heart, oesophagus, etc. are protected from radiation. A comprensive radiotherapy and diagnostic system which integrates x-ray, CT, radiotherapy planning and radiotherapy systems into a unitary concept is shown in Picture 25.

Three areas or steps in the treatment process may be identified: (a) initial evaluation of the patient and development of a treatment plan; (b) determination of the work load related to the type of treatment plan; and (c) the application of radiation therapy as prescribed in the treatment plan. Treatment may include radiation therapy or chemotherapy or a combination of both. It may also include surgery and implantation of radioisotopes. The patients are first evaluated by their physicians and the oncologist. The radiation physicist, an important member of the team, then monitors the radiation dose being delivered to protect the normal tissues near the tumour.

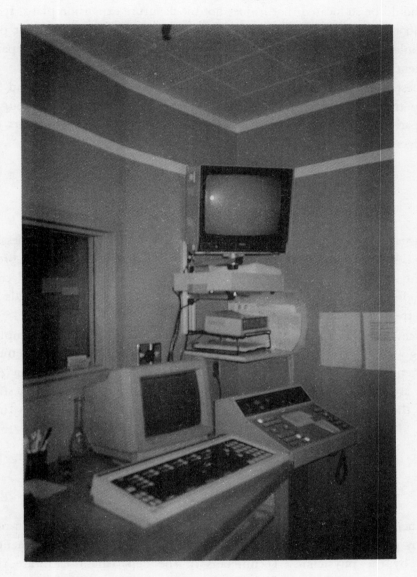

➢ **PICTURE 28** Simulator Control with Image Intensifier

➤LOCATION

The department must be carefully located in view of the need for dense shielding as a protection against radiation as mandatorily required by Bhabha Atomic Research Centre which is the regulatory body. The major factors that should be borne in mind in determining the location are the requirement of three-foot-thick walls and ceilings and the required access for the placement or removal of equipment. It should be so located that it does not block future expansion plans. It should be ideally on the ground floor in close proximity to the outpatient department. Since most patients are outpatients and are ambulatory, proximity to or easy accessibility from the outpatient department is important. The department should also be close to vertical transport facilities.

Radiation therapy is best located where it will adjoin the earth on several sides and has no department(s) below. From this point of view, a basement floor may be suitable. Although it is desirable to locate the department adjacent to diagnostic radiology, the two departments can be separated as they are not dependent on each other.

➤DESIGN

In order to overcome the usual grimness associated with a hospital set-up and the seriousness of the diseases treated in this area, the department of radiation therapy should be designed to present a warm, cheerful and pleasant atmosphere which concentrates on the patients' and visitors' comfort. Here, more than anywhere else in the hospital, the needs of the patients, families and friends extend beyond the actual medical treatment.

Radiation therapy services are not common to many facilities. Besides, they make a complex and specialized department (Fig. 4.9). We do not propose to cover all the mandatory requirements in the design of the department in accordance with the regulations laid down by Bhabha Atomic Research Centre. Hospitals are strongly advised to study these requirements at the earliest stages of planning and designing of the hospital. As in the case of diagnostic radiology, experts in the field should be consulted.

➤ORGANIZATION

The head of the radiation therapy department is a radiation oncology physician. In addition to other oncologists, he is assisted by physicist(s) whose responsibility is to develop detailed treatment plans for patients as prescribed by the physicians. Technicians in the department calibrate therapy equipment, verify all doses, position the patients before therapy, operate the radiation therapy units and deliver the prescribed doses to the patients.

Nurses assist the physician during examination of the patients and help the patients during treatment. They also deliver drugs and monitor patients during chemotherapy. A dietitian or a qualified nurse plans the nutritional needs of the patients and provides counselling.

Other staff in the department are a psychiatric social worker (who helps the patients and their families to cope and deal with the treatment and its anticipated outcome, secretary, clerk-typist for secretarial and clerical work and other support staff like cashier and housekeeping personnel.

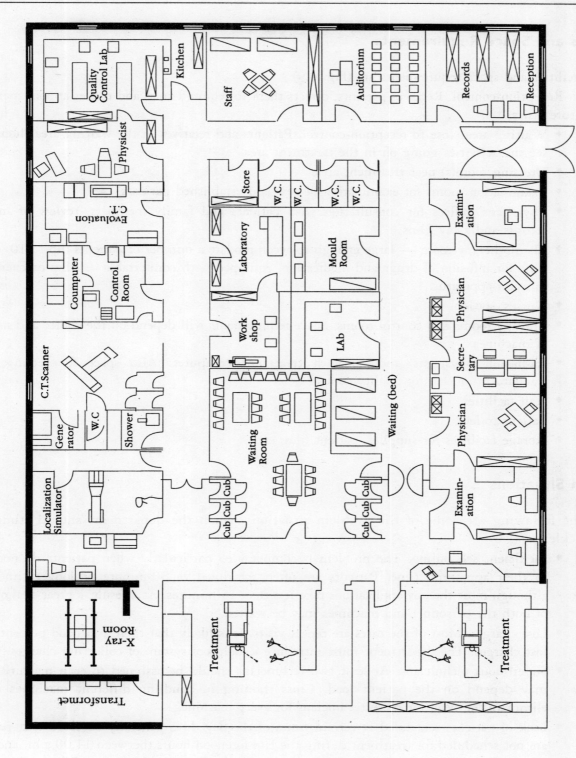

∧ **FIG. 4.9** Layout of Radiation Therapy Department

Facilities and Space Requirements

Facilities and space requirements are (Fig. 4.9):

Reception-control. Receives patients, directs them to required areas and makes appointments for future visits.

- Waiting area close to reception-control. Patients and relatives in the waiting area should not see the activities going on in the treatment area.
- Dressing room(s) near treatment area.
- Examination rooms for examination of new and established patients.
- Physicians' offices for consultation with patients and family members, review of medical records and x-ray films.
- Chemotherapy room — large enough to accommodate a number of patients (up to 10) at one time for infusion of drugs and medication, equipped with comfortable couches or chairs and infusion apparatus.
- Nurses' station.
- Therapy rooms and control rooms. Space requirements will depend on the choice and number of machines.
- Radiation physicists' room(s) with treatment computer. Also space for setting up a standardization lab for quality control .
- Staff facilities.
- Patient toilets.
- Storage facilities for supplies, patient files, etc.

Problem Situations

The following are some of the problem situations which the department should study and tackle:

- Inefficient scheduling: The problem is compounded particularly when patients do not keep to their appointed time. Patients are often scheduled in 30-minute or longer time slots although most therapy applications take less than 15 minutes. As a result, a great deal of time of both the personnel and machines may be wasted.
- Lost charges: Most of the cases are repeat visits. It is likely that there is a good percentage of lost charges. The department must introduce a foolproof system of collecting charges.
- Superfluous technicians: At least two technicians should be assigned to each machine. This may depend on the patient load. Cross training in handling different machines might eliminate the need for more technicians for every machine.
- Lack of patients scheduled at certain times of the day: It is generally observed that patients are not scheduled for treatment during the late forenoon hours (between 11.00 a.m. and 1.00

p.m.) and late afternoons (after 3.00 p.m.). This results in personnel and machines remaining idle and loss of revenue to the department.

- One way of organizing a more efficient schedule is to treat the inpatients in the early morning hours, and the outpatients during rest of the day. Treatment should be scheduled for six days of the week, and on Sundays in special cases. Emergency treatments should be on seven days of the week and 24 hours of the day with physicians, technicians and physicists being on call.

- High percentage of down time of radiation therapy machines. Machines not in working order creates problems in patient scheduling and revenue. The department may have signed an annual maintenance contract with an outside firm which may or may not have service personnel or office in the same town. Large hospitals can have their own maintenance personnel on the hospital's payroll.

NUCLEAR MEDICINE

➤ OVERVIEW

The introduction of nuclear techniques in medicine has led to the evolution of a new branch of medicine called Nuclear Medicine. This new branch utilizes radio pharmaceuticals for the diagnosis of certain diseases, their follow-up and detecting recurrence. It also deals with the treatment of certain diseases using radio pharmaceuticals.

Nuclear medicine procedures are of two types: (1) In-vivo procedures in which radioisotopes are administered to patients, and (2) In-vitro procedures where radioactivity is added to the samples collected from the patient. In-vivo tests are classified into organ imaging procedures and non-imaging procedures.

With the availability of many short-lived radioisotopes, nuclear medicine has now become an important medical specialty with wide applications in various branches of medical science. Some of the important applications of nuclear madicine are — imaging of various organs such as heart, thyroid, liver, brain, bone, kidney, etc., thyroid function studies, investigations of the central nervous system, absorption studies in gastroenterology, nuclear haematology and renal function studies like radioimmunoassay of various hormones. With the help of an on-line computer coupled to a gamma camera, a variety of dynamic function studies can also be performed using appropriate radio-pharmaceuticals (Picture 29).

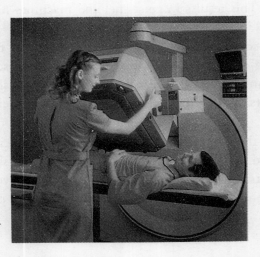

> **PICTURE 29**   Nuclear Medicine System: A large rectangular field gammacamera which revolutionized nuclear medicine with what is called Emission C.T.

➤ LOCATION

Since the nuclear medicine department caters to the needs of all clinical departments, it should be located at a central place. At the same time, because of radiation hazards associated with the use of radionuclides, planning of the department should be done in such a way that there is no radiation exposure to non-radiation employees and the general public, and also that radiation workers handling radioisotopes receive minimum exposure. A major problem always is the safe and proper disposal of radioactive material.

The department should be close to diagnostic radiology, outpatient services, social services, laboratory and medical records. Many patients are ambulatory. In some hospitals, nuclear medicine is a part of diagnostic radiology using the common facilities of the department.

➤ DESIGN

When a hospital decides to set up a nuclear medicine department, the authorities are faced with a number of questions regarding location, planning the facilities, equipment and availability of trained personnel. More importantly, they would want to know the procedure for obtaining clearance from various regulatory bodies. The first step in this process is to appoint at the earliest stage an expert or a team of experts who will guide them in the right direction.

With a view to encouraging the growth of nuclear medicine in the country so that benefits of nuclear medicine techniques can be made available to patients in all parts of the country, the Division of Radiological Protection (DRP) of the Bhabha Atomic Research Centre has brought out publications providing guidelines to hospitals wishing to set up a nuclear medicine department and help them in obtaining information regarding location, planning, equipment and procedural details. Hospitals are encouraged to contact the Board of Radiation and Isotope Technology, VN Purav Marg, Deonar, Bombay—400094.

SURGICAL DEPARTMENT

➤ OVERVIEW

The operating rooms, scopy room(s) — cystoscopy, gastroscopy, laparoscopy, etc. — pre-operative holding and preparation area and post-operative recovery room(s) make up the surgical facilities. The pre-eminent position of the surgical department in the hospital can be appreciated when one realizes that in a typical general hospital, surgical patients represent 50 to 60 per cent of the admissions, and a good percentage of them have an operation performed. They also account for an appreciable quantum of the work of and revenue from ancillary departments.

Surgical patients are of three types — inpatients, outpatients and ambulatory surgery patients. Inpatients, as the name indicates, are those who are hospitalized for surgery, the length of their stay before surgery depending on their condition and the time needed to complete pre-surgery tests. Outpatient surgery is generally limited to minor surgical procedures requiring local anaesthesia. Patients are admitted for surgery and discharged the same day. The terms ambulatory surgery (also called the same day surgery) and outpatient surgery, are often but erroneously used interchangeably. Ambulatory surgery generally involves more extensive procedures than outpatient procedures and may require general anaesthesia. The patient has his pre-operative tests done earlier as an outpatient, and gets admitted in the afternoon of the previous day to be prepared for surgery the next day, or arrives very early on the same day for surgery — this is called A M admit — recovers in the recovery room, is observed in a post-recovery area and is discharged on the same day. The present trend is increasingly for ambulatory surgery. As much as one-third of all surgery can be performed as same day surgery.

The advantage of ambulatory and outpatient surgeries is that they do not require hospitalization, and consequently minor procedures become less expensive.

Most hospitals have minor operating rooms in their outpatient departments. Depending on the degree of sophistication, emergency rooms may have either minor or full-fledged operating rooms and recovery rooms.

➤LOCATION

The best location for the surgical department is the one which permits a convenient and uncomplicated flow of patients, staff and clean supplies traffic. Surgery receives inpatients from the floors through non-public corridors, elevators and ramps. In most cases, they are returned after surgery through the same route. Convenient access to elevators is, therefore, essential. In regard to services, ideal adjacencies would be emergency, radiology, clinical laboratories, intensive care units, central sterile supply department and delivery suite, particularly if it does not have a room for Caesarean section.

➤DESIGN

The surgical suite of a modern general hospital and everything that go with it make a very complex workshop (Figs 4.10, 4.11, 4.12). The surgical procedures of the present day, involving more people and highly sophisticated equipment, have rendered ideas of planning of operating rooms of the past somewhat obsolete. Even more complicated is the question as to how to make the functioning of the operating rooms over long and tense hours, smooth and comfortable. To the architect and decision-makers, there is one other vexing problem which makes planning and designing of this area more difficult. This stems from the diversities of opinion and experience of the many persons who are directly or indirectly involved in the work of operating rooms and in policy making (Plate 6: Picture 30).

A great deal of intelligent planning is necessary. There must be a meeting of the minds among the administrators, surgeons, anaesthesiologists, surgical nurses and others concerned on such important questions as the number and type of operating rooms and the work pattern to be followed in the supportive areas. The architect must have had experience in designing hospitals. He must have a good understanding of the complexities of the operating rooms and of various management procedures, and must ensure that all the questions are discussed with every group involved before reaching conclusions.

The major decision centres round the number and type of operating rooms. While planning and equipping each operating room, a series of questions need to be answered. They are — the size, usage, lighting (surgical and general illumination), intercommunication and signal systems, electronic equipment and monitoring system, medical gas system (suction, oxygen, nitrous oxide and compressed air), and other service lines, safety precautions such as grounding for x-ray, TV camera, and against static electricity, storage, supply cabinets, clocks, film illuminators, environmental control, etc.

PLATE 6

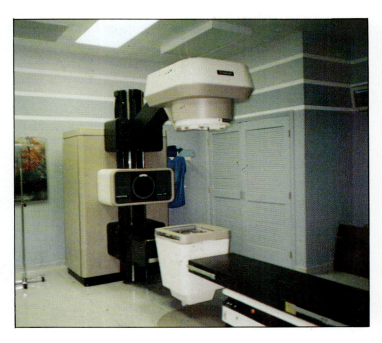

➢ **PICTURE 26** Therapy Simulator

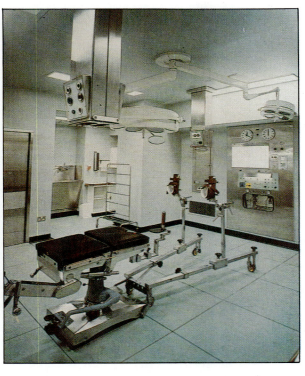

➢ **PICTURE 30** View of the Operating Room with control panel, operating table, ceiling pendant, etc.

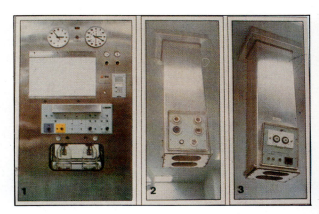

➢ **PICTURE 31** 1. A typical Fixed Services Control Panel.
2 & 3. Ceiling Pendants which provide electrical and medical gas outlets

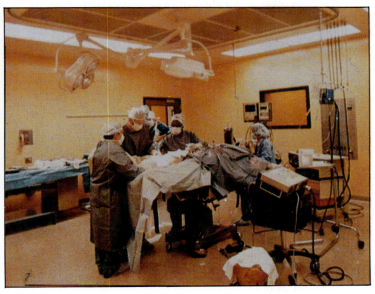

➢ **PICTURE 36** An operation in progress

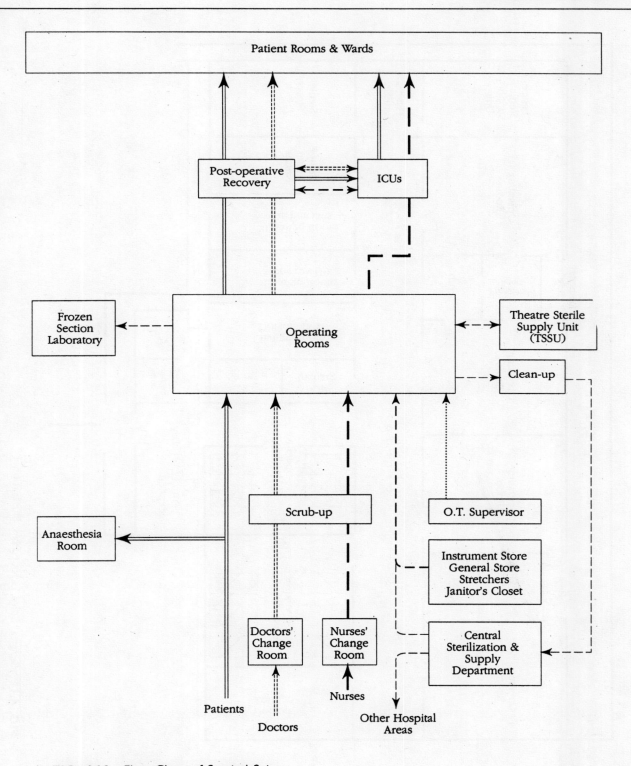

➢ **FIG. 4.10** Flow Chart of Surgical Suite

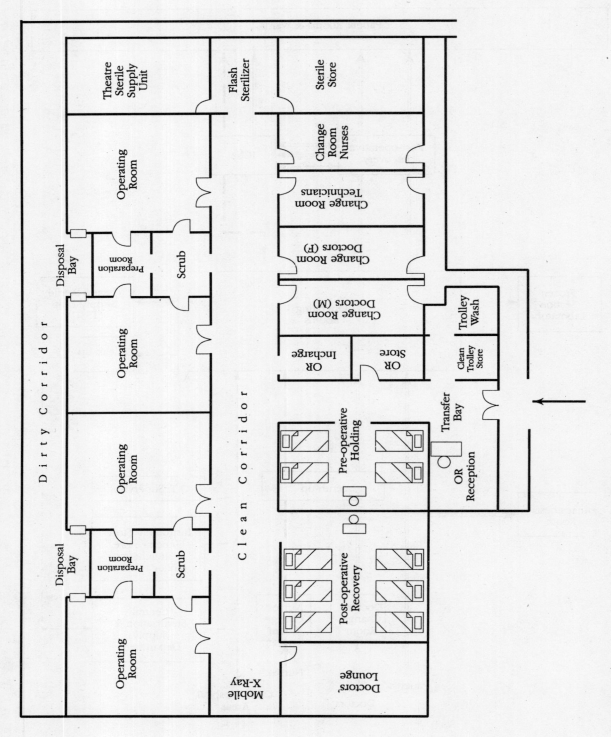

∧ **FIG. 4.11** Surgical Suite Layout Plan: Illustration I

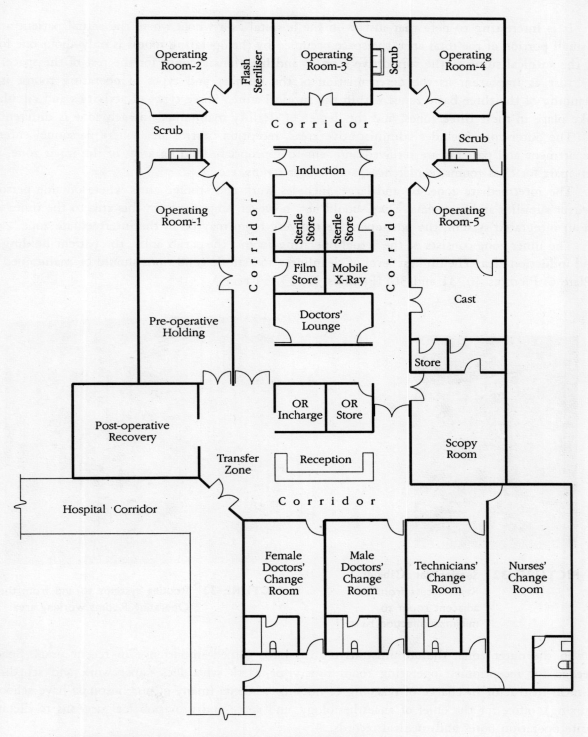

> **FIG. 4.12** Surgical Suite Layout Plan: Illustration 2

It is interesting to note that just as in the hospital as a whole where the actual patient area is a small portion of the total area, the space occupied by the operating rooms is only about one fourth of the surgical suite — the supportive services and functions account for the rest of the space.

Just as important as the determination of the number and types of operating rooms is the planning of the three basic zones within the surgical suite. Three types of activities and circulation take place in these three zones, and the degree of sterility maintained in each zone is different.

The outer zone includes administrative areas, reception-control area where personnel enter the department and patients are received and sent to appropriate holding areas of the inner zone, areas set apart for class rooms, conferences, staff locker rooms, etc.

The intermediate zone, by and large, includes work and storage areas where outside personnel deliver supplies and materials. They should not, however, step out from this area to the inner zone. In an integrated system, the post-operative recovery room may be in the intermediate zone.

The inner zone consists of the actual operating rooms, the scrub areas, the patient holding area and induction area. The highest level of cleanliness and sterile conditions should be maintained here (Plate 6: Pictures 30, 31 and 36) (Pictures 32, 33, 34 & 35).

➤ **PICTURE 32** Suction Jar Cabinet: Rear Access from an adjacent room to minimize infection risks

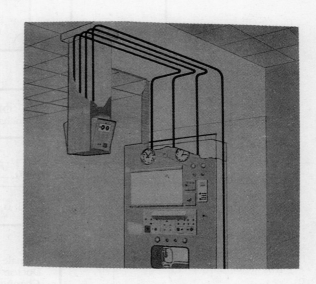

➤ **PICTURE 33** Routing systems to and from the Operating Rooms working area

In the outer zone, the administrative areas have gained importance in recent years. Space is needed for receptionist, operating room supervisor, clerk who does paper work and scheduling, clinical instructor in charge of teaching of nursing students (many of our hospitals have schools of nursing), office for the chief of anaesthesiology, and space and provision for surgeons to dictate or write operation notes and medical records.

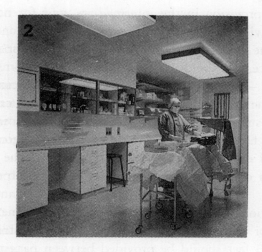

> **PICTURE 34** Preparation Room

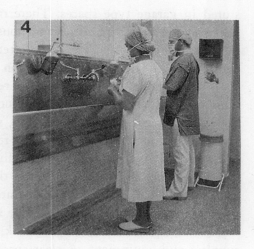

> **PICTURE 35** Scrub Room

Various personnel have to be provided with lounges, lockers and toilet facilities (Picture 37). There should be a coffee room. Operating rooms have not fixed lunch hours. Quite often, personnel need to have their lunch or snacks in between operations. Arrangements must be made for their supply. A conference or class room for meetings and in-service training programmes is necessary. All these areas should be removed from the strictly surgical areas and persons entering and leaving them in street clothes should not wander or penetrate into sterile areas without changing clothes and footwear.

Some attention should be given to planning, designing and equipping the intermediate zone and to processing and storing of a large number of items in that area. In most hospitals, the central sterilization & supply department (CSSD), which is usually located elsewhere in the hospital,

> **PICTURE 37** Doctors' Lounge

is responsible for the preparation and autoclaving of all surgical instruments, linen packs, gloves, syringes and needles. However, it is the responsibility of the surgical department to store them. In some hospitals, the surgical packs, etc. are washed, packed and made ready for autoclaving in the packing room in the intermediate zone and then sent to the CSSD for autoclaving. There are a host of other things that need storage space — clean surgical supplies such as linen, tapes, bandages, parenteral solutions, other fluids, sterile water, essential drugs and narcotics, blood supplies, and in some cases, radium and isotopes used in surgery which need special storage. In addition, equipment used in the department, some of them bulky items such as 'C' arm and portable x-rays need to be stored in garage type spaces. There must also be provision in the intermediate zone for handling waste, soiled linen, janitorial and housekeeping services.

Anaesthesia services which extend to all the zones of the surgical suite call for special consideration. The anaesthesia section requires office, work and storage areas.

The surgical or post-anaesthesia recovery room — not to be confused with intensive care units which are situated elsewhere — located near the operating rooms, permits medical care and observation by the anaesthesiologist, surgeon and surgical resident in the first hours following an operation through the use of special recovery room facilities. Nursing is directed by a surgical head nurse assisted by staff nurses specially trained and experienced in post-operative care. The size of this section will vary depending on the number of operating rooms — the rule of thumb is from one and a half to two beds per operating room.

Each recovery room should have a nurses' station with charting facilities, medication cabinet, hand washing facilities, clinical sink, provision for bedpan cleaning, storage space for stretchers, supplies and equipment. A clear space of at least 3 feet should be provided between patient beds, and between the patient bed and the adjacent wall. Provision should be made for isolation of infectious patients. Recovery rooms as well as the pre-operative holding area should have medical gas outlets.

If there is a frozen section laboratory in the surgical department, it should be located near the entrance of the surgical suite. That way laboratory personnel need not enter the inner zone. There should be an intercommunication system between the laboratory and the operating which should be operable without the use of hands or hand-held receiver sets.

There should be a darkroom which should be located close to cystoscopic, urological and orthopaedic services which usually generate the greatest load of films. The darkroom should be accessible from a corridor that will prevent outside traffic coming through the sterile area.

The staff clothing change area for men and women should be arranged for one-way traffic so that personnel entering from outside will move directly into the operating rooms after changing, without retracing steps. It should also be planned in such a manner as to make it unnecessary for doctors, nurses, etc. to enter the clean area in street clothes.

The planning of the inner zone includes the operating rooms and their supportive elements like scrub up sinks or troughs. The scrubbing area should be so located as to provide minimum travel to the operating room so that chances of contamination after the scrubbing procedure are eliminated. The type of scrub sinks and their positions should be such that there is a minimum of water splashing on nearby personnel, equipment or supply carts. Two scrub positions should be provided at the entrance to each operating room. Three positions may serve two operating rooms if they are located adjacent to each other.

The scrub sinks should be equipped with hot and cold water mixer taps which are foot-controlled. Automatic taps which are activated by infra-red sensors are now available in the market. These can be conveniently used.

Another major decision that should be made in the planning stage is whether the operating rooms should have windows. Opinion is divided. One school of thought strongly feels that outside light interferes with proper light control in the operating room, acts as an annoying distraction to the surgeons, adds extra load to the cooling system and presents problems with double sets of windows. One reason for opposing this idea is psychological. If there are windows, arrangements must be made to darken the rooms for certain procedures by providing blinds or some such means.

It is recommended that surgical corridors are not less than 10 to 12 feet wide. They are often found lined up with occupied stretchers for want of adequate holding, preparation or induction areas. This should not be the case. Hospital planners must be aware of this problem. Adequate space must be provided. An efficient scheduling system will obviate this problem to some extent.

Another important factor that should be borne in mind while planning, designing and deciding the location of the surgical department is the potential for future expansion. The design and location must permit growth in an orderly fashion without upsetting relationship in the internal organization. The surgical suite is one of the most expensive units to construct in the general hospital. It is frequently under-designed in a way that does not lend itself for future expansion. Surgery and anaesthesiology have expanded so fast in technique and extent that surgical suites are rapidly becoming obsolete.

In general hospitals, the tendency is to have all major operating rooms as nearly identical as possible so that scheduling of various kinds of surgery is possible. Operating rooms must have a minimum clear area of 360 sq. ft. ($18' \times 20'$) excluding fixed cabinets and built-in shelves. Many surgeons, however, recommend larger space — 20 ft. by 24 ft. for major operating rooms and 24 ft. by 30 ft. for special procedure rooms. Each operating room should have x-ray film illuminators which will handle at least two films at the same time, and an emergency communication system that can be activated without the use of hands for contact with the surgical suite control station and the frozen section laboratory.

An orthopaedic surgery room should have, in addition, an enclosed storage space for splints and tractions. If this storage is outside the operating room, it should be easily accessible. If plaster of Paris is used for cast work, a plaster sink should be provided.

The rapidly advancing cardiac surgery and neurosurgery units in specialty hospitals require extra large operating rooms as these types of surgery need a larger team of surgeons, nurses and technicians in addition to a great deal of extra equipment such as heart-lung machine. They also call for electronic devices like ECG, EEG, etc. for measuring bodily functions. One way of accommodating such equipment is by providing an instrumentation room adjacent to or between two extra large operating rooms with a floor approximately three feet higher than the floor of the operating rooms. Glass panels permit vision into the operating rooms.

The question that is often asked is how many operating rooms are required in a general hospital. There is no satisfactory, scientific answer to this question. The old rule of thumb is one major room for each 50 beds. In addition to major rooms, the general hospital will require one or more minor operating room(s), a scopy room and a fracture room. In specialty hospitals, there is need for special rooms for cardiac, neuro and orthopaedic surgery.

►ORGANIZATION

The surgical department may be organized around five main groups of staff — the surgeons, anaesthesiologists, nurses, technicians and aides/orderlies. Surgeons and anaesthesiologists may be full-time, hospital-based, salaried staff. However, it is not unusual to find them, not being regular staff of the hospital, operating and administering anaesthesia on a fee-for-service basis.

The surgeons' group consists of surgeons from all specialties — general, orthopaedic, neuro, cardiac, vascular, plastic and reconstructive surgery, ophthalmology, otolaryngology, gynaecology, urinogenital surgery, oral surgery, etc.

Anaesthesiologists are medical doctors who administer anaesthesia and provide medical direction in the recovery room. Anaesthetists in developed countries are certified nurse anaesthetists who work as assistants under the direction of anaesthesiologists. Nurse anaesthetists are not popular in our country. However, some hospitals have trained anaesthesia technicians.

The third group in the operating room consists of the operating room supervisor who is generally a senior nurse, and staff nurses. The nursing team usually consists of a circulating nurse and a scrub nurse — the former functions outside the sterile field surrounding the operating table. She coordinates activities, gets and hands supplies and specimens into and out of the sterile area. The scrub nurse assists the surgeon and organizes and passes instruments. Some operations require two scrub nurses, and the circulating nurse may need assistance. Bedsides the nurse, the surgeon may have another assistant, a medical doctor, during certain procedures. He retracts the surgical opening and assists the surgeon in the operating procedure.

Technicians are commonly used in specialty and teaching hospitals where complex procedures like cardiac surgery, are performed. Thus we have technicians, for example, who operate the heart-lung machine and other equipment in the instrumentation room. Another commonly found technician is the x-ray technician.

Orderlies and attendants are generally used for patients' transportation, patient positioning, cleaning work, mopping of floors and disposal of trash. Aides clean, check and wrap instrument kits.

►FACILITIES AND SPACE REQUIREMENTS

Facilities and space are required for
  ◆ Control station. Should be so located as to permit visual observation of all traffic into and within the department.
  ◆ OR supervisor's office.
  ◆ Operating rooms as required.
  ◆ Pre-operative holding and preparation area.
  ◆ Induction area.

- ◆ Recovery room(s).
- ◆ Sterilizing facilities. High speed autoclaves/flash sterilizers conveniently located to serve all operating rooms.
- ◆ Medication storage with refrigeration facilities.
- ◆ Scrub facilities as described earlier (Picture 35).
- ◆ An enclosed soiled work room with a clinical sink, work counter, waste receptacle and linen receptacle. This room can also be used for soiled holding.
- ◆ Fluid waste disposal facilities.
- ◆ Clean work room/clean supply room.
- ◆ Medical gas storage facilities for reserve gas cylinders.
- ◆ Anaesthesia work room for cleaning, testing and storing anaesthesia equipment with work counter and sink.
- ◆ Equipment storage room.
- ◆ Staff clothing change area for all categories of staff — doctors, nurses, technicians and orderlies — separate for men and women — with lockers, toilets, hand washing facilities, shower and place for changing.
- ◆ Staff lounge and toilet facilities. May be combined for male and female personnel, but preferably separate for doctors and nurses. Should be so located that the doctors and nurses can have access to recovery rooms without leaving the surgical suite.
- ◆ Dictation and report preparation area — accessible from the lounge area.
- ◆ Storage area for portable x-ray equipment, stretchers, other items of equipment and materials.
- ◆ Janitor's closet with service sink and storage of housekeeping equipment, supplies, etc.
- ◆ Laboratory for preparation and examination of frozen sections. This can be done in the main laboratory if the distance and time needed do not delay the completion of surgery.
- ◆ Provision for refrigerated blood bank.
- ◆ Consultation and conference room.
- ◆ Administrative and clerical area for clerical work, scheduling, etc.
- ◆ Family waiting room(s) outside the operating rooms complex — privacy essential.

If ambulatory surgery is performed in the main operating rooms, the following additional facilities should be provided.

- ◆ Ambulatory surgery change area where patients change from street clothing into hospital gowns. Should include lockers, toilets, and waiting area.
- ◆ Place for preparation, testing and obtaining vital signs of patients for ambulatory surgery.
- ◆ Ambulatory recovery. It is recommended that ambulatory surgery patients not subjected to general anaesthesia are separated from inpatients.
- ◆ Recovery lounge for patients who are able to leave or do not require post-anaesthesia recovery, nevertheless, need additional time for vital signs to stabilize before the patient may safely leave the hospital. A control station, space for family members and toilet should be included.

LABOUR AND DELIVERY SUITES

➤OVERVIEW

Gynaecological and obstetrical services have been customarily combined in general hospitals, and this practice continues. While strictly speaking, obstetrics deals with the process of childbirth, it is practically impossible to separate it from the problems and complications of pregnancy which is natural and without complications in a majority of women. Gynaecology treats disturbances of the function and diseases peculiar to women. This does not include surgery of the breast. Nor does it deal with any disease not directly concerned with the process of reproduction.

Routine gynaecology needs no special accommodation and may be treated in the gynaecology section of surgical wards. However, modern gynaecology, specially laparoscopic surgery (keyhole surgery), requires special equipment and is performed in the operating rooms.

Obstetrical services require special facilities. Hospitals have a responsibility to furnish safe and efficient obstetrical care which should include the utmost safety and comfort of the mother and the newborn child. Usually patients are kept in a separate wing of the hospital to avoid infection.

Two specific areas make up the obstetrical department — (a) patient accommodation which may consist of private and semi-private rooms and general wards, and (b) clinical facilities which consist of a preparation room, pre-delivery or labour room(s), birth or delivery room(s) and nursery.

In addition to the above mentioned facilities, the obstetrical department requires the services of a host of adjunct and ancillary departments such as clinical laboratory, x-ray, ultrasound, etc. For this reason, the development of a separate maternity hospital is not encouraged. It should be a part of the general hospital.

In this section, we discuss planning for facilities in the different units that go to make the clinical part of the obstetrical department, except the nursery for which a separate section has been devoted later in this book.

➤LOCATION

The clinical section of the obstetrical department should be located in a convenient but segregated area. Within the clinical facilities, the labour and delivery suites should be as remote as practicable but easily accessible from the entrance to the department so as to avoid non-related and unnecessary traffic through the suites, and to provide privacy to the patients. The facility should be close to the nursery, obstetrical nursing unit, and to vertical transport so that it is easy of access to patients. The department should be in close proximity to operating rooms. This is particularly important if it does not have an operating room of its own. The different units within the clinical section of the department should be located in such a manner that they facilitate movement of patients between them and observation of patients by the unit personnel.

Some hospitals, especially those housed in small buildings, locate the delivery suite adjacent to the surgical suite. There are obvious advantages, and some disadvantages. The advantages are that the two departments require the same isolation, the same type of nursing service, same cleaning, air-conditioning, sterile supplies, and so on. However, inter-traffic between the two is not desirable, and there is always a possibility of cross-contamination. If the two suites are close to each other, an effective control system becomes absolutely necessary.

➤ DESIGN

The delivery suite is very much like the operating rooms, and the same general considerations of location and control hold here. In designing the labour and delivery suite, attention must be focused on four or five major functional areas where much of the activities take place. These are — preparation room, labour room, delivery room, recovery room and support services area (Figs 4.13, 4.14, 4.15).

Preparation Room

Many of the patients are admitted in labour. They have to be prepared for labour and delivery. The patient receives a cleansing bath, is shaved and given an enema before being sent to the labour room. Although this preparation can be done in the patient's room, particularly in the case of private room patients, it is preferable to have a special room for this purpose. The preparation room should be located outside although adjacent to the labour and delivery suite. The room must have facilities for carrying out the procedures in connection with the preparation, such as an examining table, shower bath, wash basin, equipment for giving an enema and preparation tray. A locker to keep the patient's clothes is desirable.

If the hospital has an adequate number of single labour rooms, they eliminate the need for a patient preparation room and an observation room for suspected infectious patients.

Labour Room(s)

This is the room in which the patient remains during the first stage of labour, that is, from the time the pains commence until she is ready to be moved to the delivery room.

Labour rooms must be designed in such a way that they can serve as emergency delivery rooms.

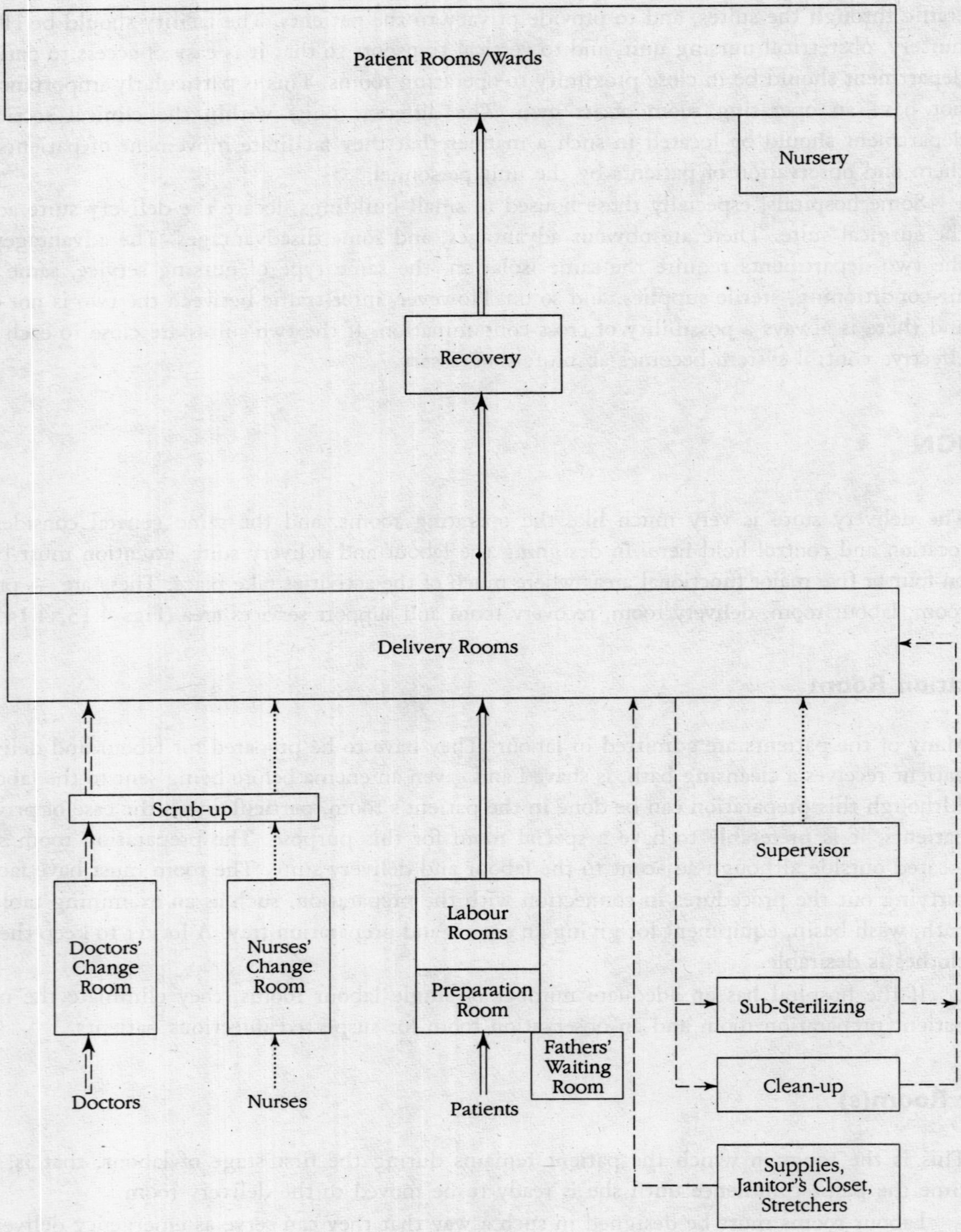

Patient Rooms/Wards

Nursery

Recovery

Delivery Rooms

Scrub-up

Supervisor

Doctors'
Change
Room

Nurses'
Change
Room

Labour
Rooms

Preparation
Room

Fathers'
Waiting
Room

Sub-Sterilizing

Clean-up

Doctors

Nurses

Patients

Supplies,
Janitor's Closet,
Stretchers

▲ FIG. 4.13 Flow Chart of Obstetrics Department

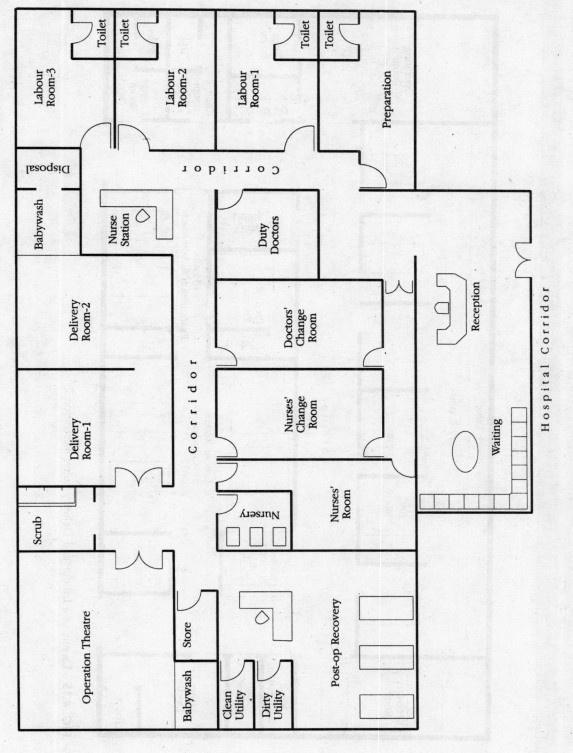

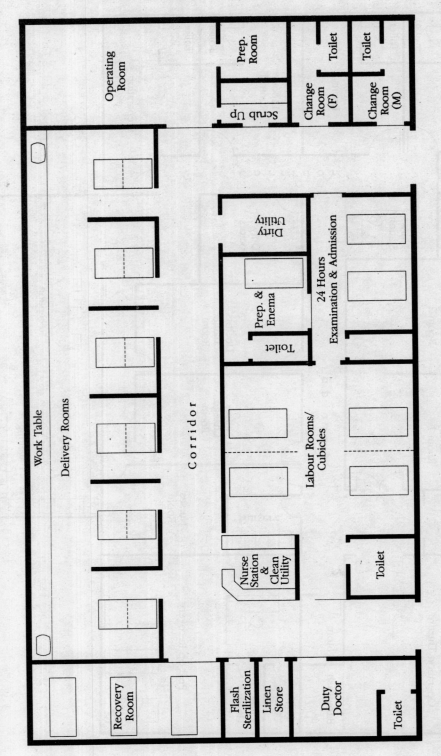

➤ **FIG. 4.15** Layout of a Labour and Delivery Suite: Illustration 2

To that end, they must be of adequate size, preferably, 18 ft. by 18 ft. Single rooms are recommended. Single rooms provide greater privacy for the patient and permit the father to visit during labour. The labour rooms must provide maximum comfort and relaxation to the patient, and should have facilities for examination, preparation and observation. They should be equipped with electronic foetal heart monitors for monitoring foetal heart beat.

The labour rooms should be close to the delivery rooms, but not so close that the two areas are intermixed or that the patients can overhear or view delivery room procedures. Both rooms are noisy — as a matter of fact, there will be more noise here than in any other part of the hospital — and so should be sound-proofed. Lighting and colour should be conducive to patients' relaxation.

The furniture in the labour rooms is much the same as in any other patient room but there should be a good extension light so that the patient can be properly examined by the obstetrician or the nurse. Besides, there should be toilet and bedpan flushing facilities, a wash basin with gooseneck type spout and foot or wrist operated controls for hand washing by the patients, doctor and nurse and a clock with a second timer.

Doors should be four feet wide to permit passage of the bed or stretcher with attendants. The bed must be furnished with oxygen, suction and compressed air outlets, nurse call system, and lighting controls. Piped—in music is desirable, as also a lockable locker for patient's personal belongings.

Delivery Room(s)

Delivery rooms should be similar to operating rooms in their design with finishes that promote maximum aseptic conditions. Facilities provided in the delivery rooms are essentially the same as for the surgical suite, such as scrub-up area with windows for view into the delivery room, lighting and operating lights, oxygen, suction and air, a clock with a second timer, built-in protection against explosion hazards, equipment and supplies.

A delivery room should accommodate only one patient at a time. It is difficult to maintain aseptic conditions in a room where more than one patient is being delivered. With two or more patients in the same room, there is a high risk of babies getting mixed up.

The delivery room should also contain a designated area to receive the newborn baby immediately after birth. This area should have a baby receiving tray, warmer, suction, oxygen and other resuscitating facilities like Ambu bag. There should be adequate lighting around the area.

Each delivery room should have a foot or elbow-operated emergency call system with a dome light and buzzer on the corridor over each delivery room. Similar light signals and buzzers must be installed in the lounge and at the nurses' station. There should also be a nurses' intercom system.

The minimum size of a delivery room should be 18 ft. by 18 ft. It should be larger if it is used for Caesarian section.

Caesarian Section Room
It is strongly recommended that there should be an operating room in the department where major

obstetrical surgery is performed. If this is not possible, one of the delivery rooms may be completely so equipped. It should, however, be remembered that maintaining aseptic conditions comparable to those of the operating room may not be possible in a delivery room set-up. The operating room should have the facilities required for the care of the newborn as found in the delivery room.

➤ Scrub Facilities

Two scrub positions should be provided near the entrance to each delivery room. One area with three positions may serve two delivery rooms if they are located adjacent to each other. It is desirable to have a viewing window at the scrub station to permit observation of delivery room interiors.

Recovery Rooms

Recovery rooms should be designed for close observation and special care of the mother by the labour-delivery staff. Every patient who has had general anaesthesia should be given constant nursing care until fully conscious. Modern maternity departments tend to have a recovery room comparable to the surgical post-anaesthesia recovery room where all post-partum patients are watched by the nurse or the doctor at least during the first six hours after delivery for evidence of delayed haemorrhage. Although recovery can be in the delivery room, the labour room or in the obstetrical nursing unit, a room used exclusively for this purpose is recommended so that the patients can be under close observation by the nursing staff.

Each recovery room can have two or more beds. Bedsides, it should have a nurses' position with charting facilities and visual control of all beds. Provision must be made for medicine dispensing, hand washing, clinical sink with bedpan flushing device and storage for supplies. Suction, oxygen and compressed air outlets are necessary.

There is some merit in what some advocate as labour-delivery room concept whereby the patient is placed in a labour room, is moved to delivery, and then returned to the same labour room for recovery care. The advantages are, among others, more privacy for the patients, minimum of utility outlets (such as nurse call, suction, etc.) and less of cleaning up procedures for the staff. Placed close to the entrance of the delivery suite, they allow the fathers to visit during labour and recovery without penetrating into the delivery suite. The disadvantage is that this requires more labour rooms.

Fathers' Waiting Room

This should be conveniently located near the labour room(s) with toilet facilities, pay telephone and drinking water fountains. Provision for personal communication between fathers and staff by means of an intercommunication system is desirable.

➤BIRTHING ROOMS AND LDRP SUITES

The LDRP suites which are an improved version of the birthing rooms of olden times are becoming increasingly popular in the west and are slowly catching up in our country. These elegantly designed suites offer the mothers-to-be the option of a family-oriented maternity care system which accommodates the entire birthing process — labour, delivery, recovery and post-partum care (LDRP) — in one room except when the mother-to-be is a high risk case. The facility offers best of both the worlds — the comforts of the home in the security of the hospital setting with the best of medical care.

LDRP suites are equipped for labour and delivery. A stylish bed opens to become a birthing bed. Examination lights are neatly placed in position; if they are portable, they are readily available. Sophisticated equipment is discreetly hidden behind screens.

Each LDRP suite is cheerfully decorated with comfortable furniture, bed, drapes and spread, and has cradle, telephone, music and TV to provide the warmth and convenience of home. Family bonding means that the spouse or the support person can be present in the delivery room. Brothers, sisters and grandparents can enjoy holding their newborn family member shortly after delivery.

The greatest benefit is that the patient is discharged from the hospital in less than 36 hours, in most cases in 24 hours.

Organization

Medical Staff The chief or head of the department is a qualified, competent and experienced obstetrician and gynaecologist. There are several other qualified obstetricians and resident doctors to assist him(her). There is a close relationship with the paediatrics department. In some hospitals, a paediatrician or a neonatologist is on call 24 hours of the day.

Nursing Staff The maternity department calls for the services of specially trained nurses. Nurses should be caring people who have personalized approach to childbirth. Besides rendering nursing care, every nurse in the maternity department should be a trained teacher. Care of the infant is carried out at the bedside of the mother by the nurse who makes each procedure a lesson for the mother. The nurse provides all care during the first few days after delivery. Later, the mother participates and finally takes over the care of the infant.

Pre-natal and Lamaze Classes

Every hospital must organize regular prenatal classes for expectant mothers — fathers should be encouraged to join too — to learn about the development of the foetus, labour, delivery, nutrition, parenting and common problems facing new parents. Hospitals may also organize Lamaze classes at which relaxation and breathing techniques are taught as these will be used during labour and delivery.

Identification of Babies

The identification of babies have serious clinical, moral, legal and public relations implications. Every hospital must establish written policies and procedures pertaining to the identification of infants and enforce these procedures.

- Every newborn infant should be properly identified in the delivery room immediately at the time of birth. This should be done before either the mother or the infant leaves the room. They must be moved from the delivery room before another patient is brought in.
- The identification item should show the mother's name and hospital number, sex of the infant, date and time of birth.
- The identification item which can be a tape or an adhesive sticker with the data written in indelible ink is placed around the wrist of the newborn.

Admission

The admission of maternity patients should not be delayed by prolonged registration and admitting formalities. Provision must be made for them to proceed straight to the department without passing through admitting or emergency departments. In a computer-assisted system in which patient's data and information are available on line, admission can be effected from any part of the hospital.

Blood Bank

Every hospital should have a blood bank with facilities to type and cross match donor's and patient's blood and dispense emergency blood when needed. It should also have facilities to screen donors for blood group, Rh factor, Hepatitis B virus, AIDS, STD, etc. The blood bank should be as close to the obstetrical unit as possible.

Facilities and Space Requirements

Space and facilities are required for the following:

Clinical Areas

- Preparation room
- Delivery room(s)
- Labour room(s)
- Recovery room(s)
- Operating room

Service Areas

- Control/nursing station. Should be so located as to permit observation of all traffic which enters and leaves the obstetrical suite.
- Supervisor's office.
- Fathers' waiting room.
- Sterilizing facilities with high speed autoclave(s) convenient to Caesarian section room and delivery rooms.
- Recessed scrub area with sink equipped with gooseneck spouts, foot-operated controls, thermostatically controlled temperature valves, space for nail brushes, sterile caps, and masks, and an easily visible clock with a second timer.
- Controlled storage for drugs.
- Enclosed soiled work room/storage room with sink, counter, etc.
- Clean work room/supply room.
- Anaesthesia storage facilities.
- Equipment storage.
- Staff clothing change area.
- Lounge and toilet facilities for obstetrical staff.
- Janitor's closet.
- Alcove for stretchers.
- A recessed place for film illuminator, and a desk and chair for chart work.
- Duty rooms with sleeping accommodation, toilet, shower, for resident duty and on call doctors, separate for men and women. Bunk beds may be provided to accommodate more doctors.

PHYSICAL MEDICINE AND REHABILITATION

► INTRODUCTION

Physical medicine and rehabilitation is a medical specialty concerned with the diagnosis and treatment of certain musculo-skeletal defects and neuromuscular diseases and problems. In their efforts to deliver comprehensive care, modern hospitals have included rehabilitative medicine as part of their inpatient and outpatient care. The physical medicine and rehabilitation department is organized to provide continued specialized treatment of a variety of prolonged, often reversible, physical and mental disabilities.

Physical medicine and rehabilitation is often wrongly considered as an adjunct to orthopaedic surgery. Every service in the hospital should be considered to be dealing with the rehabilitation and reconditioning of the convalescent and of the patients suffering from chronic diseases.

Rehabilitative medicine encompasses such fields as physical therapy, occupational therapy, recreational therapy, speech and hearing therapy, bracing and prosthetics and pulmonary medicine.

Very few hospitals offer comprehensive physical medicine and rehabilitation services although there are institutions providing physical therapy services in varying degrees of scope and specialization. The reasons for this are many, the most important being the lack of proper indoctrination on the part of hospital authorities and medical staff to the importance of rehabilitative medicine, acute dearth of qualified staff and lack of funds.

It is important that these various branches of rehabilitative medicine establish common goals and approach to treatment so that none of them works at cross purposes with the others. Effective coordination, co-operation and mutual dependence — which come from working in tandem — are, therefore, the hallmark of the department.

PHYSICAL THERAPY

➤ OVERVIEW

The physical therapy department provides a specialized rehabilitative service. Therapy is necessitated as a result of surgery, trauma, stroke and other functional impairment. Although traditionally therapy was provided by the nursing personnel, with improved technology and sophisticated equipment, physical therapy has evolved into a specialized field. A well developed department may now offer specialty programmes like cardiac programme, chest therapy (in cooperation with respiratory therapy) and sports medicine.

The objectives of the physical therapy department may be stated thus:
* to render suitable physical therapy service to patients in order to prevent, correct or alleviate physical disability by the use of physical measures like heat, cold, light, water, electricity, sound, massage, therapeutic exercises and rehabilitative procedures;
* to assist each patient in reaching his maximum functional level, to teach him to live within the limits of his capabilities and assist him in assuming his place in society;
* to extend the patient's activities of daily living beyond those required at the time of discharge: and

◆ to contribute to the comfort and well being of the patient.

The department is relatively small in most hospitals. The size of the department depends on such factors as the amount of orthopaedic surgery done in and the various other rehabilitative services offered by the hospital.

➤FUNCTIONS

The following are the major functions of physical therapy:
- ◆ to restore body functions;
- ◆ to hasten convalescence and shorten the patient's stay in the hospital;
- ◆ to prevent and minimize residual physical disabilities;
- ◆ to return the individual to optimum living;
- ◆ to provide an aid to the physician in implementing early ambulation programmes;
- ◆ to start self-care activities before a patient becomes too dependent on others for help;
- ◆ to provide information derived from special techniques, for example, electrical muscle testing; and
- ◆ to help plan follow-up rehabilitation programme for the patient after discharge.

➤LOCATION

Since the department provides therapy services for outpatients, inpatients and emergency patients, it should be centrally located to accommodate all categories of patients. It should be close to the elevators with easy access to both outpatients and inpatients. It should be adjacent to other rehabilitative services such as occupational, recreational and speech therapy as well as social and outpatient services. The orthopaedic surgery, one of the major users of physical therapy, should be as close as possible. Many hospitals find it advantageous to cluster the service departments such as radiology, laboratory and pharmacy round or close to physical therapy. In finding an ideal location for physical therapy, the overriding consideration should be the fact that a majority of the patients who seek the service of physical therapy are physically handicapped to various degrees, whether temporarily or permanently. This reason makes it necessary that the department is located on the ground floor. The scope and popularity of the department can be enhanced tremendously if it is located in a conspicuous area.

➤DESIGN

An assessment of the physical facilities required for the department should be made in the earliest planning and design stage. An estimate was once made that not less than 10 per cent of the inpatients of a general hospital required physical therapy services. This percentage varies from hospital to hospital, and will actually be much higher depending on the facilities available in any given set-up, how well the medical staff have been indoctrinated to the use of physical therapy, and how effective and efficacious the programme is. Hospitals which plan for a good set-up must naturally expect a far greater number of patients than estimated.

➤ FUNCTIONAL AREAS

In designing the department, the hospital planners and designers should pay attention to three major functional areas (Fig. 4.16). They are (a) a hydrotherapy area, (b) a gymnasium, and (c) a cubicled treatment area. In addition, there should be designated areas for waiting patients and support services such as administration, staff facilities, storage and toilets.

The hydrotherapy area should provide for whirlpools and a work station. There may be a small whirlpool, an extremity whirlpool, and a lowboy whirlpool. If a hubbard tank (Plate 7: Picture 38) is to be installed, a separate room of 16′ × 18′ is required. The hydrotherapy area requires plumbing and water supply. Attention should be given to reinforce ceilings of the areas below to provide special drains. The ceiling of the hubbard tank room should be at least 10 feet high in order to accommodate the electric hoist apparatus. Space must be provided in the room for examination and exercise outside the tank. Other facilities needed are a special area for preparation of hot packs and storage for clean linen and for disposal of wet linen after treatment. Physical therapy requires a large amount of linen. Whirlpools and tanks must be located in such a manner that wheelchairs can move about without interfering with other patients receiving treatment in that area.

The gymnasium is a large rectangular room for individual and group exercises (Picture 39).

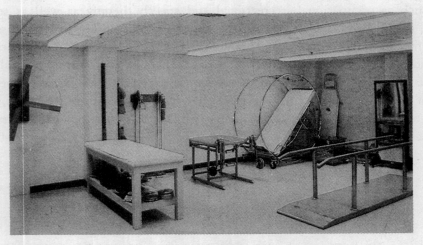

➤ **PICTURE 39** Physical Therapy Department: A partial view of the gymnasium with parallel bars, wheels, etc.

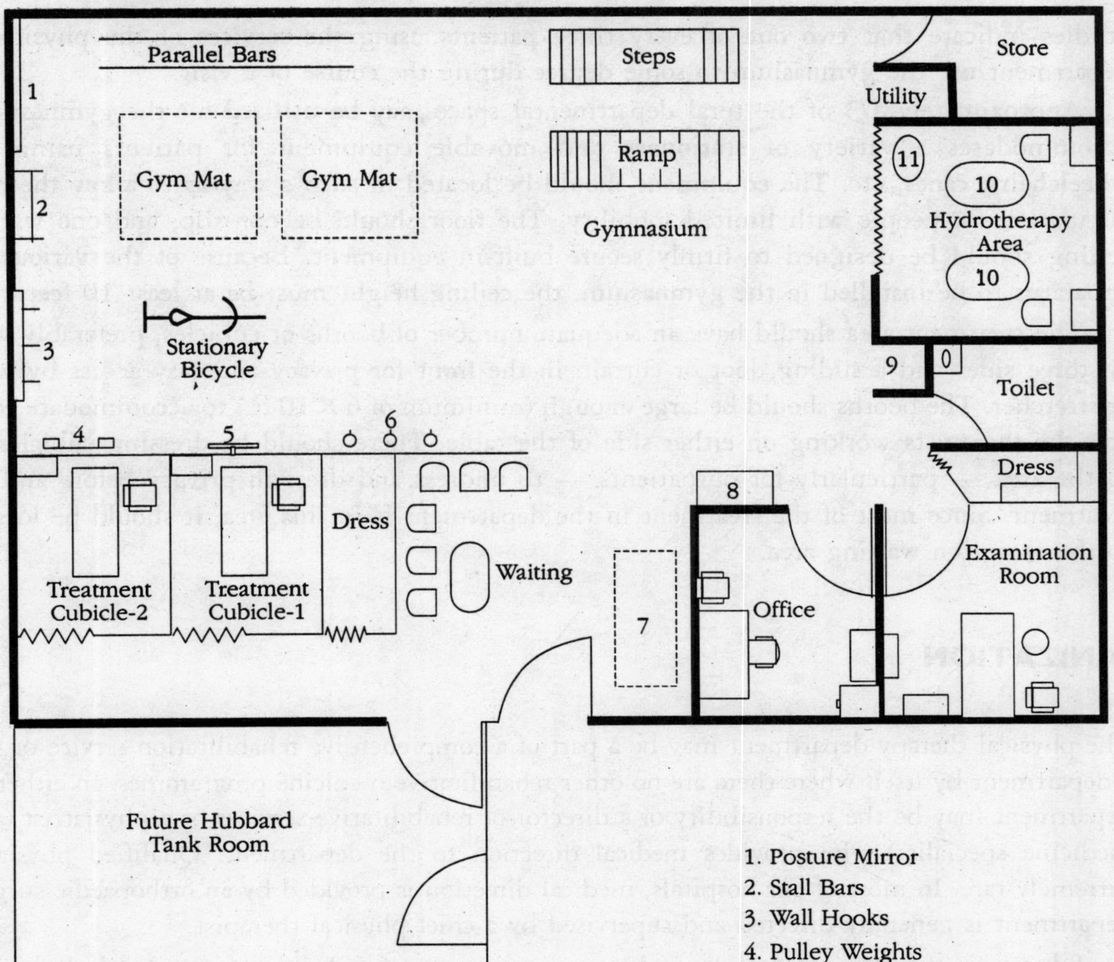

> **FIG. 4.16** Layout of Physical Therapy Department in a Medium Sized Hospital

Labels in figure:

Parallel Bars

Gym Mat | Gym Mat

Stationary Bicycle

Treatment Cubicle-2 | Treatment Cubicle-1

Dress

Waiting

Future Hubbard Tank Room

Steps

Ramp

Gymnasium

Store

Utility

Hydrotherapy Area

Toilet

Dress

Examination Room

Office

1. Posture Mirror
2. Stall Bars
3. Wall Hooks
4. Pulley Weights
5. Shoulder Wheel
6. Hooks
7. Stretcher
8. Work Top with Shelves Below, & Wall Mounted Cabinets Above
9. Paraffin Bath
10. Whirlpool with Adjustable Chair
11. Linen Hamper

Studies indicate that two out of every three patients using the services of the physical therapy department use the gymnasium to some degree during the course of a visit.

Approximately 1/3 of the total departmental space may be utilized for the gymnasium which accommodates a variety of stationary and movable equipment for patients using crutches, wheelchairs, canes, etc. The equipment should be located in such a way as to allow the maximum circulation for people with limited mobility. The floor should be non-slip, and one wall and the ceiling should be designed to firmly secure built-in equipment. Because of the various kinds of apparatus to be installed in the gymnasium, the ceiling height must be at least 10 feet.

The treatment area should have an adequate number of booths or cubicles, preferably with walls on three sides and a sliding door or curtain in the front for privacy and easy access by wheelchair or stretcher. The booths should be large enough (minimum of 6 × 10 ft.) to accommodate equipment and the therapists working on either side of the table. There should be dressing cubicles adjacent to the area — particularly for outpatients — to undress and dress in privacy before and after the treatment. Since most of the treatment in the department is in this area, it should be located close to the reception waiting area.

➤ORGANIZATION

The physical therapy department may be a part of a comprehensive rehabilitation service or it may be a department by itself where there are no other rehabilitative medicine programmes. In either case, the department may be the responsibility of a director of rehabilitative services or a physiatrist (a physical medicine specialist) who provides medical direction to the department. Qualified physiatrists are extremely rare. In most of our hospitals, medical direction is provided by an orthopaedic surgeon. The department is generally directed and supervised by a chief physical therapist.

Education and training for physical therapists consist of a three to four year diploma or degree programme. Many colleges and universities offer degree programmes. The physical therapists are registered under their professional organization, although at present, this registration is not mandatory to practise in the way it is for doctors, nurses and pharmacists.

The following are the staff of the department:
- ◆ Chief physical therapist with a degree or diploma in physical therapy and experience;
- ◆ Staff physical therapists with diploma or degree;
- ◆ Physical therapy assistants, technicians or aides who have received on-the-job training; and
- ◆ Secretary, clerk and receptionist for clerical duties, billing, cashiering, patient scheduling, etc.

Facilities and Space Requirements

Space and facilities are required for the following:

- Reception and control station — so located that visual control of waiting area and activities is possible;
- Office and clerical area with space for clerical work and storage of files. In smaller hospitals, office and clerical area may be combined with reception-control station.
- Offices for physiatrist and chief physical therapist with facilities for consultation;
- Waiting area out of traffic for patients;
- Patients' toilets with wash basins and accessible to wheelchairs;
- Storage space for wheelchairs and stretchers out of traffic;
- Locked cabinets for storing staff's personal effects;
- Janitor's closet with service sink, and facilities for housekeeping;
- Staff lounge, lockers and toilets;
- Conference room for demonstration, teaching, etc.;
- Ramps with handrails at the main entrance and of sufficient width for wheelchairs;
- Individual treatment booths or cubicles with screens for privacy;
- Hand washing facilities close to the treatment area;
- Hydrotherapy area as described earlier;
- Work stations and counters at different areas of work;
- Space for paraffin wax bath. As this can be messy, this area should be recessed;
- Gymnasium (exercise area) with all facilities;
- Storage for clean linen and towels;
- Storage for equipment and supplies;
- Area for temporary storage of soiled linen, towels and supplies; and
- Changing/dressing room for outpatients.

Back School

It is estimated that eight out of every ten persons will have a back injury sometime during their lives. And yet, many times, back injuries can be prevented by learning correct techniques of lifting, eliminating excess body weight and adopting good posture habits at work and home. For example, it is said that 40 hours of sitting can put more strain on the back of a workaholic than 40 hours of standing or even lifting. Back injuries cost people and their employers an enormous amount of money, not to mention reduced production, absenteeism, increased turnover, staff medical bills and pain.

Many hospitals establish what they call a "Back School" in which physical therapists hold classes and impart instructions to patients suffering from back pain or injury as to how to avoid these problems.

Problem Situations

One of the problems the department faces stems from the fact that most of the physical therapy outpatients receive treatment on a routine basis. For example, a certain treatment is scheduled twice a week for two months. This may cause problems of registration and collection of charges. Does the patient have to register and get billed every time he visits the hospital? And where? At the central registration or at the department? The patients may find it inconvenient to go every time to the central registration and make payments. If they go straight to the department, charges may be missed and such visits may fail to be included in the repeat patient census.

The second problem is the lack of coordination between the department and inpatient areas as far as the scheduling of inpatient therapy is concerned. A therapist may schedule therapy and go to the patient floor only to find that the patient has either gone for an x-ray or some other procedure, or is bathing. This may happen frequently.

A different kind of problem arises when the physical therapist needs a patient's medical chart and finds that the chart is being used by a physician or a nurse.

Generally, inpatients who are scheduled to have therapy in the physical therapy department are sent by escort service which is often inefficiently handled. Wards may not have escorts for this purpose or the escorts may be busy. As a result, schedules in the department go haywire.

OCCUPATIONAL THERAPY

Occupational therapy is a professional branch of health care that evaluates and treats individuals whose abilities to cope with or perform the tasks of daily living are threatened or impaired by physical illness or injury, the ageing process, congenital or developmental disability, poverty or cultural differences or psychological and social disabilities.

Occupational therapy translates the ability to move, no matter how limited, into activities that we all do everyday: bathing, dressing, hygiene, grooming, eating, transferring (moving from the bed to a wheelchair, from a wheelchair to a toilet, tub, etc.), home making and leisure time activities. The patient may be one whose life has been affected by illness or injury such as cerebrovascular accident (stroke), arthritis, cerebral palsy, spinal cord injury, amputation, etc. The simple truth we often tend to forget is that what seems so easy for us may be extremely difficult for these patients.

The goal of the occupational therapy is to return the patient to his or her optimum level of independence. It is important to remember that occupational therapy can help decrease the patients' length of stay in the hospital. More importantly, early therapy means that the patients can retain more of their physical independence and become less psychologically dependent, and much of the attendant problems that result from injury, illness or inactivity can be reduced through early intervention.

The occupational therapist identifies and selects activities suited to the individual's physical capacity, level of intelligence and interests to assist the patient in developing maximum independence in the activities of daily living. Some of the crafts utilized in occupational therapy are basket work, artwork, printing, woodwork, weaving, sewing, etc.

➤LOCATION

The occupational therapy which is a branch of physical medicine, should be located adjacent to the physical therapy department. The therapy provided in the occupational therapy department is in one sense an extension of the work of the physical therapy department. Besides, the members of the rehabilitation team need to consult each other and coordinate therapeutic activities of individual patients. Therefore, adjacency is important.

➤DESIGN

One of the areas that should receive attention in designing the department is "Activities of Daily Living" room which is equipped with cooking range, refrigerator, oven and other gadgets needed for daily living.

➤ORGANIZATION

There may be a qualified occupational therapist who is responsible for the day-to-day work of the department. In some hospitals, the chief physical therapist may be in overall charge of the occupational therapy unit.

The staff of the department are qualified occupational therapists who have undergone a 3-year degree or diploma course in occupational therapy. Quite a few medical colleges offer this course. Besides occupational therapists, there may be occupational therapy assistants or aides who have received on-the-job experience but have no formal training.

Facilities and Space Requirements

Many of the areas needed for the occupational therapy unit are common areas among the various sections of the rehabilitative medicine, particularly physical therapy department. It would be unwise and uneconomical to duplicate them. Additionally, the occupational therapy should have the following facilities:

- ◆ Activities of daily living room;
- ◆ Work areas and counters suitable for wheelchair access;

◆ Hand washing facilities;
◆ Storage for supplies and equipment; and
◆ Crafts room — can be off site or common facilities can be used.

RECREATIONAL THERAPY

➤OVERVIEW

The underlying philosophy of recreational therapy is that recreation, an important tool in the lives of the people, offers them a chance to express their playful selves. That expression should not be suppressed due to physical limitations. The department aims at accomplishing what physical and occupational therapies achieve through the means of recreational activities which the patients are encouraged to look upon as a leisure time and fun activity rather than as a formal therapy. In that kind of an atmosphere, the patients are relaxed and enjoy what they are doing. This helps in the speedy rehabilitation process.

Goals of Recreation Therapy in the Rehabilitation Setting

◆ Diversion from the structure of other therapies;
◆ Promote socializing opportunities among the patients;
◆ Encourage patients' awareness that they can still participate in active games and hobbies, and that they can still perform tasks successfully;
◆ Teach patients new skills or adaptation to old ones; and
◆ Inform and guide patients into community recreation resources after discharge.

Outside a hospital setting, recreational therapy is used in prisons, camp sites, old age homes and activity centres.

➤LOCATION

The recreational therapy unit should have a close relationship with physical therapy for the continuity of care. Close proximity is also desirable with the psychiatric department which is one of the major users of recreational therapy. In hospitals which have a psychiatric department, recreational therapy forms an integral part of that department. Some of the outdoor activities of recreational therapy may be conducted off site.

➤ORGANIZATION

Not many hospitals have a formal recreational therapy programme, nor extensive facilities for the programme. Wherever it exists, it may be a part of the physical therapy unit under the supervision of the chief physical therapist. The work may be assigned to one of the physical therapists.

There is no formal training programme in recreational therapy, but any physical therapist with imagination can organize recreational therapy activities.

SPEECH AND HEARING THERAPY

➤OVERVIEW

This unit of rehabilitative medicine is concerned with the diagnosis and treatment of speech and hearing disorders or communicative disorders or disorders of the processes underlying speech, swallowing or chewing. The purpose is to restore or develop the patient to the highest level of communicative functioning in relation to the needs of the person and his or her environment. The unit is also concerned with the diagnosis and treatment of the hearing impaired. It carries out diagnostic evaluation of patients with auditory problems and determines effective conservation, habilitative and rehabilitative programmes.

Hearing and speech problems affect one out of every ten people. Studies indicate that more people suffer from hearing, speech and language impairments than from heart disease, paralysis, epilepsy, blindness, cerebral palsy and tuberculosis all put together. Most people with communicative disorders could be helped medically, surgically, through amplification or through rehabilitative treatment.

Goals of Speech Pathology

- ◆ to assess and diagnose speech and language disorders;
- ◆ to improve intelligibility of speech;
- ◆ to aid in developing socially appropriate speech;
- ◆ to improve muscle function of the oral mechanism;
- ◆ to provide alternative methods of communication for those who will not regain normal speech;
- ◆ to provide information to the patient and his family about the nature of the disorder, treatment, duration of therapy, etc.; and

◆ to counsel patients and their families regarding speech and language disorders and their implications on education and vocational placement.

Goals of Audiology

◆ To evaluate the degree and type of hearing impairment.
◆ Based on the diagnosis, make appropriate referrals of patients for whom habilitation is indicated.
◆ To recommend appropriate rehabilitation programmes.
◆ To provide professional evaluation procedures and guidance for the selection and use of suitable hearing aids.
◆ To counsel patients and their families with regard to hearing impairment, education and vocational placement.

The hearing loss is the inability to hear what in technical terms are expressed as auditory signals less than 25 dB in the frequency range of 250 Hz to 8000 Hz. In layman's language this means inability to hear low pitched sounds or soft words. It occurs when the ear is not as sensitive to sound as it used to be or as it should be, and is produced by any abnormality in the ear. Environmental sounds are not as loud to the patient as they are to others, nor speech as clear as it used to be. Examples: the TV needs to be adjusted to a louder pitch or the affected person has difficulty in hearing over the telephone.

Basic hearing tests using tones of different frequencies and speech are conducted to determine the type and degree of hearing loss. From this test, a record of the patient's hearing, called an audiogram, is taken. Most modern audiology clinics are equipped with impedance audiometers, brain stem evoked response audiometers, electrocochleography, etc. to provide thorough and advanced audiological assessments. Nearly everyone with a hearing problem can be helped by medical treatment, surgery or hearing aids.

Hearing and speech have a close relationship because hearing is the natural and normal way to understand speech.

➤ LOCATION

The speech and hearing therapy unit need not be in a prime location. Traditionally this unit is a part of the ENT department. However, speech and hearing therapy has advanced and become so sophisticated that in most modern hospitals the unit is accorded an independent status under trained professionals. This is largely because of the vast clinical services the unit provides and the constant interaction it has to have with many departments such as medicine, neurosciences, paediatrics, ENT, physical therapy and occupational therapy. Since most of the patients who need the services of the unit for testing and treatment are ambulatory, the unit should be easily accessible to the outpatient department. Besides, many patients who suffer from aphasia (total or partial loss of the power to use or understand words) require outpatient services initially.

➤DESIGN

There is one particular area of this unit in the design of which, as in the case of all specialized areas of the hospital, experts should be consulted. This is sound-treated or sound-proofed room for audiological assessment where sensitive tests are carried out. If there is a specially trained and experienced speech and hearing therapist or consultant, he/she may be able to provide specifications for the room and other inputs; or he/she may summon the help of outside specialist(s).

➤ORGANIZATION

The key persons in the unit are the specially trained speech pathologist and audiologist. The speech pathologist is concerned with the diagnosis and treatment of those people who have speech disorders like stuttering, voice disorders, aphasia, articulation disorders, delayed language, and disorders of swallowing and chewing. The audiologist is concerned with the diagnosis, referrals and treatment of those who are hearing impaired. Disorders may be those affecting the external ear, middle ear, inner ear, nerve pathways or central auditory cortex.

Institutions like the All India Institute of Speech and Hearing in Mysore train speech pathologists and audiologists. Students receive wide and in-depth training in various branches of speech pathology, audiology, psychology, linguistics, electronics, acoustics, anatomy, physiology, pathology, paediatrics, neurology and ENT insofar as they relate to speech and hearing therapy. Their professional potentials include independent assessment and treatment of speech and hearing disorders.

Facilities and Space Requirements

The following facilities are required:
- ◆ Reception-control room
- ◆ Waiting area
- ◆ Speech pathologist's and speech therapist's offices
- ◆ Audiologist's office
- ◆ Sound-treated room(s) for audiological assessment
- ◆ Therapy rooms
- ◆ Storage room.

PULMONARY MEDICINE

➤ OVERVIEW

Pulmonary medicine, a branch of internal medicine, is a relatively new specialty. A department of pulmonary medicine providing a variety of specialized services may be found in most of the modern hospitals. The department is headed by a physician who has specialized in pulmonary medicine. The services provided by the department are respiratory therapy, pulmonary function test and arterial blood gas, bronchoscopy, pulmonary rehabilitation and sleep laboratory. Respiratory therapists provide such services as the maintenance of mechanical ventilators, administration of aerosols, oxygen, nebulisers, humidifiers, intermittent positive pressure breathing (IPPB), performing pulmonary function tests, arterial blood gas studies, etc. Respiratory therapists carry out various therapies as per the orders of the pulmonary medicine specialist under whom they work.

➤ LOCATION

Although both inpatients and outpatients receive therapy for a variety of conditions, the inpatients account for 90 per cent of the care given by the department. Besides, almost all therapeutic work is performed at the bedside. Diagnostic functions are undertaken in the department. The department should, therefore, be conveniently located for service to inpatients. Because much of the work is performed away from the department, breakage of small but expensive mobile equipment while transporting is a common problem. The location should not add to this problem.

In addition to the inpatient areas, the other major users of pulmonary medicine are intensive care units, post–operative area, and emergency room. The department should have easy access to these areas. Close proximity to vertical transport is necessary.

➤ DESIGN

Pulmonary medicine requires a wide range of expensive but fragile equipment including ventilators, flow meters, humidifiers, nebulisers, etc. These need a fairly large storage area.

After use, the equipment must be disassembled, washed, cleaned, reassembled, serviced, packaged, separated and sterilized. Facilities for these activities are essential as also a good distribution system. The services required are oxygen, suction and air. A good communication system is very important.

➤ORGANIZATION

The chief of the department is a pulmonary specialist or a physician with an interest and training in pulmonary medicine.

There are no formal training programmes for therapists. In the absence of such a training, the therapists require extensive on-the-job training under a physician specialist in an established department.

It is advisable to cross train therapist staff to rotate between diagnostic and therapeutic sections.

In many hospitals pulmonary medicine units have been combined with cardiology departments to form cardiopulmonary service departments. Cardiology and pulmonary diagnostic testing have been combined to make a cardiopulmonary laboratory unit. This integrated unit has proved beneficial as it has provided improved and integrated cardiopulmonary services and better staff utilization.

Facilities and Space Requirements

The following facilities are required:
- Physician's office cum examination room.
- Chief respiratory therapist's office.
- Room for bronchoscopy. Should have a cardiac monitor and provision for oxygen and suction.
- A fairly large room for sleep laboratory. The lab must be equipped with many pieces of equipment to monitor heart, EEG, blood pressure, oxygen saturation, carbon dioxide, respiration, etc.
- Space for technicians with charting facilities.
- Treatment room(s) — can be partitioned into cubicles for examination and treatment.
- Clerical area.
- Storage area for movable equipment, small processed equipment and supplies. To enable the therapist to respond to calls quickly, it is recommended that the processed equipment and supplies are stored in emergency carts, and the movable equipment is stored on the opposite side.
- A small repair room where therapist dismantles, cleans, maintains, and repairs small items of equipment routinely. The room should have utilities, a work bench, and wall mounted cabinets.

CHAPTER 5

Planning and Designing Nursing Services

CONTENTS

GENERAL NURSING UNIT

➤ OVERVIEW

While designing a nursing unit, as in any other area of hospital operation, it is of utmost importance to remember that functional planning always precedes the architectural drawings. During the initial stages of planning, the architect should consult representatives of all the departments and incorporate in his plans all possible economies and efficiencies. A nurse may not know how to read a blueprint or understand its intricacies and complexities, but she certainly knows the functions of nursing and can interpret them to the architect to give him a clear idea of the operations of her department.

Nursing units should be designed to serve certain functional goals. The design should result in:
- building at the lowest possible capital cost,
- the most economical operation,
- provision of the highest quality patient care,
- providing the most desirable patient comfort and environment,
- the most efficient operation of the unit,
- the greatest degree of job satisfaction for nursing and medical staff, and
- meeting the needs of visitors.

Good nursing care is difficult to provide if facilities are inappropriate. There is a close relationship between physical facilities and safe patient care. Inadequate planning precludes efficient service. For example, failure to provide basic necessities like handwashing facilities may lead to physicians and nurses spreading an infection from one patient to another. It is important to remember that even minor defects in designing can make the operation of the hospital inefficient and uneconomical. Circulation and flow of traffic also enhance or adversely affect the efficiency and economy of operation (Fig. 5.1).

Efficiency of operations, economy of space and comfort of the patients are the considerations while designing a nursing unit, which ideally should be so compact that a nurse need not travel more than 25 feet to serve any of the patients. Travelling within the nursing unit has been estimated to occupy one-sixth of a nurse's time. The shortest possible traffic routes assist in maintaining aseptic conditions and save steps for everybody — nurses, doctors, patients and other hospital personnel. Time wasted on unnecessary steps costs money besides leaving people fatigued at the end of the day.

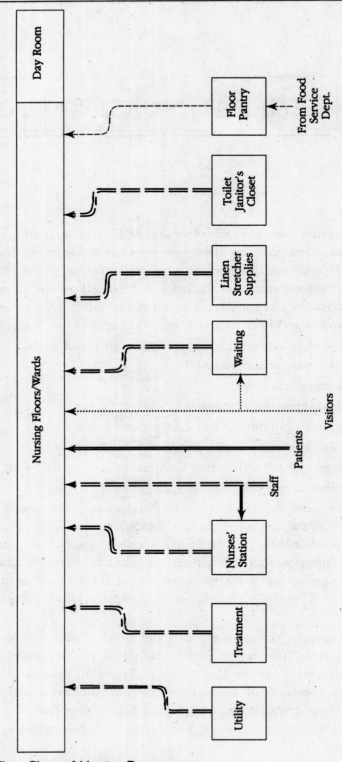

▷ **FIG. 5.1** Flow Chart of Nursing Department

➤FUNCTIONS

The functions of a nursing unit are better understood by looking at the three primary components that constitute the unit, namely, the patient rooms. the nurse control station, and the work areas.

The patient area which may consist of private and semi-private rooms and multi-bed general wards is designed to be a safe and aesthetically pleasing treatment area which is conducive to fast recuperation of the patients. It must contain space for equipment, staff and the various needs of the patient. The nurse control station provides work space for the nursing staff. It is located and designed in such a way that the nurses can observe patient rooms and direct the traffic entering and leaving the unit and at the same time carry on the activities associated with the care and safety of the patients. The functions of the work area relate to handling materials necessary for patient care, handling and maintaining communications and patient records, social and physical needs of the patients, and the specific needs of the staff.

➤LOCATION

Nursing units have a close relationship with the operating rooms, pharmacy, central stores, laboratory and the dietary. In maintaining this relationship, they are highly dependent on vertical transportation and an efficient communication system. The location of these facilities must be considered from the point of view of their relationship to the nursing units.

➤DESIGN

The size of the nursing unit and the distribution of different categories of beds — private, semi-private and multi-bed — should be decided during the planning stage. Whether or not the unit should be a unitary ward serving one clinical unit under one consultant should also be decided at that time. Consideration should be given to the cost of construction of the unit, staffing requirement, and the distance between the nursing station and the patient rooms and supply points. Any duplication of facilities and equipment should be avoided. In short, the unit should function efficiently in which to work.

The size of the nursing units varies. They may be very small or very large. Both are uneconomical. Nursing units which are too small are more expensive to build and maintain. From the service point of view, they are as unsatisfactory as the units which are too large. For instance, a 10-bed unit requires the same type of utilities, types of equipment and supervisory staff as a 20-bed unit. The most common size is between 20 and 40-beds (Figs 5.2 and 5.3). In larger units, the nursing station, telephones, utilities, etc. need to be duplicated.

The optimum size of a general nursing unit is governed primarily by the number of patients and nursing staff that the nurse in charge of the unit can manage efficiently. Other factors are the activities of the unit, bed utilization and the degree or acuity of patient illness. The nursing service

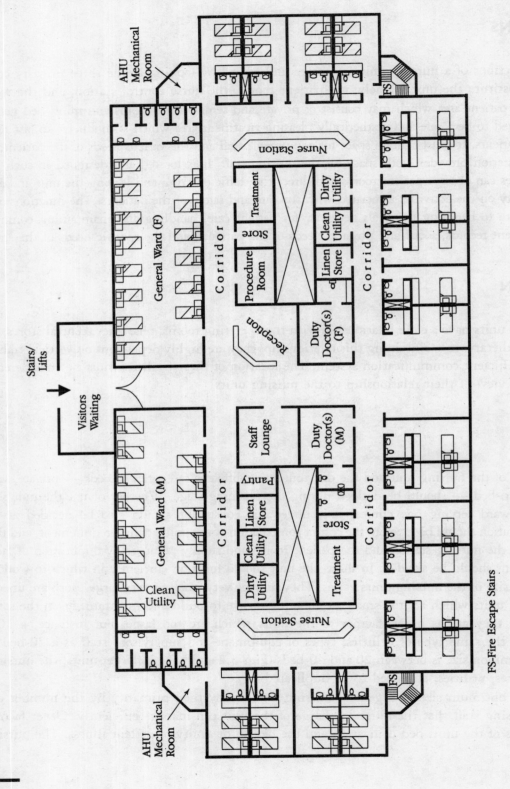

➤ **FIG. 5.2** Layout of a Nursing Unit: Illustration I

FS–Fire Escape Stairs

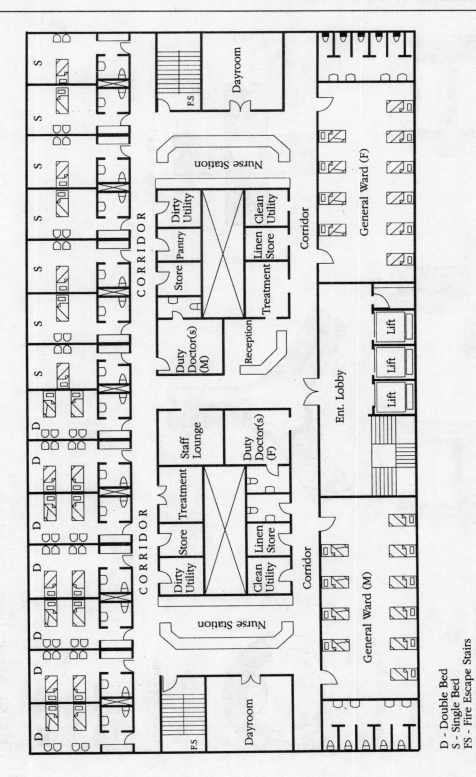

D - Double Bed
S - Single Bed
FS - Fire Escape Stairs

➤ **FIG. 5.3** Layout of a Nursing Unit: Illustration 2

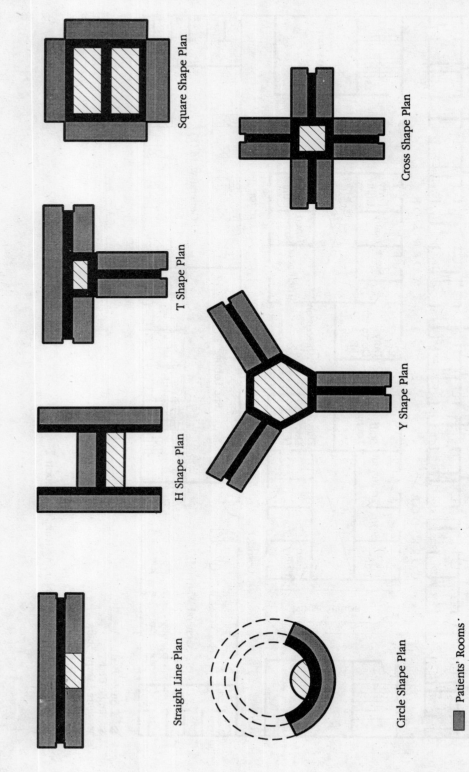

> **FIG. 5.4** Schematic Designs of Nursing Units

Square Shape Plan

Cross Shape Plan

T Shape Plan

Y Shape Plan

H Shape Plan

Straight Line Plan

Circle Shape Plan

Patients' Rooms

Nurse Station

Corridor

operates on three shifts to cover the 24-hour period. If there are two units on each floor, during the most active hours of 7.00 a.m. to 3.00 p.m., each unit would function independently; while in the less active shifts, staffing of the two units may be combined.

Nursing units can be designed in a variety of ways — square shape plan, straight line plan, Y shape plan, H shape plan, T shape plan, cross plan, and circle shape plan (Fig 5.4). These can be with a single corridor or with double corridors. There may be other variations as well.

The most common floor plan resembles a racetrack. The nurses' station, utility and supply rooms, treatment room and conference or class rooms are located centrally. A single long corridor with patient rooms on both sides extends in two directions from the nurses' station.

Patient Rooms

It is recommended that the minimum size of a one-bed room should not be less than 125 sq.ft.,

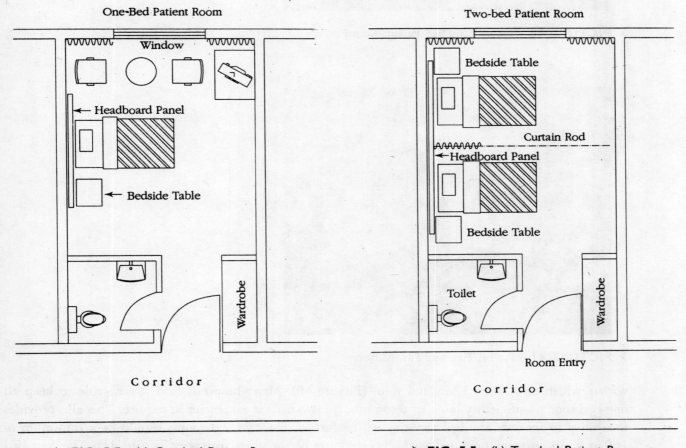

➤ **FIG. 5.5** (a) One-bed Patient Room   ➤ **FIG. 5.5** (b) Two-bed Patient Room

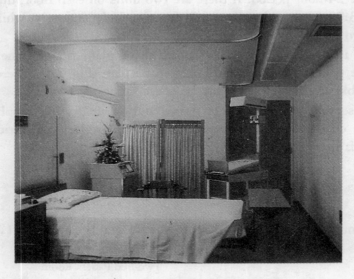

> **PICTURE 40**   Typical One-bed Patient Room

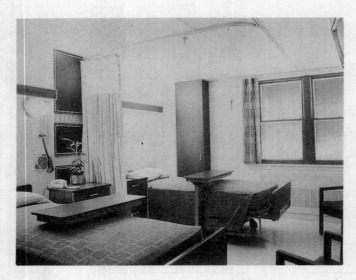

> **PICTURE 41**   Typical Two-bed Patient Room

with a width of at least 12 ft. and 6 in. (Picture 40). Many hospitals find it advisable to keep all one-bed rooms sufficiently large to accommodate two beds if exigencies so require. This also provides flexibility to increase the bed capacity in the future. The two-bed rooms should be a minimum of 160 sq.ft. in size and provided with cubicle curtains for visual privacy (Picture 41). The four-bed

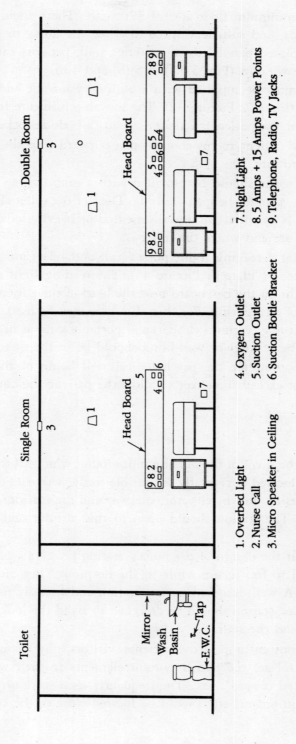

Single Room

Double Room

Toilet

Mirror →
Wash
Basin →
Tap
E.W.C.

Head Board

➤ **FIG. 5.6** Patient Rooms: Wall Elevations

1. Overbed Light
2. Nurse Call
3. Micro Speaker in Ceiling
4. Oxygen Outlet
5. Suction Outlet
6. Suction Bottle Bracket
7. Night Light
8. 5 Amps + 15 Amps Power Points
9. Telephone, Radio, TV Jacks

rooms should have a minimum floor area of 320 sq. ft. There should be no less than four feet of space between the beds, and sufficient space to allow the nurse to pass between the bed and the wall. If the beds are placed parallel to the exterior wall, patients can avoid facing the wall or the outside glare from the window (Fig. 5.5(a), 5.5 (b) and 5.6).

Each patient bed must be supplied with a built-in wardrobe and an outlet on the wall for the nurse call signal cord (Plate 7: Picture 42). The movable furniture in each private room includes a bed, a lounge chair (for the patient), a visitor's chair, a bedside locker and an overbed table.

As a rule, doors of patient rooms should open outward into the corridor with an outside lock only and no inside hardware.

The patient room toilet should be provided with a grab bar, an emergency call button within easy reach, and a flush valve for bedpan cleaning. Door of the toilet should open towards the patient room. Night lighting is essential because disoriented or heavily-sedated patients frequently do not remember where they are and where the toilet is.

Electrical outlets for a reading light, nurses' call and television should be at the head of bed, and so also the telephone (Plate 7: Picture 43). The reading light should be fixed at the head of the bed, and night light on the baseboard near the head of the patient's bed. The patient should be able to control levels of all lighting. For this, dimmers may be used. An additional electrical outlet for cleaning equipment (e.g. vacuum cleaner) and portable x-ray is needed on the opposite wall. The television which may be ceiling- or wall-hung should be in the direct line of sight from the bed.

Windows are important for the psychological well being of the patients. The height of the window sill should not exceed three feet to allow the patient the outside view.

The Nurses' Station

The nurses' station is the pivot of the nursing unit around which all the activities of the unit revolve. It should therefore be located as centrally as possible to the activities of the unit. It should command the entrance to the nursing unit by elevator, stairway and the corridor, and provide optimal visibility of the patient wings. The station should open to the corridor and also have a counter (Plate 7: Picture 44).

It is important that the design of the nurses' station projects a positive and reassuring image to the patients who need to feel secure while in the hospital. They often wake up and are fearful or anxious in the night. A well-designed nurses' station and friendly nurses provide them reassurance and comfort. The nurses' station should be designed to avoid the look of a cage which is undesirable for both the patients and the staff.

The nurses' functions encompass the patients, visitors, physicians and other personnel. General coordination and control are the two important elements in their work; the design of the nurses' station should be geared toward these. The frequently used work areas like medical area, the linen store, clean supply, equipment, etc. should be located close to the control station directly in front of the viewing area.

The following facilities are required in this area.

* A desk for secretarial activities of nurses and ward secretary. If the ward operation is computerized, the computer terminal may be kept here.
* A separated area, if possible, for physicians with dictating equipment if that is provided.
* A charting area with counters and chairs.
* Nurses call system panel (Plate 8: Picture 45).
* Wall clock and bulletin board.
* Office space for the head nurse.
* Telephone, paging and intercommunication systems.
* A chart desk/rack with chart holders.
* Lockable cabinet for storage of drugs.
* Storage for stationery, forms, etc.
* Handwashing facilities.
* Small refrigerator for drugs, etc.
* Place to conveniently store emergency cart, medication cart and general and utility cart (Pictures 46, 47, 48).

➢ **PICTURE 46** Emergency Cart, also called Blue Cart, Red Cart or Crash Cart

➤ **PICTURE 47** Medication Cart

➤ **PICTURE 48** General and Utility Cart

➤ORGANIZATION

This is discussed under the nursing administration unit.

➤FACILITIES AND SPACE REQUIREMENTS

In addition to the facilities like the patients' rooms, nurses' station, etc., already dealt with in detail, additional facilities as mentioned below are also required.

- ◆ Examination and treatment room with wash basin, counter, shelf, etc.
- ◆ Clean work room or clean holding room.
- ◆ Soiled work room with clinical sink, work top, waste receptacle, etc.
- ◆ Clean linen store.
- ◆ Equipment storage room (for IV stands, inhalators, cots, walkers, mattresses, etc.)
- ◆ Recessed space for storage of stretchers and wheelchairs.
- ◆ Staff facilities like lockers, etc. for personal effects.

- Staff toilet(s).
- Pantry with sink, work top, storage cabinets and space for temporary storage of dietary trays.
- Small laboratory, where desirable, for residents and interns.
- A multipurpose room for staff and patient conference, education, demonstration, etc.
- Solarium — a day room located at the end of the nursing unit. The room should have large windows, a table, individual seating, a television set, magazine and book racks, storage cabinets (for cards, occupational therapy materials, etc.), emergency call button connected to nurses' station, etc.
- Janitor's closet.

PAEDIATRIC NURSING UNIT

➤ OVERVIEW

The paediatric nursing service encompasses the care of children. Care of the newborn is usually in a separate unit located in the obstetric unit. The paediatric care calls for understanding of the unique needs, fears and behaviour of the children. This is reflected in the type and degree of nursing care given to them. Some cases may require protracted convalescence and educational and occupational therapy. Such cases become the concern of paediatric nurses. An added responsibility of the nursing staff, which sometimes turns out to be a problem, is the relationship with the parents of the convalescing children.

As many as 12 to 15 per cent of beds in a general hospital may be required for paediatric patients depending on the character of the community the hospital serves. A paediatric unit should not be as large as an adult unit since children need more care and attention.

Parents, siblings and friends are of paramount importance in the recovery and rehabilitation of the sick child. It is necessary therefore that the parents live with, assist in the care of, and maintain continuous contact with the child. The psychological benefit to a paediatric patient of seeing siblings and friends is well recognized. It is also generally accepted that children adjust to hospitalization better when they have companionship of other children in the same room. These points should be remembered in designing the paediatric unit.

➤ LOCATION

The paediatric unit is generally noisy. It should therefore be located in a quiet area away from the main stream of hospital traffic. If possible, it should be located adjacent to an enclosed outside terrace which may be used as a play area. It is recommended that, wherever possible, natural sunlight should be available in all the rooms. The unit should also be in close proximity to the vertical transportation so as to be conveniently accessible to the other services.

➤DESIGN

Many of the facilities discussed under the general nursing unit are also required for the paediatric nursing unit. However, this unit has some special requirements, as follows.
- ◆ In view of the size variation and the need to change from cribs to beds, and vice versa, space provided for cribs should be the same as for the beds (Picture 49).

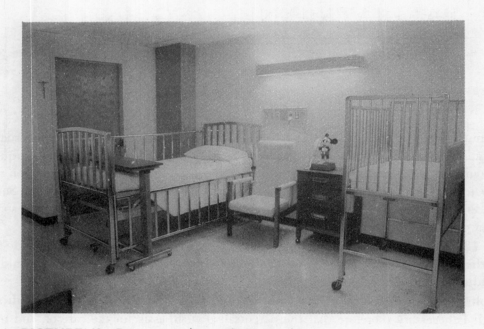

➢ **PICTURE 49** Private room in paediatric ward

- ◆ Where parents are allowed to remain with the young children, provision must be made for toilets, sleeping and storage of personal effects.
- ◆ Separate provision for infant examination and treatment may be necessary.
- ◆ It is essential that each paediatric unit is provided with one or more isolation rooms. These rooms should have an anteroom with gowning and masking facilities, sink with handwashing facilities without the use of hands, foot operated waste receptacle, and small cabinet for clean gowns, masks, etc. The isolation rooms should be remote from other rooms but convenient to the nurses' station.

- At least 25 per cent of the beds should be in single rooms for the critically ill patients and for those who disturb others.
- A recreation/play room should be provided for group activities and recreation. This room would serve as a playroom for the younger children (Plate 8: Picture 50); for games, occupational therapy and school work for the older children; and a social room and library for the adolescents. The room should be so designed structurally as to minimize the transmission of noise and impact through the floor, walls and the ceilings. The recreation room makes an ideal place for ambulatory children to dine together.
- Storage space for linen, supplies, toys, educational and recreational materials, and provision of bookcases, bulletin board and chalkboard.
- Special needs and interests of adolescents in recreation and companionship with their own age group should be borne in mind.
- There should be a formula preparation room with a counter, sink and a small refrigerator. This is necessary regardless of whether the formulas are primarily prepared in that room or purchased from outside.
- Walls between patient rooms and the corridors should have clear glass panels so that children can see activities around them, and nurses can see the children easily.
- If piped medical gases are supplied, the isolation rooms and single rooms for the critically ill patients should be provided with outlets, oxygen, suction and air.
- Sick children require an appropriate and cheerful physical environment. Therefore, in the design of a paediatric unit, the decor, lighting, colour, equipment and recreational facilities are all vitally important.

OBSTETRICAL NURSING UNIT

➤OVERVIEW

The responsibilities of obstetrical nursing service include prenatal care, observation and comforting of patients in labour, providing assistance in the delivery room, care of the mother after delivery and care of the newborn. Obstetric nurses participate in instructing new mothers in postnatal care and care of the newborn. Care of the infant is carried out at the bedside of the mother by the nurse who makes each procedure a lesson for the mother. The nurse provides all the care during the first few days after delivery. Later the mother participates in the learning process and finally takes over the care of the infant. Care of the newborn, particularly the premature, requires special nursing skills.

➤LOCATION

Ideally the obstetrical unit should be located on the same floor as the labour-delivery suites and in close proximity to them. This enhances operational efficiency. The unit should also be adjacent or close to the nursery to minimize the distance of travel and consequent exposure of babies between the nursery and the mother's room. Although in keeping with the Indian culture, babies generally room-in with the mothers, there are occasions when babies are kept in the nursery as, for example, when the mother does not wish to assume care of the infant rightaway, or is too sick or exhausted to do so. Infants are kept in the nursery when they are suspected of infection, when they need to be under observation, or when they are born premature. The obstetrical unit should also be close to the vertical transport.

➤FACILITIES AND SPACE REQUIREMENTS

Facilities for obstetrical nursing unit are the same as for the general nursing unit. A conference room is necessary for patient education programmes and prenatal classes. Since ultrasound scanning has become an almost routine procedure during the antenatal period, a scanning room should be provided if that has not been provided elsewhere in the hospital such as in the obstetric and gynaecology outpatient clinic. A dayroom is highly desirable in the obstetrical unit. Such a room should be cheerfully decorated. Since most obstetric patients are ambulatory, the dayroom will be utilized more than in the other units. Besides, many of these women like to socialize; the day room provides an excellent facility for this. For the same reason, many patients prefer double rooms because of the social interaction they provide.

PSYCHIATRIC NURSING UNIT

➤OVERVIEW

Mental illness is one of nation's foremost health problems. While the cases of temporary alcoholic-dementia, emotionally disturbed patients, etc. are treated in specialized mental hospitals, a large number of people also receive psychiatric diagnosis and treatment for varying degrees of mental illness outside them. Many general hospitals recognize a responsibility for the mentally ill and provide facilities to treat them. Nursing care of psychiatric patients requires a knowledge of their various behavioural patterns and how to cope with them.

The system of delivery of psychiatric service of the past is changing, in that it now takes into account the philosophy and concept that recognize the increased awareness of the community towards mental illness as well as the new methods of treatment and medications. As a result, the isolated facility of the past for the treatment of mental illness — the "asylum" with a stigma attached to it — is no longer considered appropriate.

➤ FUNCTIONS

The basic functions of a psychiatric unit are to provide
  * diagnostic and treatment services,
  * full or partial hospitalization,
  * consultation for physicians, and
  * education for staff, patients and members of the public.

➤ LOCATION

The psychiatric unit should be located on a floor which has separate entrance from the rest of the hospital.

➤ DESIGN

The unit should be designed as an open unit, with a non-institutional atmosphere and requiring minimum security, nevertheless, providing a safe environment for patients and the staff. The design should also avoid the typical general hospital look. Light, paint, decor, etc. should be thoughtfully chosen to provide a desirable therapeutic effect as these things easily affect the mood and attitude of the psychiatric patients.

While designing the unit, a factor that should be kept in mind is the distinction between the patients brought by relatives who are ready to stay with the patients throughout the period of their hospitalization and those whose total responsibility is taken up by the staff. In the latter case, relatives are not immediately available to report about the patient's condition and exigencies that might arise from time to time. It then behoves the staff to continually keep them under observation. The unit should be designed to provide for this. There should be a single exit, tamper-proof fittings, precautions against suicide and escape and 24-hour security.

Another differentiation is made among three groups of patients:
1. acutely disturbed patients who require temporary physical or chemical restraints;
2. chronic, deteriorated a-motivated patients; and

3. patients who do not belong to either of the above mentioned groups — those who come for diagnosis or for a relapse and do not pose management problems.

For the first group, specific isolation rooms are needed with a provision to physically restrain them and to prevent risks of suicide, self-harm and homicide.

The psychiatric unit has four broad sub-units:

1. The treatment-consultation area composed of staff offices for individual or family care sessions and for administrative purposes;
2. Conference-therapy area for group therapy sessions and for observation of these sessions by staff;
3. Inpatient area to accommodate hospitalized patients; and
4. Activities area for therapeutic activities such as occupational therapy, recreational therapy, etc.

The design of the unit should provide space for various functions; at the same time the various sub-units should be able to share as many common facilities as possible.

Ideally, the psychiatric nursing unit should have 20 to 24 beds. This provides for close observation of patients by the medical and nursing staff. It is also good from the therapeutic point of view. Private accommodations are desirable.

The unit should have one or two seclusion rooms, as described earlier, for patients requiring security and protection, or for management of acute behavioural disturbances. These rooms, meant for short-term occupancy by patients who have become violent or suicidal, should be specially designed to prevent hiding, escape, injury or suicide. They should be so located as to enable direct observation by the staff. They should have padded beds, high ceilings, tamper-proof fittings and an absence of such objects in the vicinity as could be made instrumental for hurting oneself or the others, like glass, bottles, rope, knife, etc. Isolation rooms for these patients should have a separate entry. Doors should open out and should permit observation by the staff.

Additionally, there should be two rooms for the physically ill patients or patients receiving treatment requiring bed rest or skilled nursing care. These should be adjacent to the nursing station so located as to permit close observation of the patients by the nursing staff.

➤ FACILITIES AND SPACE REQUIREMENTS

Facilities in the psychiatric unit are by and large similar to those in any general nursing unit except for the following special features and requirements.

* Windows should be operable by keys or some such devices which are under the control of the staff. Safety glazing may be used instead of regular glass. This security measure is intended to inhibit any possible escape or suicide.
* Drug storage should have provision for security against unauthorized access.
* Space for storage of stretchers and wheelchairs should be outside the unit. They should, however, be readily available.
* There should be interview rooms where the therapist can work up and see the patient.

Individual cubicles are necessary. While acoustical privacy is necessary, visual privacy is not so essential.

- ◆ Recreational and occupational therapy areas (both indoors and outdoors) should be provided, particularly for a-motivated and chronically ill patients. These should be close to the unit, if possible. The areas can be used for group discussion, family interaction session and other group sessions.
- ◆ Medical examination and procedure room or rooms close to the nurses' station.
- ◆ Consultation room(s). There should be at least one consultation room for every 30 psychiatric beds. Rooms should be designed to provide both visual and acoustical privacy.
- ◆ It is recommended that toilets and showers are in a central location. The object is to eliminate individual facilities adjacent to rooms.

ISOLATION ROOMS

➤OVERVIEW

All hospitals knowingly or unknowingly admit patients with communicable diseases. It is the responsibility of the hospital to protect other patients and hospital personnel from these diseases. Barrier nursing and other techniques alone, or a single room which could constitute an improvised isolation unit, are not enough. When the unit depends entirely on techniques alone, lapses are likely to occur in case of overworked or insufficiently trained staff. Physical barriers are therefore necessary.

➤LOCATION

Isolation rooms may be located within the individual nursing units and placed at the end of the corridor. They should be available for normal acute care when not required for isolation cases. Isolation rooms may be grouped as a separate isolation unit.

➤DESIGN

Each isolation room should contain only one bed and should have all the facilities of the acute care patient room. In addition, it should have the following facilities.

- Entry into the room should be through an ante-room, work area, or a vestibule that has provision for aseptic control which includes handwashing facilities with a gooseneck spout, gowning and storage of clean and soiled materials. These facilities in the work area are separate from those in the patient's room. Where strict isolation is envisaged, a vestibule may not serve the purpose; an enclosed ante-room will then be necessary.
- Viewing panel(s) may be provided for observation of the patient by staff from the ante-room.
- One ante-room may serve several isolation rooms.
- Each isolation room should have toilet, bath and handwashing facilities so arranged as to permit access from the bed without passing through the ante-room or the work area.
- Isolation room is similar to other patient rooms, but requires a hook strip for gowns near the door.

The design features described here refer to general isolation rooms for communicable diseases. Facilities required for specialized procedures such as organ transplants, bone marrow transplant and burn cases which call for special design provisions to meet the needs of functional programmes are not covered here.

INTENSIVE CARE UNITS (ICUs)

➤OVERVIEW

Intensive care units (ICUs) are specialty nursing units designed, equipped and staffed with specially skilled personnel for treating very critical patients or those requiring specialized care and equipment. Centralizing the acutely ill patients, as is often done, in contiguous units in an intensive care complex consisting of surgical-medical intensive care unit, coronary care unit and specialty units such as renal and burn units, results in multidisciplinary care and economical use of the space and equipment. ICUs use sophisticated electronic instruments for observation, signaling, recording and measuring physiological functions besides monitoring temperature, blood pressure and respiration rates. More nurses are required per number of patients in the ICU, sometimes on a one to one ratio for each shift — 3 : 1 ratio for three shifts of the day is considered ideal — to give close attention to the critically ill or post-operative patients.

Establishing and maintaining an ICU complex is costly because it requires special space, equipment and staff. Not every hospital can provide all types of intensive care. Some hospitals may have a small combined unit; others may have separate, sophisticated units for highly specialized treatments; still others may not be able to provide even the most basic care. In this section the ICUs and the CCU are discussed comprehensively so that hospitals can use this information to set up their intensive care units in the scale of their choice.

➤ FUNCTIONS

The functions of the intensive care complex are as follows.

- ◆ To concentrate in one centralized area the critically ill and the post-operative patients for close observation and skilled nursing care by specially trained personnel.
- ◆ To enhance the physician's ability to treat the acutely ill patients through the use of centralized and highly skilled support personnel and specialized equipment.
- ◆ To provide close personal and monitor-assisted surveillance of the critically ill patients so that the readings and data relating to their physiological functions are instantly available to the professional staff for facilitating timely diagnosis, treatment and evaluation of care.
- ◆ To utilize equipment and highly trained personnel more effectively and economically.
- ◆ To improve the overall patient care on the patient floors by moving to the ICU the acutely ill patient whose treatment is often carried out at the expense of other patients so that nurses can give more time to the less critically ill patients in the wards.
- ◆ The surgical intensive care unit, in particular, provides care of the post-surgical patients who develop complications and require close nursing observation and care.
- ◆ The medical intensive care unit provides care for emergency patients suffering from coma, shock, haemorrhage, convulsions and respiratory and other medical problems.
- ◆ The coronary intensive care unit cares for patients with acute cardiac conditions and those who require continuous, individual observation and care utilizing electronic monitoring and therapy equipment.

➤ LOCATION

There is no unanimity among the medical and nursing experts as to where ICU should be located. There are two schools of thought. One suggests that the ICUs should be in a centralized place and be contiguous with, or readily accessible to, one another. The argument is that patients admitted to the medical-surgical intensive care unit may have, or suddenly develop, cardiac complications. Having intensive care facilities in a centralized place allows the specially trained professionals and equipment an almost instant access to patients in all clinical services when an emergency develops. Such an arrangement also eliminates the need for duplication of costly equipment and personnel.

The second school of thought favours that the location should be dependent on the type of patients. For example, the surgical ICU should be close to the operating rooms while the medical ICU should be in close proximity to the medical ward to facilitate follow the concept of progressive care, i.e., the patient is moved from the intensive care unit to intermediate care or step-down unit, and then to the general patient care area.

We recommend that each hospital weigh the advantages and disadvantages of these two systems carefully and decide which one would most suitably meet its needs.

Intensive care units should be close to the emergency department, operating rooms, recovery rooms, respiratory therapy, laboratory and radiology, and so located that the specialized cardiac team is able to respond promptly to the ICU emergency calls with a minimum travel time. Most admissions to ICUs are either through the emergency department or from the operating rooms following major surgery.

The ICUs should not be too far away from the general nursing units. This will reduce to a minimum the movement and time required to transfer patients from one unit to the other, especially from the nursing unit to the ICU, in an emergency. For this reason, ICUs should be close to the vertical transportation to facilitate rapid transport of patients and personnel.

The ICUs should be located away from heavy traffic and noise. In deciding the location, the electrical influence of equipment like elevator motors and x-ray equipment on the displays of monitors should be kept in mind.

➤SOME POINTS TO CONSIDER BEFORE DESIGNING

Defining Intensive Care and Intensive Care Units

It is important that the planners and medical specialists clearly define what constitutes intensive care, and interpret it to the architect by means of a written manual with details of work activities, equipment, finishes and space requirements.

Determining the Number of ICU Beds

In the earlier times when ICU consisted largely of coronary care beds, five per cent of the total number of beds was the "rule of thumb" for determining the number of beds in the ICU. Intensive care now encompasses many disciplines; it is recommended that 10 per cent of the total beds be designated to ICUs.

Size of Units

It is generally recognized that for effective operation, there should be no more than twelve to sixteen beds per intensive care unit. A six-bed unit is probably the most economical unit. It requires approximately the same number of staff as smaller units.

➤DESIGN

Functions, accessibility and direct visual contact between the patient beds or rooms and the nurses' station are of paramount importance in designing the ICUs. The design should provide for maximum visibility of the patient, not only of the monitor. The patient room should be close enough to permit observation of respiration, facial colour and other revealing symptoms.

Proximity of the nurses' station to the patient room saves steps for the nurse and serves to boost patients' morale.

Differences between the ICU and the CCU are few. The environment and the staff are the more important ones. Patients in the CCU are generally alert and often ambulant. They need windows for visual contact with the outside world. Besides, interior design should be planned to avoid depressing effects or overstimulation from certain colours and lighting. Live plants provide freshness and cheer to the patients. As for the staff, they must be specially trained for quick action in any emergency.

Each cardiac patient should have a separate room or enclosure for acoustical and visual privacy. He should also have access to a toilet directly from his room or cubicle. Portable commodes may be used, but then provision should be made for their storage, servicing and odour control. Monitoring equipment should have provision for display both at the bedside and at the nurses' station.

Two sample ICU layouts are given in Figs 5.7 and 5.8.

➤ SOME DESIGN ELEMENTS FOR PLANNING

Facilities and Space Requirements

- Individual rooms with full height glass walls between the rooms and the corridor are recommended. This will allow visibility of patient's face and minimize cross infection. Where individual rooms cannot be provided, curtain or cubicle screening of beds may be provided.
- Privacy is required for all adult beds since the unit will be serving both male and female patients. Curtains over the glass walls may be drawn when necessary. Doors may have view panels with curtains to allow total privacy when needed.
- The entrance door to rooms should be at least 3 ft. 8 in. preferably 4 ft. wide for easy movement of beds and large equipment.
- Rooms should be at least 120 sq. ft. with free unencumbered movable space surrounding the bed.
- In order to increase free floor space, there should be a minimum amount of movable furniture such as a bed, a chair and overbed table. Rest of the necessary articles should be fixed equipment like ceiling mounted monitoring console on retractable swivel post, a support column with provision for blood pressure equipment, etc.
- Mechanically or electrically operated beds which can be adjusted to various positions are recommended. The base and mattress should be firm to enable effective cardiac compression. A detachable rail at the head end allows easy access for endotracheal intubation. Beds, movable on wheels and provided with a locking device and detachable side railings are preferable.
- The following items should be placed on the wall at the head end of each bed or on a free standing column (Plate 8: Picture 51):
  1. Medical gas outlets — two for oxygen, one for compressed air and two for suction;

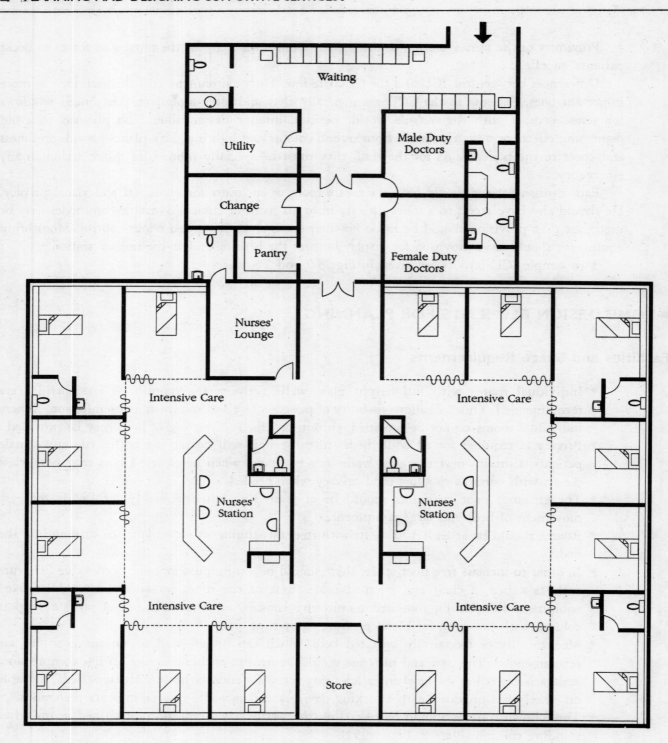

➤ **FIG. 5.7** Layout of an Intensive Care Unit: Illustration I

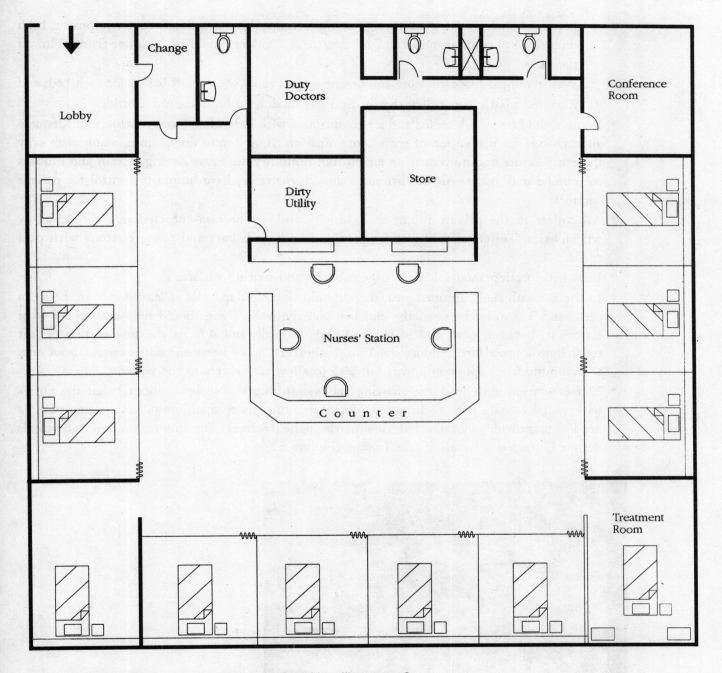

> **FIG. 5.8** Layout of an Intensive Care Unit: Illustration 2

2. Nurses' call button;
3. Telephone outlet — optional;

4. Electrical outlets — for ceiling lights, night lights, dimmers, fluoroscopy equipment, high intensity lights for examination and treatment. Outlets should be no lower than 30 in. off the floor;

5. Wall mounted blood pressure equipment (swivel type) with a cuff basket for each bed; and

6. Recessed plastic pan below the medical gas module to hold vacuum bottles.

♦ Room should be soundproofed and air-conditioned with individual room controls and adequate air exchange for prevention of cross contamination. If air-conditioning units manipulate only the temperature and movement of air but not humidity, excessive drying of skin and mucous membrane may occur. Air-conditioning should, therefore, have humidity control for patient comfort.

♦ All toilets in the private rooms and cubicles and on the non-cubicled side should allow wheelchairs. Toilets should also be provided with grab bars and panic buttons with pull cords.

♦ Each bed location should have storage cabinets and writing surface.

♦ In the non-cubicled curtained area, there should be a clear space of at least 4 ft. 6 in. between beds, and 3 ft. 6 in. between the end bed and the wall. There should be a clear space of at least 3 ft. between head end of the bed and the wall, and 4 ft. at the foot end to permit resuscitation procedures, endotracheal intubation, etc. Space between beds is essential not only to accommodate bulky equipment but also to allow easy access to the patient.

♦ Nurses' station with space for charting and monitoring should be so located that the nurses have visual contact with each patient. In larger units, more than one nurses' station may be needed to provide unobstructed view of the patients' faces. For this reason, some hospitals design U-shaped or semi-circular layouts (Picture 52).

➤ **PICTURE 52** Coronary Care rooms arranged in semi-circle around a nursing station

- The central control station should be designed to seat two or three persons only. This will discourage staff from frequenting or idling in that area.

- Each unit should have equipment for continuous monitoring with visual display for each patient. Monitors should have in-built high/low alarm and capability to provide hard copy of wave forms needed for patient care.

- Central monitoring alone is almost a thing of the past. Most hospitals now have sophisticated, fully-computerized individual monitors at the bedside where patient care takes place (Picture 53). Data entered on bedside terminal is automatically communicated to the PC located at the nurses' station where it is formatted and printed on a printer for insertion into the patient chart. The bedside monitor should be so placed that it is easily visible to the staff but not to the patient. Bedside computers are discussed in detail in the chapter "Hospital Information System".

- Advanced computers have the capability to display laboratory data along with the physiological data. Small bedside computers can be used to calculate the fluid balance.

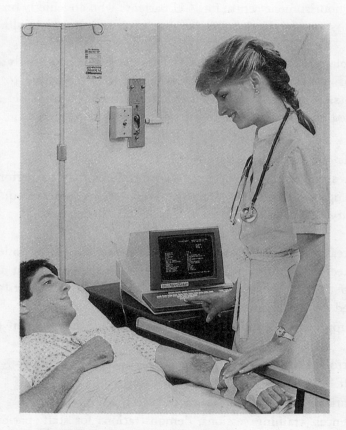

➢ **PICTURE 53** Bedside Computer System

- All units need resuscitation equipment like ventilators, blood gas analysers, infusion pumps, a defibrillator with built-in oscilloscope, etc.
- One or more emergency carts, also called "crash carts", "red carts", or "blue carts", depending on local preference (Picture 46), should be kept in readiness in each ICU immediately accessible to the ICU nursing station. Equipment and medical supplies are normally assembled on carts and kept ready to roll at any time. The carts should be replenished immediately after use. All the supplies necessary to re-stock the carts should be in a clean utility room and drug cabinet of the unit.
- Considerable storage space is required to maintain in stock a large number of essential items like tracheostomy trays, sterile pacemaker catheter, dressing trays, Foley's catheters, catheterization kits, urinary drainage bags, sterile gloves, needles, syringes, and intravenous solutions.
- Respiratory therapist is an important member of the ICU staff. His office should be readily accessible so that he can be promptly summoned to attend to patients in a respiratory crisis. Space for storage of respiratory equipment is required.
- Provision should be made for a nourishment centre for ICU patients who are chiefly on liquids. Nourishment for staff who miss regular food hours because of emergencies should also be available.
- It is essential to provide for isolation room(s) in each ICU for infected or potentially infected and critically ill patients. This is particularly necessary when it is not possible to provide single rooms for all the patients. A critically ill road accident victim with multiple injuries, for example, cannot be admitted and treated next to a fresh post-operative patient.

Other needed facilities are:

- A treatment room. Medical gas outlets are necessary.
- One room for special procedures where emergency invasive procedures may be performed. It should be designed for fluoroscopy, and be large enough to hold various items of equipment. Medical gas outlets are necessary.
- Clean linen storage area.
- A soiled holding and work area with clinical sink and handwashing facilities. Soiled linen should be bagged — if it is infected linen preferably in bags distinguished by colour and with sufficient space left at the top to make complete closure — and stored temporarily in the soiled holding area until picked up by laundry personnel.
- A storage for housekeeping supplies with janitor's closet, service sink and bedpan flushing facilities.
- Lockable drug storage for controlled drugs.
- Handwashing facilities for staff.
- Facilities for physicians for charting and dictation (with dictating equipment.)
- Multipurpose room for conferences, training sessions, demonstrations for staff, patients, and patients' families.

- Staff lounges, preferably separate for male and female staff.
- Duty doctors' room containing sleeping accommodation with bath, toilet, etc. There should be separate rooms for male and female doctors who are on regular night or on-call duty.
- Family waiting room with toilet and pay telephone facilities. There should be a gowning area and storage for clean gowns, masks, etc.
  Casual visiting should be discouraged in order to maintain high standards of cleanliness. Most ICUs, however, allow family members to visit, one at a time, for 5 to 10 minutes every one or two hours. A sign board which says patient needs rest usually helps the family to understand this rule.
- One or two beds in the medical-surgical ICU may be supplied with wall mounted dialysis machines.

►CODE BLUE TEAM AND CODE BLUE ALARM

'Code Blue' is a term used in hospitals to announce an emergency of serious nature such as a cardiac arrest. In advanced countries, in patient rooms and other strategic areas, there is a button marked "CODE BLUE" which when activated sets off a distinguished emergency alarm signal both at the nurses' station and at the telephone operator's room. While the nurse attends to the patient instantly, the telephone operator goes on the public address system announcing 'Code Blue' three times giving the location of the emergency. There is a pre-written procedure to deal with such an emergency and members of a pre-appointed code blue team respond to the call immediately for wherever they are.

To avoid panic among patients and visitors, emergencies in hospitals are announced using codes — "Code Blue" for medical emergency, "Doctor Red" for fire, "Code Black" for bomb threat, "Code White" for security emergency, "Doctor Major" for disaster, and "Code Green" for all clear.

►OTHER SPECIALTY ICUs

This chapter does not include other specialty ICUs like neuro ICU, oncology ICU, respiratory ICU, paediatric and neonatal ICU, etc. because of their unique requirements of specialty intensive care. General provisions and standards discussed in the foregoing sections should be used in them insofar as they are applicable. Adaptations, adjustments and additions should be made to meet the functional needs of staff and patients with special consideration for ancillary services needed in their operation.

NEWBORN NURSERIES

➤OVERVIEW

Nurseries are one of the areas of the hospital where patients are most vulnerable to infections. They should therefore be planned and designed to provide the best possible care, safety, and welfare of the infants. There has been a rapid evolution in the concept and principles of care in this area as a result of which the location, design, guiding concepts and principles governing nurseries have changed markedly. Traditionally, what was only a nursery has now given place to the concept of nurseries like the newborn, observation, premature, isolation, high-risk and intensive care nurseries.

Location of nurseries is influenced by many factors such as specially trained staff and professional philosophy. This is particularly true of isolation and premature nurseries. The contemporary trend is to locate these in the paediatric service with the premature nursery as part of the neonatal intensive care unit.

➤LOCATION

The nurseries should be located in the obstetrical nursing unit as close to the mothers as possible. It should also be close to the labour-delivery suites so as to minimize travel distance and exposure of the newborn infants. It is desirable that they are in close proximity to the paediatric service, particularly if premature nursery and neonatal intensive care unit are also located there.

➤DESIGN

Although the plan, location and size of nurseries differ from hospital to hospital, the underlying principles in their design are the same and apply to all of them. The following are some of the basic principles that should be remembered:
   ◆ Limit the number of infants in each nursery;
   ◆ Allow wide spacing of bassinets within each nursery;

* As far as possible, separate bassinets by cubicle partitions to prevent crowding;
* Promote the use of aseptic techniques and individual care and provide handwashing facilities;
* Limit the number of bassinets served by one nurses' station;
* Separate premature infants from infected infants or those suspected of infection; and
* Provide optimum conditions of temperature, relative humidity and ventilation.

The major nurseries or components of nurseries are generally a full term nursery for normal infants, a premature nursery for the care of infants with a low birth weight, an observation nursery for infants suspected of infection, an isolation nursery designed for the care of infants which have an infectious disease, and a nursery for intensive care (Fig. 5.9 and 5.10).

The need for a close, natural adaptation of mother and the newborn infant to each other right from birth is ingrained in the Indian culture. The basic physical and emotional needs of both the mother and the child are best satisfied by 'rooming-in', that is, by placing the infant and mother together soon after birth. So in the Indian setting, it is hardly necessary to have a large nursery for full-term infants as it is in the West. However, the full-term nursery may not be entirely done away with for the following reasons:

* Some mothers do not wish to assume the care of infant rightaway; some may be too sick or exhausted and may not be able to do so.
* Mothers who have girl babies are sometimes under great stress and pressure from their families. They may take some time to room-in with the infants. They would, meanwhile, need to be helped.
* In some hospitals, the infant is removed from the room during the medical staff rounds and examination, cleaning periods, mother's meal time, and in the presence of any infection in the mother.

In the light of the observations noted above, each hospital should plan the size of its full-term nursery. If a full scale nursery is decided upon, the maximum number of bassinets should be 12. Staffing requirement is generally one nurse to care for every six infants. So larger nurseries do not conserve staff. On the other hand, smaller nurseries provide better conditions for the care of infants. They are easier to clean. They also minimize the risk of one infant incubating an infectious disease and passing it on to other infants.

The nurses' station serves as a control point. It also provides work space for the nurses. The nurses' desk should be so placed as to provide her a view and supervision of the nurseries and the entrance from the corridor. The nurseries should be visible through large view panels in the partitions. Each nurses' station should be provided with a rack for charts, a waste receptacle with foot-controlled cover and chairs at the nurses' desk. A clean workspace on one side should have a counter with a sink with gooseneck spout and knee or foot operated controls, wall-mounted cabinets above and a refrigerator to hold a 24-hour supply of mixtures. The soiled workspace with similar facilities should be on the other side.

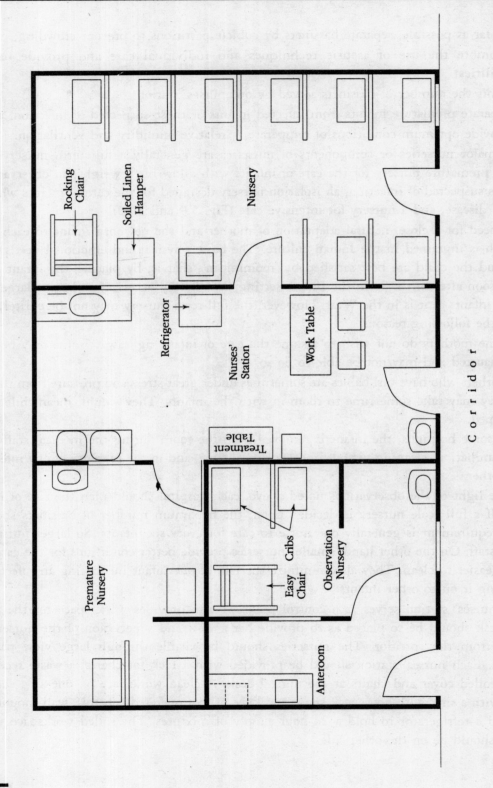

FIG. 5.9 Layout of a Nursery

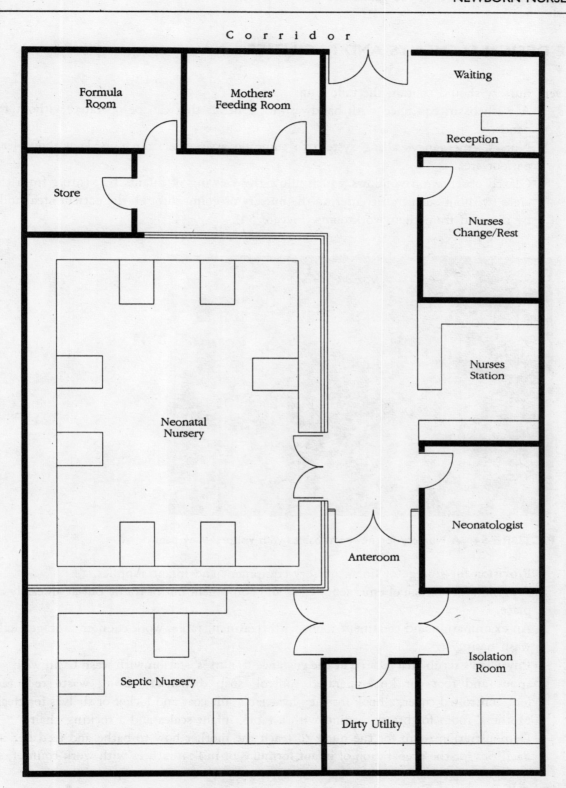

Corridor

Formula Room

Mothers' Feeding Room

Waiting

Reception

Store

Nurses Change/Rest

Neonatal Nursery

Nurses Station

Neonatologist

Anteroom

Septic Nursery

Isolation Room

Dirty Utility

➤ **FIG. 5.10** A Typical Neonatal Unit

➤ SOME DESIGN ELEMENTS AND FACILITIES

Every nursery should contain the following:
* A wash basin equipped with handwashing controls that can be operated without the use of hands.
* Nurses' emergency calling system to summon assistance from outside without leaving the patient area.
* Glazed observation windows which allow the viewing of infants by visitors from the public areas (without such traffic entering the nursery or going through the patient area) and viewing by the staff from their workrooms (Picture 54).

PICTURE 54 A Nursery for newborn babies with visitors' view panel

* Provision for storage of linen, nursery equipment and infant supplies.
* Provision for storage of emergency carts and equipment out of traffic but at an easily accessible place.
* An examination and treatment room with treatment table, work counter, storage facilities and wash basin.
* Physician's scrub area located at the entrance to nurses' station with wash basin with gooseneck spout and foot- or knee-operated control, soap dispenser, towel, waste receptacle with foot-controlled cover, a hook for the physician's suit coat and locker or shelves for clean gloves.
* Mothers' room for breast feeding with a table, infant scales and a rocking chair.
* Demonstration room for the nurse to teach the mother how to bathe and feed her infant.
* Facilities for the preparation of infant formulas or milk mixtures with work counter, sink, and

a small refrigerator, regardless of whether commercial formulas are used or they are primarily prepared in the hospital. Also, clean-up facilities for bottle washing and sterilizing supplies and equipment should be provided. Refrigerated storage and provision for warming infant formula must be available for use by the nursery personnel at all times.

- In the nurseries, the peak workload comes during the morning hours when there may be one nurse for every five or six babies. This ratio may be reduced during the afternoon and night. There should be adequate work and moving space for the nurses in the nurses' station.
- Each nursery should have a connecting workroom or anteroom with scrubbing and gowning facilities at the entrance for physicians, nurses and housekeeping personnel, work counter, wash basin and storage of supplies. The anteroom is particularly important in the case of observation nursery.

➤FULL-TERM NURSERY

The full-term nursery is intended for normal infants, and should be located in the post-partum nursing unit, as close to the mothers as possible. An area of 30 sq.ft. per infant or bassinet is recommended exclusive of the nurses' station, with a provision for at least 3 ft. between bassinets and on all sides. If cubicles are used, they should be large enough to permit bedside care. All partitions must be of clear glass to permit the nursing staff to observe the infants. In the absence of cubicle partitions, the bassinets tend to be crowded. The optimum number of full-term infants that any member of the nursing staff can care for is in the range of 8–10.

Furnishings in each full-term nursery include a bassinet, a bedside cabinet, incubators, a scale, a utility table (for infant scales although modern incubators may have built-in scales), a wash basin with gooseneck spout and foot- or knee-operated controls, soap dispenser, a waste receptacle with foot-controlled cover, a metal can with foot-controlled cover for soiled linen, outlets for piped oxygen — one double outlet for every four bassinets and outlet for suction from the central system (or a mechanical suction device) — a rocking chair for use of the nurse for feeding infants, facilities for examination, soiled holding area, etc.

➤OBSERVATION NURSERY

An observation nursery should be provided for infants suspected of infection. If the diagnosis proves to be positive, the infant is moved to other facilities and placed in isolation. If the diagnosis is not positive, the infant is returned to the regular nursery or to the mother's room provided he has not been exposed to any infection from any other infant in the observation nursery.

A minimum of 40 sq. ft. per infant or bassinet is recommended.

The observation nursery should be located adjacent to the full-term nursery, but it should be a completely separate unit with glazed partitions to permit observation of the infants by the nursing staff.

There should be a connecting anteroom for the observation nursery between the nursery and the corridor with provision for facilities mentioned earlier. The anteroom should have the same facilities as the work and treatment areas of a full-term nursery which include, among other things, a work counter, a sink, hook strip and shelves or cabinets for clean gowns. The furnishings in the observation nursery should be similar to those in a full-term nursery.

➤ PREMATURE NURSERY

The premature nursery is used for infants with a low birth weight and is generally located in paediatric service. If it is located in the post-partum unit, as is sometimes done, common facilities of the full-term nursery can be shared.

Premature infants need more specialized care (Plate 8: Picture 55). So a premature nursery room should accommodate no more than five infants and should have minimum of 40 sq. ft. of space per incubator or bassinet.

Not many hospitals can maintain a separate premature nursery. If the number of premature babies are less than five, there may not be any justification for a separate nursery for them. In such cases, they may be accommodated in the full-term nursery provided that space is not a restricting problem, and the combined complement of infants does not exceed the optimum number recommended for one nursery. Also, during the periods of low census, it may be necessary to care for premature infants in a full-term nursery.

In the premature nursery, suitable environmental temperature and humidity are important factors. When these are properly maintained, only 50–70 per cent of the infants may require incubators. Apart from the incubators, other furnishings here are the same as in a full-term nursery. A rocking chair is provided for tending to the infants.

Piped oxygen and suction should be supplied from the central system.

If the premature nursery is located in the paediatric department, premature infants born outside the hospital may be accepted for treatment. If the unit is located in the post-partum unit, many hospitals do not, as a matter of policy, admit infants born outside the hospital.

➤ ISOLATION NURSERY

Isolation nursery is designed for the care of infants who have infectious diseases. It requires 50–60 sq. ft. of space per bassinet. Infants in the isolation nursery may not have been transferred from other nurseries in the hospital. It may be that they acquired the infection after they were discharged from the hospital.

Cubicles could be used to separate infants from each other. Strict isolation techniques should be practised, by all the staff working in this nursery. Physical facilities are the same as in the premature nursery and the same elements of care of the infants are required here.

➤UTILITIES

Attention should be given to special requirements in the matter of air-conditioning, ventilation and electrical services. Air-conditioning is required for nurseries to ensure constant temperature and humidity conditions so beneficial to the care of newborns. Air-conditioning system also removes odour, and materially reduces bacterial contamination of the environment through its ventilating features. In this way, air-conditioning supplements aseptic techniques aimed at reducing the hazard of infection and cross-infection.

Early in the planning stage, experts should be called in to advise, and the architect should be given a detailed brief regarding the services required in the nurseries. The same thing holds good with regard to finishes and colour schemes.

CHAPTER 6

Planning and Designing
Supportive Services

CONTENTS

6 CHAPTER

- ❑ **Admitting Department**
- ❑ **Medical Records Department**
- ❑ **Central Sterilization and Supply Department (CSSD)**
- ❑ **Pharmacy**
- ❑ **Materials Management**
- ❑ **Food Service Department**
- ❑ **Laundry and Linen Services**
- ❑ **Housekeeping**
- ❑ **Volunteer Department**

ADMITTING DEPARTMENT

➤OVERVIEW

Functions of the admitting department revolve largely around admitting, transfer and discharge of patients. In some progressive hospitals, the functions of this department are enlarged to include reception, round-the-clock enquiry or information and what is called centralized patient service which is a new concept in telemarketing and public relations . This department coordinates patient's arrival, registration, medical records and initial tests. It also makes pre-admission reservation of beds and the follow-up of patients' well-being and return visits after they are discharged.

The importance of the role of admitting process cannot be overemphasized. The data collected during the admitting process is vital to the quality of care the patient receives. This information is transmitted to the medical staff, nursing units, business office and the ancillary services such as the laboratory, which need accurate and timely information to function well. The efficiency of the department is measured by such factors as the length of the time patients must wait during admitting, confidentiality of information, simplified forms and procedures in the admitting process, system of escorting patients through service departments and to their rooms, and finally the demonstrated concern and courtesy of the staff. In some cases, the admitting department may be the patient's first point of contact with the hospital; in that case, it may influence or establish the patient's first and, perhaps, the lasting opinion of the hospital. The same is true of the opinion formed by the patients' families and the general public.

Many of the patients who pass through the admitting process are ill and worried. Some of them are apprehensive or physically incapacitated. The patients and those accompanying them are in a state of mental stress. Delay in admission can cause them emotional trauma. It is imperative, therefore, that there should be a minimum of delay between the patient's arrival at the hospital and his being established in his room. The admitting staff are expected to exercise the utmost care, consideration and courtesy in dealing with patients and their relatives, and in meeting their responsibilities with composure, grace and resourcefulness. Similarly, discharge of the patient, which should be a pleasant occasion for the patients and their relatives, often turns sour, even traumatic, because of poor planning, lack of information, delay in the preparation of bills and inefficient working. This often results on a poor last impression of the hospital.

Admission sets the stage for an uninterrupted programme of patient care. It provides the medical staff, nursing unit and service areas with essential information consisting of the patient's background and, in most cases, a preliminary diagnosis.

Admission as well as discharge calls for cooperation between many departments of the hospital. The trend is towards pre-admission procedures to make the patient's transition from his home to the hospital smooth and brief. Patients are usually admitted to the hospital through the outpatient department, emergency service and from their homes, by and large, as elective cases. Some patients may be admitted directly by doctors who are given the privileges to practise in the hospital. The priority of admissions is determined by the degree of urgency which usually classifies cases into three types, namely, emergency cases which require immediate admission, urgent cases which require admission as soon as possible (usually within 72 hours) and elective cases which are scheduled in advance and are generally admitted in order of their reservations.

➤FUNCTIONS

The following are the main functions of the admitting department.

- ◆ Admit, transfer and discharge patients. Check master patient index file for any previous file of the patient. If there is a file, get the medical record from medical record department pre-released for coordination. If the patient is to be admitted for the first time, arrange for the number to be assigned by the medical record department.
- ◆ Interview the patient to collect information needed by the various hospital departments. Generate appropriate patient records. Coordinate admitting with the other hospital functions.
- ◆ Compute and arrange for payment of pre-admission deposit according to the policy of the institution.
- ◆ Accept reservations for scheduling future admissions.
- ◆ Disseminate information relating to admission, transfer, discharge, death, etc. to the departments concerned.
- ◆ Maintain a bed index showing current occupancy status of all the patient rooms in order to schedule and assign beds for admissions and transfers.
- ◆ Arrange for the safekeeping of patients' valuables, either with the cashier or in the department's safety vault.
- ◆ Prepare admission and discharge lists and midnight census.
- ◆ Coordinate pre-admission tests in the case of elective admissions in order to reduce the actual time needed to complete the admitting process.
- ◆ Arrange for patient escort service to the diagnostic or test areas, if necessary, and then to the patient rooms. Some hospitals make it mandatory that on discharge, all patients are escorted to the main entrance of the hospital by wheelchair.

In many hospitals, particularly if the hospital is computerized, emergency service and obstetric department perform admitting functions for their patients in their own departments using the common database. Patients are pre-instructed to proceed to those departments directly.

➤ LOCATION

Admitting department should be strategically located. Since some of the patients seeking admission are physically incapacitated, confused or even disoriented, the department should be (a) situated on the same level as the hospital's main entrance, (b) be readily identifiable, and (c) provided with a sign that can be seen without difficulty from the hospital's main entrance reception area and the information desk. It should be adjacent to or in close proximity with the emergency service, outpatient department, medical records, laboratory, radiology and the cashier's booth. It must have convenient access to or communication with the business office, operating rooms and the administration. Because of the physical condition of some patients — cardiac and obstetrical patients, for example — who require immediate care in areas quickly reached by elevators, the department should be located near the main vertical circulation of the hospital.

The admitting department should be contiguous with the hospital's main lobby as well as the main waiting area of the outpatient department. It is economical to have all the admitting functions along with those of enquiry/information and centralized patient service in one unit rather than in a number of units. It results in simplified administration, closer supervision and savings in terms of personnel, space, and equipment. Time and effort can be saved, errors eliminated and operating efficiency of both the departments increased if admitting and medical records are located adjacent to each other. As described earlier, the admitting department needs to coordinate the patient's medical records during admitting.

➤ DESIGN

Admitting and discharge should take place in a pleasant and comfortable environment in which patient is assured of privacy and individual attention. Cubicles in which the hospital representative can interact with the patient on a one-to-one basis are highly recommended. Special consideration should be given to decor, furniture, etc. which should be made pleasing to the eye.

Admitting and discharge of patients should not be through the main lobby of the hospital or in the midst of heavy traffic. Although the lounge seating in the main lobby can be used by relatives and others accompanying the patient to wait comfortably while the patient is being admitted, a subsidiary waiting area — a screened alcove or sublounge off the main lobby — adjacent to the admitting department should be provided exclusively for admitting. The reason for this is the distracting noise and crowd in the lobby during the peak hours, whereas admitting requires a quiet and peaceful environment (Fig. 6.1 & Picture 56).

The main public entrance to the hospital, public telephones, toilets, drinking fountains, information desk, elevators, wheelchair storage area and cashier's booth should be conveniently located for the use of patients, their relatives and visitors. (Fig. 6.1 & Fig. 7.1) So also, an appropriate space should be provided for keeping the patients' suitcases and other personal belongings.

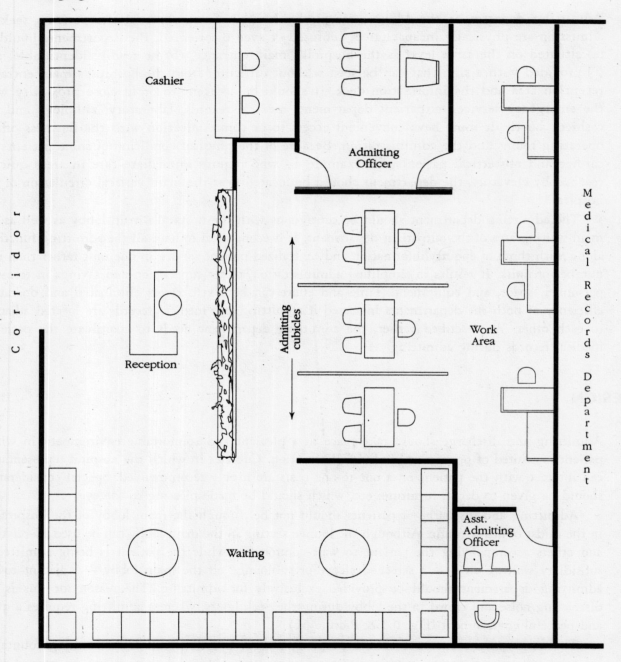

➤ **FIG. 6.1** Layout of Central Admitting Department

➤ORGANIZATION

The reporting relationships of the admitting department vary considerably, but generally the chief of admitting reports either to the administrative services or the financial services. However, with registration usually being a part of the medical records, a close working relationship between admitting and registration and medical records is necessary.

There is no particular training or specific credentials required for working in the admitting department. Most of the skills are learned on the job. However, the staff, consisting mostly of clerks possessing general office skills, should have orientation towards accounting, medical terms and records, nursing, and ancillary services.

The admitting process is a seven-day a week, round-the-clock function. In many hospitals, late night shift coverage (11.00 pm to 7.00 am) is performed by nursing or emergency department staff.

➤SPACE REQUIREMENTS

The admitting department requires space and facilities for the following.
- ◆ Admitting patients' waiting area, large enough to accommodate patients and their relatives.
- ◆ Admitting reception area for clerks who receive patients, record their arrival, assign them to interview offices (cubicles), coordinate records, follow their admissions and arrange escort services. Records are delivered by escorts to the nursing units along with the patients.

➢ **PICTURE 56** Admitting Cubicles

- ◆ Admitting office(s) or cubicle(s) (Picture 56) to serve as individual work spaces for the admitting officer, patient and the responsible patient relative(s) who need to be there. It should be just large enough to avoid unnecessary crowding. Privacy is essential.
- ◆ Office of the chief/supervisor of admitting.

* Clerical work area for maintaining hospital occupancy directories, records, forms, notices, etc. related to admitting.
* A small safe or vault if the admitting department is responsible for receiving prepayment and patients' valuables during the evening and night hours when the cashier's booth is closed.
* Wheelchair and stretcher alcove. This is not necessary if the main wheelchair holding area is readily accessible.
* Supplies room.
* Space for computers and office equipment.
* Toilets.

➤OTHER CONSIDERATIONS

Admitting and Computer

Admitting is one of those departments of the hospital which uses computer extensively. The extent of the use of computers ranges from a stand-alone personal computer to the one linked to the hospitalwide computer network. For example, in an on line, real time information system with a terminal located in the admitting department, information such as bed availability and status of beds, and whether a particular room is occupied, being cleaned, or vacant is obtained instantly. Additionally, the patient has to register only once and is not required to repeat information at various points.

A common database is used by all the hospital departments. Other applications of computers are entry, maintenance and retrieval of census data, capturing charges, generating requisition slips, posting charges, printing forms, communication with other departments, etc. The systems can be expanded hospitalwide with computer terminals covering nursing stations and other departments.

Day Surgery and AM Admit

With increasing number of surgeries being performed as day surgery (also called the same day surgery) and with some hospitals performing as much as one-third of all their surgeries as the same day surgery, the admitting department has come to assume a greater responsibility. Patients have pre-surgery tests done days ahead of surgery as outpatients and get themselves admitted on the previous day to be prepared for surgery the next morning. Admitting department coordinates all activities relating to their admission and surgery.

In the case of AM ADMIT, the patient does not stay in the hospital the previous day or night. He gets admitted early in the morning on the day of surgery — hence the name — and is sent to a special preparation area to be prepared for surgery after the necessary paper work in the admitting department which opens its doors to such patients as early as 5.30 or 6.00 a m.

MEDICAL RECORD DEPARTMENT

➤OVERVIEW

The medical record department maintains records and documents relating to patient care. Among a host of activities, its main functions are filing, indexing and retrieving of medical records. The primary purpose of establishing a medical record department is to render service to patients, medical staff and hospital administration in support of good patient care. The quality of care rendered depends on the accuracy of information contained in the medical records, its timely availability to and the extent of utilization by the professional staff. To achieve economy, accuracy of information, and good communication which are of vital importance to medical records system, all information should be concentrated in the original medical records of patients. This should then be indexed and filed in the department. The three basic principles of medical records are: they must be accurately written, properly filed and easily accessible.

Medical records are used as primary tools for evaluating the quality of patient care rendered by the medical staff. For effective implementation of this, the medical staff must adopt and self-enforce rules and regulations for the production of timely, accurate and complete medical records. In well established institutions, there is a medical records committee which oversees this function and appraises the quality of medical records.

These records are widely used for teaching and research purposes. In the context of increasing malpractice liability suits against hospitals and physicians, well documented medical records are a good legal protection.

The physician is primarily responsible for the quality of his patients' medical records. It is his duty to review, correct and countersign records that are written by residents and junior doctors working under him. Each entry in the medical record must be signed by the person making the entry, and the signature should be identifiable so that responsibility for accuracy and authenticity can be fixed. The language used in writing medical records should be clear and concise and should not lend itself to misinterpretation. Abbreviations, symbols, etc. should be of acceptable standard. The medical record department should maintain a list of acceptable abbreviations and symbols for every one to follow.

Every hospital should formulate policies, rules and regulations for the production, completion and maintenance of medical records.

In many hospitals, registration is an integral part of medical records. The front office which registers all patients, assigns each new patient a unique number, collects patient demographics and other necessary data, assigns patients to physicians, and creates records. In the case of returning patients, it retrieves their records and updates them. It maintains a master patient index for all the patients. Registration is the starting point for outpatient visits and all patient related activities.

➤FUNCTIONS

The following are the important functions of the department.

- Planning, developing and directing a medical records system which includes patients' original clinical records and also the primary and secondary records and indexes. These may be in the central record room, the clinical service area, adjunct departments, or the outpatient department of the hospital.
- Maintaining proper facilities and services for accurate and timely production, processing, checking, indexing, filing and retrieval of medical records.
- Development and direction of procedure for the proper flow of records and reports among the various services and departments, including clinical services and the outpatient clinics where they are needed.
- Development of a statistical reporting system which includes ward census, consolidated daily census, outpatient department activities, and statistics in relation to services such as radiology, clinical laboratories and pharmacy.
- Preparation of vital records of births, deaths, reports of communicable diseases, etc. for mandatory and regulatory agencies, and statistical reports such as number of admissions, discharges by major clinical services, discharge diagnoses and length of stay by diagnoses, types and numbers of surgeries performed, etc. for use by the administration, medical staff committees and the education and research departments.
- Coding all diagnoses and operations according to international classification of diseases for statistical purposes.
- Safeguarding the information in the medical records against theft, loss, defacement, tampering or use by unauthorized persons.
- Determining in coordination with medical staff and administration the action to be taken in medico-legal cases relating to the release of medical records in a variety of situations, and determining the legality and ethical appropriateness of such actions in conformity with the laws of the land.

➤LOCATION

In order to provide prompt medical record service for the care of all patients (outpatient, inpatient and emergency patients) at all hours, and to foster close working relationship and continuous communication which is so essential among the related departments, the medical record department should be located close to the admitting area, the outpatient department, emergency room and the business office. It should also be located close to or on the corridor leading to the doctors' lounge so that the medical staff can conveniently stop by and complete their records and study cases. Proximity to admitting, outpatient and emergency departments eliminates unnecessary walking and consequent delay in procuring medical records. It also permits a skeleton staff to manage the work of medical record department during the evening and night shifts. While carrying on with their normal duties like filing, etc., the night crew can also furnish records to the emergency department. Location is important particularly in small hospitals where medical record department may remain closed during the night. In that case, it should be within easy walking distance for the authorized admitting or emergency department staff to enter the department and retrieve records for emergency patients. The need for security surveillance to safeguard medical record information also has a bearing on the location.

➤DESIGN

The front office of medical records — the registration together with the enquiry — is often the patient's first point of contact with the hospital. It is here that public relations plays a vital role. In addition to courteous and helpful staff, the physical design should be one that projects a warm and welcome feeling. Good functional design, logical placement of work areas, and a good system of communication among the various sections of the department and between the other departments are vitally important. The department should also be designed with the best possible means of transportation of medical records through all stages of their use and processing (Fig. 6.2).

➤ORGANIZATION

Staffing

Medical record department may be headed by a medical record administrator or officer who reports to the director for professional (medical) or administrative services. He should be a graduate with a degree or diploma in medical record administration. Besides him, the remaining staff in the department consists of medical record technicians and medical record clerks. The Christian Medical Association of India and various medical colleges offer degree and diploma courses in medical record administration. In large hospitals, there may be an assistant medical record officer and supervisors for major functional areas such as filing and indexing, coding and abstracting, transcription, discharge analysis, medical audit, utilization review and registration.

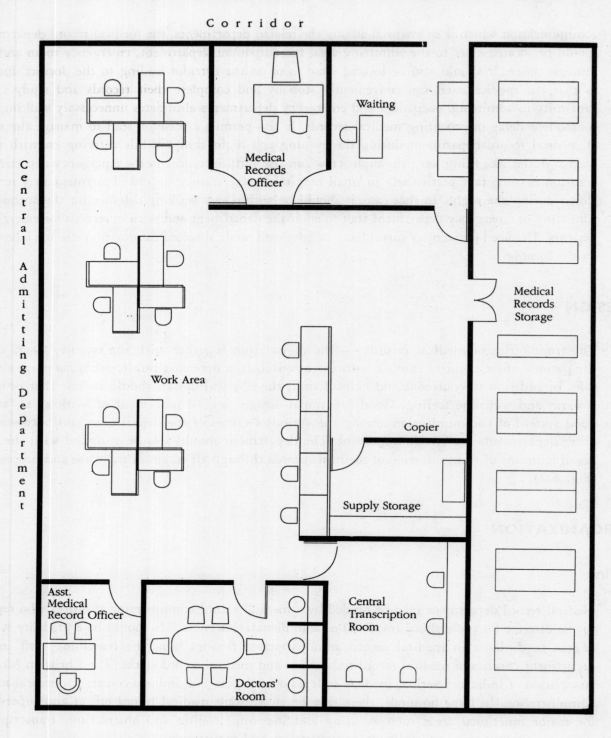

➢ **FIG. 6.2** Layout of a Medical Records Department

Unit Record

The unit record is a single record that documents the entire medical care provided to an individual in all the services of the hospital, namely, in the inpatient and outpatient sections and the emergency room. The single unit consolidates and retains all the records in a chronological order, that is, it maintains patient care information in the order of occurrence of events and findings. This way the record provides the doctors with the necessary references to a patient's current and past conditions, all tests and procedures performed on him, and his response to therapy.

Some hospitals maintain separate records for inpatient care and outpatient visits. The disadvantage of this system is that the patient's complete history cannot be reviewed quickly and easily. Other methods of assembling medical records are:

(a) chronological by source of information or section (physicians's notes, nurse's notes, laboratory reports, etc.), and

(b) what is called problem-oriented medical record.

These methods are, however, not popular.

Numbering System

The most widely used method for numbering is the unit numbering, used in conjunction with the unit record system mentioned above. In this system, a single, permanent number is assigned for each patient (as against different numbers each time a patient is admitted). The unit number ensures accurate identification of the patient and complete information about his investigation and tests and the accounting records.

Filing System

The most popular method of filing is the straight numerical filing, starting with the lowest number and ending with the highest. Activities relating to filing and retrieving are most concentrated in the area where records with the highest numbers are stored because they are the most recent and active files. This is the easiest method of filing as the staff is familiar and comfortable with this principle of filing. However, the chances of misfiling and not finding the misfiled charts are high in this system.

The other method of filing is the terminal digit filing which provides equal distribution of medical records in the storage area and therefore allows the staff to be evenly spread within the area. The filing is based on the last two digits of the medical record number. The entire file is divided into hundred sections from 00 to 99 and the records are stored in these sections according to their last two digits. For example, all records ending with 40 are filed together. In an advanced system, the terminal digits are colour coded. (Plate 9: Pictures 57 and 58) The greatest advantage of this system is that the filing clerks can visualize the actual location of the records. It also speeds up filing and retrieval of files and virtually eliminates any chance of misfiling.

Dictating and Transcription System

Various dictating and transcribing systems are available. In an advanced system, doctors dictate their notes or discharge summaries from various locations in the hospital — from the wards, operating rooms, ICU & CCU complex, emergency room, etc. — using either a remote dictating equipment or the telephones which are linked to the central transcription room in the medical record department where the dictation is tape recorded. The medical secretaries then transcribe the recorded dictation.

➤SPACE REQUIREMENTS

The medical record department requires space and facilities for the following.
- Reception-cum-registration area.
- Offices for medical record officer and assistant medical record officer.
- Space for sectional supervisors.
- Work area for personnel, computers and equipment for record processing, assembling, numbering, indexing, utilization review, discharge analysis, correspondence, word processing, quality assurance, tumour registry, etc.
- Record storage for active and inactive files. Active files are the files where the date of discharge or last visit is within three to five years of the current date. These files should be readily accessible. Inactive record storage should also be located near the active files area as far as possible. Inactive records may be microfilmed or stored in computer assisted system.
- If microfilming is done, space for storage container, viewer, copier and camera equipment for in-house filming is required. Microfilming is a formidable job which is also very expensive. Besides, in a computer age who would want to microfilm thousands of records at such an enormous cost?
- Space for copier which is used to a considerable degree.
- A room for medical staff to complete records, study cases, and review and abstract records with tables, chairs, dictating equipment, etc.
- An area with book cases or shelves to temporarily house medical records pending completion or temporarily used by the medical staff.
- Transcription area with space for central recording equipment, tables, typewriters, etc. for medical secretaries to transcribe dictation. Should be close to the doctors' records completion room to clear any doubts in dictation.
- Space for master patient index depending on the kind of system used, for immediate identification of current and past patients. Computer assisted system is now widely used.

PLATE 9

➤ **PICTURE 57**   Colour Coded Filing System

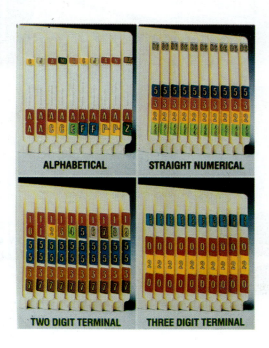

➤ **PICTURE 58**   Various kinds of Colour-
Coded Filing Systems—
Alphabetical Straight
Numerical,
Two-Digit Terminal and
Three-Digit Terminal

PLATE 10

➤ **PICTURE 64** Dining Hall of Cafeteria

➤ **PICTURE 65** Double-Door, Pass-through Washing Machines

➤ **PICTURE 67** A View of the Main Lobby

- ◆ Medical record carts storage area.
- ◆ Supplies storage area, and storage space for unused medical record file folders, forms, etc.
- ◆ Staff facilities.
- ◆ If the medical record department is on two floors with the record storage area on a lower floor, an electrically operated dumbwaiter may be necessary. This must be planned at the design stage.

➤OTHER CONSIDERATIONS

Ownership of Medical Records

Medical records are created and maintained for the benefit of patients, medical staff and the hospital. It should, however, be remembered that they are the property of the hospital which has the responsibility for their safe custody and to preserve the confidentiality of the information contained in them. The hospital has the right to restrict removal of medical records from the records room or from the hospital premises, determine who may have access to them, and lay down as a policy the kind of information that may be taken from them. Except for authorized patient care purposes within the hospital, medical records may be removed from the department only on the order of a court of law and with the prior permission of the chief executive officer. Even when the records are given out, it is a wise policy not to part with the original records; only a photostat copy should be given.

Confidentiality of Information

While the information contained in the identification section of the medical record is not confidential, the clinical data obtained professionally is, and it should be safeguarded. Employees are obligated to safeguard the confidential information of patients. In one hospital where he worked, the principal author of this book, G D Kunders, introduced an undertaking to be signed by all employees having access to patient records, not to divulge any patient information that may have come to their knowledge in the course of their work. A great deal of harm can be done to people by employees who divulge patient information.

Confidential information may be released with appropriate authorization. However, the information acquired by a physician in a doctor-patient relationship is privileged information which the physician may not disclose even in a court of law.

The question is often asked whether the patient has a right to see or review his own medical record. This is a delicate issue. While in a straightforward case, he may have access to his record, a psychiatric patient, for example, may not be competent either to ask to see his record or to review it. Special care should be exercised in the release of information to patients, their employers, news media and the insurance companies.

Record Retention

The length of time the medical records should be retained for and in what form (original or microfilmed) is a complex issue, if not a controversial one. Records are retained for various purposes in addition to patient care such as for legal and research purposes. It is not necessary to retain records permanently for any purpose, and certainly not for the purpose of proving birth, age, residence, etc. It is generally accepted that hospitals are seldom required or requested to produce medical records older than ten years for clinical, research, legal or audit purposes.

It is, therefore, recommended that patient records be retained for ten years. This may either be in the original or in the reproduced form.

Computerization

Computers are widely used in the areas of registration and medical records. In many hospitals, they are extensively used in registration — for maintaining information and patients' personal data (demographics), assigning patient numbers, making appointments and assigning to physicians, creating records, etc.

In medical records, computers can be used for: (a) patient records, and (b) medical records administration. For the most part, however, computers have not made much inroads into the patient records area, but in the records administration area, they are used for chart abstracting, medical record indexing, diagnosis coding, chart location, master patient index, statistics, etc.

CENTRAL STERILIZATION AND SUPPLY DEPARTMENT

➤OVERVIEW

Despite all the advancements in medical science, hospital-acquired infection remains a serious problem in health care today. To combat this, hospitals must have a scientific and effective method of disinfection and sterilization. In modern hospitals, this process is centralized and takes place in what is called the central sterilization and supply department (CSSD), (Picture 59). From various parts of the hospital like operating rooms, wards, outpatient clinics and other departments, all soiled items are collected in the CSSD for processing, and then transported back to the end users. In the CSSD, the process of cleaning, disinfecting, inspecting, packing, sterilizing, storing and distributing

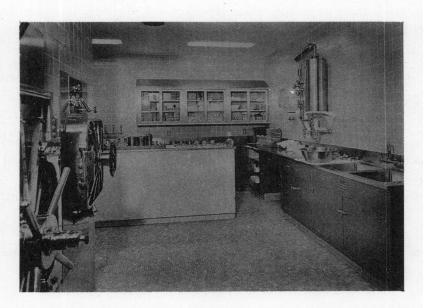

➤ **PICTURE 59** A View of Central Sterilization and Supply Department

is carried out by trained personnel. This ensures better control and reliable results, and consequently reduced risk of infection.

Sterilization of instruments, operating packs, trays, etc. is performed by heating them with pressurized steam or by gas sterilization. Steam sterilization is called autoclaving. However, certain items such as rubber, plastic and delicate instruments cannot be autoclaved and so have to be sterilized by using ethylene oxide or similar gases. Gas sterilization requires certain safety precautions such as aeration prior to use and special exhaust ventilation. Under both systems, sterilization is performed on cleaned instruments wrapped in special linen.

In the decentralized system, the sterilization facility or what is called substerile department is located near the area where the sterilized items are used, for example, the operating rooms. This is called the theatre sterile supply unit (TSSU). The advantages of this system is that it allows for direct communication, the number of instruments is small and transportation is more or less eliminated. However, with so much of traffic in and around the sterile area of the operating room complex and the open window in most places through which sterilized instruments and packs are passed on to the operating rooms, maintaining sterile condition is difficult.

The CSSD services the nursing units, the operating rooms including anaesthesia and recovery rooms, intensive care units, labour-delivery suites, the nursery, outpatient department, radiology, pharmacy and the clinical laboratories. The primary activities of the department are sterilizing, storing and distributing the dressings, needles and syringes, rubber goods (gloves, catheters, tubing), instruments, treatment trays and sets, sterile linen packs, etc.

Disposable sterile supplies are being increasingly used in hospitals. These need only to be stored and not processed for reuse. Since these disposable items are expensive, their use in Indian hospitals has not significantly affected the workload of the CSSD.

➤ OBJECTIVES

The objectives of the CSSD department are to:
1. perform processing and sterilizing of equipment and materials under controlled conditions by trained and experienced personnel thereby contributing to total environmental control in the hospital;
2. effect greater economy by keeping and operating the expensive processing equipment in one central area;
3. achieve greater uniformity by standardizing techniques of operation; and
4. gain a higher level of efficiency in the operations by training personnel in correct processing procedures.

➤ FUNCTIONS

The functions of the CSSD are as follows.
- Receiving and sorting the soiled material used in the hospital.
- Determining whether the items should be reused or discarded.
- Carrying out the process of decontamination or disinfection prior to sterilizing.
- Carrying out specialized cleaning of equipment and supplies.
- Inspecting and testing instruments, equipment and linen.
- Assembling treatment trays, instrument sets, linen packs, etc.
- Packaging all materials for sterilizing.
- Sterilizing.
- Labelling and dating materials.
- Storing and controlling inventory.
- Issuing and distributing.

➤ LOCATION

Accessibility to elevators, dumbwaiters and stairs is of utmost importance in determining the location of CSSD. It should also be centrally located in relation to the departments which use its services the most. Generally, the largest users are the surgical department, including the recovery room, and the nursing units. The facility should ideally be located in the "service core" area adjoining the departments from which it receives materials such as general store, linen store and laundry.

➤DESIGN

The workflow pattern should be planned in such a manner that the personnel traffic and the movement of supplies and equipment is accomplished in an efficient manner, the flow of work is continuous from receiving to issuing without retracting steps, and the receiving and clean up areas are physically separated from the rest of the department. Work flow must be so planned as to allow a separate entrance to receive soiled and contaminated material from the departments, and another for issuing clean and sterile supplies and equipment. There should be a third entrance, if necessary, to receive materials from general stores and laundry.

In a well designed, state-of-the-art CSSD, there are three organized zones: soiled area, clean area and sterile area (Fig. 6.3). Unclean items from the various user departments of the hospital are received at the soiled reception area in the same trolleys, instrument trays, baskets or containers as they are delivered in. Most of them are loaded straight onto the pass-through washer-disinfectors. Trolleys and some instruments are cleaned and disinfected manually. Steam and hot water are the most common disinfection agents used in hospitals.

In the clean area, sorting, inspection and packing of the clean disinfected materials take place. It is interesting to note that in an advanced system, the instruments leave the trays only a couple of times — once during inspection at the packing tables, and then at the point of use. After packing, the instrument trays are put into baskets for sterilization in the pass-through autclaves. Fabrics are sorted out and packed in a separate area before sterilization.

The double-door pass-through autoclaves of the required size are built into the wall between the clean and sterile areas. Materials are loaded on the clean side and unloaded on the sterile side. Both automatic and manual loading and unloading autoclaves are available as well as autoclaves with formaldehyde and ethylene oxide for heat sensitive goods and cycles for fluid production.

After sterilization, the autoclaves are unloaded in the sterile area and the materials are stored there. The storage area should be dry and free of dust.

The concept of three-zone CSSD prevails largely in the European hospitals. However, American hospitals are increasingly veering round to the concept of two-zone CSSD. They believe three zones are superfluous and a waste of valuable space.

It is advisable to have one high-speed autoclave, preferably in the operating room complex, as a standby in the event of the CSSD breakdown. Flash sterilization is performed in the user departments, particularly the operating rooms, to resterilize the instruments needed immediately or those which have been dropped accidentally. Flash sterilization is simply autoclaving an instrument when it is unwrapped.

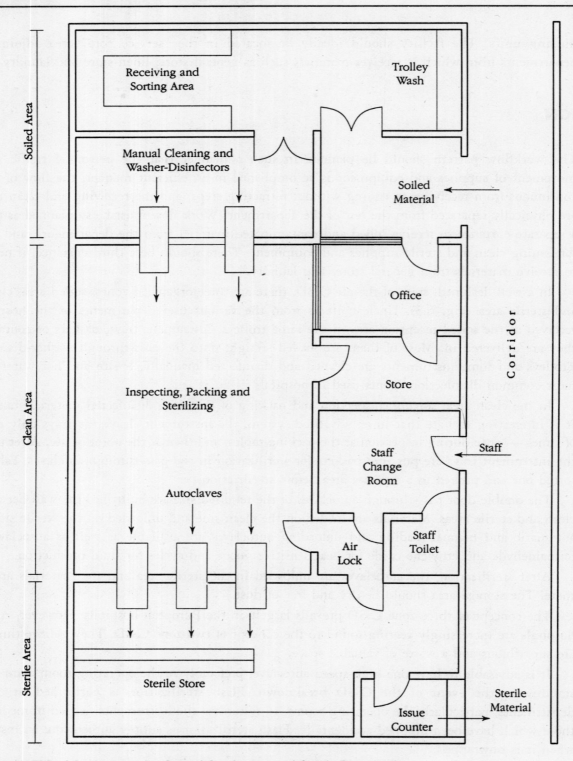

> **FIG. 6.3** Layout of a Central Sterilization and Supply Department

➤SOME PROCEDURES

Cleaning and washing of instruments, trays, etc. should be performed prior to reassembling and wrapping instrument kits. Cleaning and washing can be done either manually or by automatic washers. Ultrasonic cleaners are considered most effective in cleaning joints, hinges, etc. They, however, cause erosion of the surface of instruments and shorten their life.

Surgical linen is inspected before wrapping instruments or linen pack to check for holes, tears or rips by passing it over a light table.

Linen packs of sheets, drapes, wraps, etc. are assembled for operating rooms, labour rooms and delivery suites. Special linen packs are prepared to suit special procedures such as laparascopy, mastectomy and orthopaedic hip surgery.

Processing of instruments, one of the activities of the CSSD, includes assembling appropriate instruments and supplies into kits and wrapping the kits with sterile linen. Kits and trays may be of various types, such as surgical instrument kits for operating rooms, suture kits for nursing units and emergency department, cut down trays for nursing units, and special procedure trays for radiology.

Instruments used regularly are sometimes assembled to make pre-wrapped kits and stocked, or they are prepared when needed as per order. A combination of both the methods is common.

Sterilization is done in batches, which means that several packages are sterilized in a single load. For infection control, these packages are labelled and dated, and later reviewed periodically against test indicators. If a batch is found to be below standards, the packages are removed from the shelves. A wrapped and sterilized kit is considered sterile for a certain length of time after which it has to be resterilized. The length of time a kit remains sterile depends on the type of wrap used, that is, whether the kit is wrapped with single or double thickness surgical quality linen. Labelling and dating of package is, therefore, one of the important steps in the sterilization process.

The CSSD may also be engaged in the manufacture of parenteral solutions, normal sterile saline solutions and sterile distilled water. However, because of the risks involved, only a few hospitals today prepare parenteral solutions. Even in the case of saline solutions and sterile water, the trend is to purchase them from outside in plastic pouch containers which reduce breakage and are also convenient to handle.

➤ORGANIZATION

Traditionally, CSSD has been a part of the nursing service department supervised by a nurse or a person with paramedical training, and reporting to the director of nursing or the nursing superintendent. This pattern prevails in many hospitals. It is also not uncommon for the operating rooms to perform their own sterilization and not have much interaction with the CSSD. The sterilization room is located next to the operating rooms for ease of transporting sterile packs. However, the trend is to centralize this function. In many hospitals in developed countries, the department goes by the name of "Central Service Department" and encompasses many other functions in addition to sterilization, such as purchasing, stocking and distribution of supplies under a materials manager or an assistant administrator.

The various personnel in the CSSD comprise a supervisor who may be a nurse and one or two nurses. The remaining staff typically consists of assistants, technicians, aides, orderlies and messengers who are trained on the job. Usually in a new set-up with sophisticated equipment, the firm which supplies the equipment also trains personnel in handling it as part of a package deal. There is now a growing trend towards putting the CSSD in charge of an experienced manager. The chief of CSSD is generally a member of the hospital infection control committee.

➤FACILITIES AND SPACE REQUIREMENTS

The following facilities and space requirements are needed for the CSSD.

1. Reception-control and disinfection area. Work space and equipment are needed for cleaning and disinfecting medical and surgical equipment which are sorted, racked and passed through washer-sterilizers to the clean area.
2. Facilities for washing and sanitizing carts.
3. Staff change rooms, lockers, toilets, staff room (lounge), etc.
4. Supervisor's office. It should be out of the flow of activities but provide unobstructed view of the processing area. For this, a glass-walled office is recommended.
5. Clean work area. Space for special instruments preparation, for inspection and testing instruments, equipment and linen, for assembling treatment trays and linen packs, for glove preparation, for packaging all materials for sterilizing, and for clean store.
6. Assembling area. Requires work stations for assembling medical-surgical treatment packs, sets and trays. Work benches with multiple drawers for instruments and supplies should be provided. The linen pack area requires large work tables, and for inspection, a special inspection (light) table for examining linen wrappers for minute instrument holes.
7. Supply storage area.
8. Double-door, pass-through autoclaves. These are high-vacuum steam and gas sterilizers.
9. Adequate space for loaded sterilizer carts or trolleys prior to sterilization, for carts during the cooling period following sterilization and, where applicable, for carts for sterilized supplies for the surgical suites and labour-delivery suites prior to delivery of these supplies.
10. Sterile store.
11. Sterile disposable store.
12. Issue counter.

13. Clean cart storage area.
14. Provision for supply of steam, hot and cold water and other utilities and services.

PHARMACY

➤ **OVERVIEW**

Pharmacy is one of the most extensively used therapeutic facilities of the hospital and one of the few areas where large amounts of money are spent on purchases on a recurring basis. It is also one of the highest revenue generating centres. A fairly high percentage of the total expenditure of the hospital goes for the pharmacy services. This emphasizes the need for planning and designing the pharmacy in a manner that results in efficient clinical and administrative services. Frequently the pharmacy is not organized and managed as its importance warrants.

A good pharmacy is a blend of several things: qualified personnel, modern facilities, efficient organization and operations, sound budgeting and the support and cooperation of the medical, nursing and administrative staff of the hospital. A modern and progressive pharmacy calls for new methods of operations, new production approaches and new distribution techniques to achieve reduced costs and increased efficiency. Automation, prepackaging, unit dose drug distribution, decentralization are some of the methods that are being increasingly used these days in addition to computer-based ordering system, computer assisted pricing, billing, cashiering and checking of reorder level, stock, out-of-stock and overstock positions, expiry dates, etc and a host of other functions.

Keeping in mind this picture of the pharmacy of the future, it is imperative that planners include enough flexibility and expansion potential in initial planning of the pharmacy itself so that expansion programmes can be carried out with minimum disruption to its physical location or without having to move it to an entirely new location.

Pharmacy is a specialized area. Its operation calls for intimate knowledge of drugs and drug therapy. Because of this and the amount of drugs and supplies involved, pharmacists usually handle their own purchases and stocking of drugs rather than let the purchasing department perform these functions for them. In large hospitals, there is a pharmacy and therapeutics committee of which the chief pharmacist is a member, to oversee the activities of the pharmacy.

➤FUNCTIONS

The following are the primary functions of the pharmacy some of which are performed directly by its chief.

1. Purchase, receive, store, compound, package, label and dispense pharmaceutical items.
2. Serve as a source of drug information to physicians, pharmacists and other health care professionals, and the patients. This involves compiling, storing, retrieving and disseminating drug information and providing pharmaceutical advice and consultation regarding drug therapy.
3. Participate in the hospital's educational programmes.
4. Plan and organize pharmacy department, establish policies and procedures, and implement them in accordance with the policies of the hospital.
5. Serve as a member of the pharmacy and therapeutics committee, be actively involved in its functions and activities, and implement its decisions.
6. Carry out research and participate in the evaluation of new drugs.
7. Participate in performing therapeutic assessment of drugs and in the preparation of a hospital formulary so that equally effective but less expensive drugs may be put on the formulary. (A formulary is a list of drugs approved by the medical staff and the pharmacy committee for hospital use and kept in the inventory.)
8. Keep track of drugs and formulations or combinations banned in the country and elsewhere, and keep abreast of WHO's revision of "essential list of drugs" and other notifications.
9. Carry out quality assurance programmes to ensure quality when in doubt of the efficacy or potency of a drug by sampling and analysing it either in the hospital or through the drug inspectorate.
10. Comply with statutory regulations, initiating licenses to be obtained, and maintaining records as legally required.
11. Many hospitals are recognized to provide pharmacy students practical training which is in partial fulfilment of their course requirements.

➤LOCATION

In determining the most suitable location for the pharmacy, the following factors should be considered.

- ◆ Flow of outpatient traffic through the hospital.
- ◆ Flow of drugs and other raw materials into the pharmacy.
- ◆ Flow of drugs and services from the pharmacy to the inpatient areas and other departments.
- ◆ Needs for future expansion.

These factors make it evident that pharmacy should be conveniently accessible from the outpatient department, central receiving (or pharmacy bulk) store and the inpatient areas. A ground floor location close to the outpatient department and to elevators servicing the inpatient areas is ideal.

It is assumed that the outpatient and inpatient dispensing activities are combined. Many hospitals may, however, find that when the outpatient department is the overriding consideration in determining the location of the pharmacy, the result is a less than optimal location for the inpatient dispensing activities. They may soon find that a separate inpatient pharmacy facility needs to be established. In may of our hospitals, inpatients are required to buy their requirements of medicines directly from the pharmacy on a cash 'n' carry basis. Medicines are not supplied and billed. Every hospital sooner then later and much to its consternation discovers that its pharmacy facility has become woefully inadequate. Keeping this in mind, the pharmacy should have at least one outside wall to allow for expansion, and must be located adjacent to an area, for example, a store room which can be relocated easily.

➤ DESIGN

Since there is no such thing as a typical hospital, there cannot be a typical hospital pharmacy. Each hospital must therefore pattern its own pharmacy and solve its individual pharmacy-programming problems, and at the same time adhere to the accepted norms of good pharmacy practice and the legal requirements (Fig. 6.4).

Broadly speaking, the pharmacy has four main functional areas: the dispensing area, the production or preparation area, the administrative area and the storage area (Fig. 6.4). These areas must be designed and located for convenient access, staff control and security. Facilities required in these areas are discussed later in this section.

➤ ORGANIZATION

The head of the pharmacy services is usually a chief pharmacist who may possess a B Pharm. or M Pharm. degree and adequate experience. He is normally responsible to the medical director or the medical superintendent. In larger hospitals, he may be required to work in conjunction with the pharmacy and therapeutics committee. Every pharmacist requires registration with the pharmacy council without which he cannot practise.

Other personnel in the pharmacy department are the registered staff pharmacists, pharmacy aides or helpers, pharmacy store keeper and the pharmacy clerks.

Normal working hours of a pharmacy in most hospitals are from 7.00 a.m. to 11.00 p.m., seven days a week. Between 11.00 p.m. and 7.00 a.m., coverage is normally by the staff on an on-call basis.

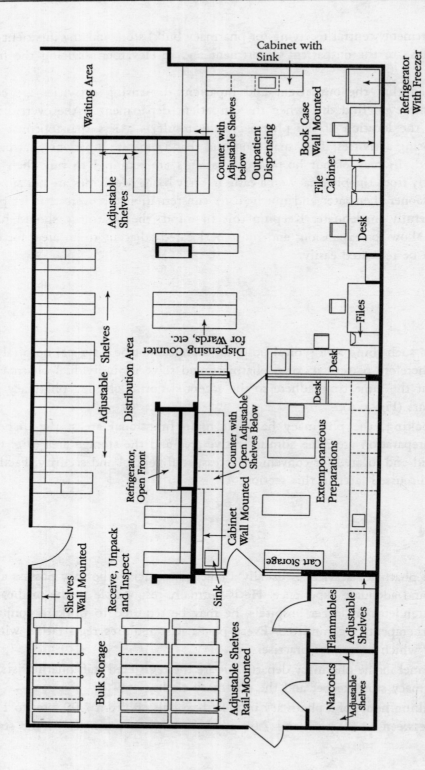

➤ **FIG. 6.4** Layout of a Pharmacy

FACILITIES AND SPACE REQUIREMENTS
The pharmacy department requires space for the following facilities.

Dispensing Area

1. Patient waiting area.

2. Patient dispensing counter, preferably glass panelled, with space for computer-assisted pricing, billing and cashiering on one side and for dispensing on the other.

3. Active storage. Adequate space for a large number of active drugs stored in routine shelves laid out efficiently (Picture 60).

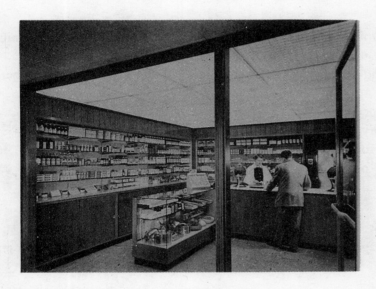

> **PICTURE 60** Dispensing Pharmacy

4. Pick up and receiving counter and space for temporary storage of carts.

5. Area for review and recording of drug orders.

6. Extemporaneous compounding area.

7. Work counters and cabinets for pharmacy activities.

8. Refrigerated storage.

9. Storage for narcotics and other controlled drugs (secured storage).

10. Storage for alcohol and for volatile and flammable substances.

11. Space for maintaining patient medication profiles and cross-checking of medication, for providing drug information, and a room for pharmacist to meet with the patient who requires extensive consultation, instructions or counselling, if these functions are performed.

Manufacturing Area

- ◆ Bulk compounding area.
- ◆ Provision for packaging and labelling.
- ◆ Provision for quality assurance.
- ◆ Clinical sinks and handwashing facilities.

The preparation of parenteral fluids comes under the mandatory regulations of the drug control act which has now been made stricter and more comprehensive. Hospitals wanting to manufacture these fluids are advised to thoroughly study the regulations and procedures.

Administrative Area

1. Reception and the clerk-typist's area for office functions including filing, communications, references, etc.
2. Chief pharmacist's office, and office space for assistant chief pharmacist and clinical pharmacist (if there is one).
3. Waiting area for visitors, medical representatives and salesmen.
4. Conference room-cum-library.
5. Staff facilities like lockers, toilets, lounge, duty room for on-call duty pharmacist(s), etc. Some or all of these can be offsite.

Storage Area

For convenience, some of the storage requirements already mentioned in other sections are repeated below.

- ◆ Bulk storage
- ◆ Active storage
- ◆ Refrigerated storage
- ◆ Volatile and alcohol storage
- ◆ Secured storage for narcotics and controlled drugs
- ◆ Storage for general supplies, equipment, files, stationery, etc.

➤ OTHER CONSIDERATIONS

The traditional pharmacy services are rapidly undergoing a change all over the world, especially in the dispensing and distribution systems, and many innovative approaches and methods have been introduced in recent years. Though not all the hospitals can implement these changes, it is our hope that some of the larger and progressive hospitals in our country would introduce and test these newer systems and be pace setters for other hospitals to follow. Some of these changes are described here.

Clinical Pharmacy

In most of our hospitals, pharmacy is engaged in its traditional activities such as drug ordering, preparation, distribution and dispensing. Of these, dispensing prescriptions as ordered by physicians is the most important. Except for monitoring drug incompatibilities occasionally, pharmacists have no role in determining what to order. But hospital pharmacists are now increasingly becoming involved in what is called "clinical pharmacy" which includes activities like taking medication history, monitoring drug use, drug selection, patient counselling and surveillance of adverse reaction of drugs. In other words, they are becoming involved in determining what to order and thus become a part of the team effort in determining treatment.

Unit Dose Dispensing System

Another important change that has taken place in the field of pharmacy is in the medication dispensing system — from the traditional pharmacy system to the unit dose system. In the traditional system, the pharmacy sends to each patient in the nursing unit, a supply of medication which may be for several days. The nursing unit then prepares the individual dose from the supply. In the unit dose system, the doses are premeasured by the pharmacy so that the nurse has only to administer the medication. The system uses a cassette mechanism which designates one drawer for each patient in the medication cart (Picture 47) or cabinet. The nurse rolls the unit dose cart to each individual patient room, removes the dose of medication to be given from the respective patient drawer in the cart, and administers it to the patient. In the emergency cart maintained in the nursing units, certain drugs are kept in single dose packages which are ready and convenient to administer.

While the unit dose system is expensive — initial one-time cost largely involves the purchase of unit dose carts and packaging equipment and increased pharmacy personnel — there are also several advantages. It reduces nursing time for pouring, counting and dispensing, reduces medication errors, and increases control and recording of medications by the pharmacy.

IV Additive System

The concept of unit dose system can be extended to intravenous (IV) solutions, for which there are two methods: the traditional method and the IV additive method. The activity relates to mixing of medications with IV solutions. In the traditional system, IV solutions are stocked in the nursing unit. Medications are sent to the unit by the pharmacy, and the nurse mixes or adds medications to the IV solution. In the additive system, the medications and the IV solutions are mixed in the pharmacy itself. The pre-mixed bottles are then sent to the nursing unit and the nurse merely administers the solution.

As in the case of unit dose system, the advantages are reduction in nurses' time as well as in wastage and medication errors.

Pharmacy and Therapeutics Committee

Every hospital should have a pharmacy and therapeutics committee consisting of physicians representing the various divisions of medical staff, pharmacist(s) and representatives of administration, to oversee the work of the pharmacy. More specifically, the following are some of the duties and responsibilities of this committee.

1. Develop a formulary of accepted drugs for use in the hospital.
2. Serve the medical staff, pharmacists and hospital administration in an advisory capacity in all matters pertaining to the use of drugs and in the selection of drugs to be stocked.
3. Evaluate clinical data concerning new drugs requested to be included in the formulary and for use in the hospital.
4. Add or delete specific drugs from the formulary.
5. Prevent unnecessary duplication of the same basic drugs to be stocked.
6. Recommend drugs to be stocked in the nursing units and other areas.
7. Study problems or reported adverse reactions to the administration of drugs.
8. Issue communication(s) to physicians, pharmacists, nurses and administrative staff regarding proposed changes in the formulary such as additions to and deletions from the list, changes in the working of the system and in the contents of the formulary.
9. Adoption of a policy that the inclusion of drugs in the formulary should be by their non-proprietary names.
10. Ensure that the labelling of medication containers be by the non-proprietary names of the contents.
11. Issue written communication to the nursing and pharmacy staff regarding the existence of a formulary in the hospital and the policies and procedures governing its operation.
12. Issue guidelines for the control, appraisal and use of drugs not included in the formulary, investigational drugs and non-formulary drugs.

Hospital Formulary

One of the major responsibilities of the pharmacy and therapeutics committee is to develop or adopt a suitable formulary of selected medications. A formulary is the official compilation of drug products that have been selected and approved for use within the hospital. The two main objectives of the formulary are: (a) it promotes rational therapeutics, and (b) it prevents unnecessary duplication, waste and confusion and thus promotes economy for both the hospital and the patient. When many brands of the same drug are stocked and prescribed, it results in a loss to the patient as well as to the hospital. However, economy in medication should not be construed to mean prescribing inferior remedies.

It should be remembered that a mere list of medications placed on the pharmacy shelves does not constitute a formulary. The drug list should be expanded to include specifications about how a medication should be used. Formularies should also include recommended daily dosage and cautions, warnings, restrictions, pharmacology and similar other information to facilitate correct use of the drugs.

A detailed discussion of the formulary system is outside the scope of this book. However, a brief mention of the administrative procedures pertaining to the development of a formulary would be in order. The following are some of the steps involved in this process.

1. Appointment of a pharmacy and therapeutics committee by the medical staff composed of physicians, pharmacist(s) and representatives of the administration.
2. Outlining the purpose, organization, functions and scope of the committee, and an organized method for this committee to evaluate the therapeutic claims of competing or suggested drug products.
3. Periodic publication of the authorized drugs.
4. Procedures for revising the list.

➤PROBLEM SITUATIONS

Theft in Pharmacy

Pharmacy is one of the most theft-prone places in the hospital and, what is worse, pharmacy thefts can be costly, difficult to check, and may go unnoticed. Theft is usually by the employees themselves or in collusion with them. The most common points where thefts take place are the dispensing area, stores, purchasing process, receiving and invoice payment and the nursing units.

Substantial losses may take place in the dispensing and purchasing areas and continue for a long time without being discovered. The chief pharmacist or the person responsible for purchasing may, in collusion with the vendors, manipulate supply or bills, and divert part of the supply to privately owned drug stores. With an incredibly large number of items kept mostly in open shelves of the dispensing pharmacy, the task of exercising any meaningful control over the drugs is a formidable one even with all the checks and balances and control measures. The problem becomes serious during evening and night shifts when there may be only one pharmacist on duty, and even more serious when, in smaller hospital, the pharmacist doubles the duties of the cashier as well.

Every hospital must recognize that it has a moral obligation to make theft and fraud as difficult as possible, if not altogether impossible, by instituting proper control systems. Too often the general climate in the hospital is such as to provide ample scope and temptation for employees to indulge in such activities without anybody taking cognizance of such offenses or punishing the offenders. A sound system of controls acts as a deterrent and creates fear in the employees that frauds and thefts will be detected and punished.

MATERIALS MANAGEMENT

➤OVERVIEW

Materials management encompasses acquisition, shipping, receiving, evaluation, warehousing and distributing of all goods, supplies and equipment for an organization. Acquisition may be through purchase, lease or rent; warehousing is storage of goods, supplies and equipment, and inventory control; and distribution is delivery or pick up of goods and supplies. Traditionally nursing units, laboratory, x-ray and certain other departments performed their own purchasing functions, negotiated prices and maintained their own inventory. There is unnecessary duplication of personnel, facilities and efforts in this system, and the best possible prices cannot be negotiated. The current trend is thus increasingly towards centralizing materials management functions. Nevertheless, we have many hospitals where pharmacy, food service, maintenance and certain ancillary services like laboratory and radiology perform their own materials management. The argument is that their purchases require a high level of technical competence or knowledge which the purchase department normally does not have. In most cases, it may be so.

Computers are now increasingly used in materials management in the areas of purchase, inventory, reorder and invoice production, and a number of other functions.

➤TEN GOLDEN RULES OF MATERIALS MANAGEMENT

There are certain cardinal rules that every materials manager should know and practise.

Rule One Successful materials management is built upon an effective management system and good supervision.

Rule Two Purchase order is often called the umbilical cord of the purchasing process. For internal control and a smooth flow of goods, the receiving store, the accounts and the requesting department should receive copies of the purchase order. The original goes to the vendor and a copy of it to the purchasing file. The procedure and documents should be simple. Initiation of every purchase order costs money. It should, therefore, be cost effective. Consider this: if the cost of initiating a purchase order is Rs 50 or 75, and the cost of the ordered commodities is Rs 100, it will not be economical.

Rule Three Centralize the purchasing system. Decentralized purchasing is contrary to all the underlying principles of good materials management. Centralized purchasing eliminates uncontrolled purchases by departments, secures a reduction in prices through improved purchasing methods and quantity buying and shipping, results in improved allocation of space, and reduces inventory costs and staff. It saves staff time.

Rule Four Negotiation is the key to sound purchasing. Every purchase officer should learn the art of hard-nosed but ethical negotiation. Here are some tips for good and successful negotiation.

- Never purchase at list price.
- Negotiate for bulk price or price for the quantity the hospital would need for the whole year, and then get the price for any quantity you want to purchase.
- Regardless of how much you buy, always ask for a discount.
- Always get a price protection on the agreed price. Start negotiating price protection for two years. Even if you end up with a short period, it is worthwhile.
- Try to get at least a month's time to pay the invoice. Whichever way you look at it, it is a positive gain.
- Remember, even at a good bargain price, the supplier makes a profit. So don't feel sorry for him.
- Always keep in mind that there are tougher negotiators than you. They may be getting a better price than you do.
- Remember, it is the quality and determination of the negotiator which always get a good price in buying and not merely the reputation and size of the hospital.

Rule Five There should be an effective receiving programme with responsibility, accountability and internal control built into the system. The fundamental rule is that the three functions of purchasing, receiving and paying of invoices should be handled by three different persons. Receiving is one of the most important functions of materials management which should be carried out professionally and ethically.

Rule Six Establish an optimum level of inventory, and a simple but effective inventory control programme. The inventory should not be so large that you can replenish it only with the aid of a computer. Remember the dictum: the secret to managing an inventory is to control it, not count it.

Rule Seven Establish effective and result-oriented requisition and distribution systems. The goal of the distribution system is: the right item to the right place at the right time in the right quantity and at the least total cost.

Rule Eight Establish written policies and procedures. Internal control in materials management starts with them.

Rule Nine If you want to buy the right supply in the right quantity at the right place and time, and simultaneously effect cost containment, standardization and evaluation of all products and services is a good place to start with.

Rule Ten Wherever possible, go for contract purchasing through prime vendors i.e., buy the entire categories of supplies from a single source on negotiated terms of quantity, quality, price and time. By being innovative in their purchasing approaches, hospitals can create a buyer's market instead of competing with each other in a seller's market. Since contract purchase encompasses all supplies of a certain category, hospitals can get the best deal in a fraction of time and effort they spend on traditional purchase. A host of items, from medical-surgical to dietary items, can be purchased through prime vendors.

➤RECEIVING

The receiving area is the entry point for all materials coming to the hospital. It is here that a careful check should be made and errors detected. Some of the most common receiving errors are: supplies not ordered are shipped; items are changed or substituted without the prior approval of the purchase department; outdated or defective materials are sent, or they are damaged, and there are shortages in the supply.

The cost of correcting any mistake committed by the purchase department, the vendor or the shipping agency after the supplies have passed through the receiving section can be very high. So great care should be taken to thoroughly check them at the entry point itself.

The receiving person should do the following:

* Check quantity — number of packages and weights against shipping or packing slip.
* Check quality — inspect the materials and condition of containers, and record any damage.
* Verify receipt of materials against the supplier's packing slip and the purchase order.
* Record in the receiving report all discrepancies between the materials ordered and the materials received, like shortages, surplus, incorrect or damaged materials, incomplete supply, etc. The unit of measure used in the receiving report should be the same as that in the purchase order. The report should carry the signature of the receiver and the date. It should state that the order received is complete.
* Enter the serial number of equipment items in the receiving report to enable the accounts department to maintain an accurate and complete assets register. There may be several pieces of equipment of the same model.
* Distribute the materials to the appropriate departments or stores, as the case may be, and obtain their signatures.

➤INVENTORY CONTROL

The central store is responsible for receiving, unpacking, storing and distributing supplies used in the hospital. Inventory control is one of its primary responsibilities.

The store generally stocks all items that have not been delivered to departments upon receipt, but have been stored for distribution and use at a later date. The most common materials that are stored are: medical-surgical items, dressings, housekeeping items, office supplies and stationery and forms. Materials usually kept in their respective departments are: drugs, other pharmaceuticals, intravenous solutions and sets, laboratory supplies, x-ray supplies, all processed sterile supplies, food items, and china and tableware. Linen may be kept in a separate linen store.

The inventory serves two important purposes.: 1. It provides maximum efficiency and optimum inventory and 2. it provides a cushion between the forecast and actual demand for materials.

To achieve the primary goal of patient care, stores should carry on optimum inventory. This should not, however, be taken to mean a minimum inventory. Inventories are a protection against unforeseen failures in supply, sudden increases in demand, or unforeseen delays in delivery.

Among the criteria for inventory control is the availability of space: Prepackaged and disposable goods, packed by similarity of service or use, are available in plenty and are being increasingly used in hospitals. But they require more floor space and general circulation area. Another criterion is that it must be more economical to maintain an item in inventory than to purchase it on demand. Among the questions hospitals should ask are:

- What are the consequences of not having an item available from the patient care point of view, both in real and perceived terms?
- What are the economic consequences if any individual department is allowed to stock this item?
- For what use is the item intended?
- Is the item used by more than one department?

➤ DISTRIBUTION

From the store to the user points, the hospital distribution is an intricate system. An effective distribution should ensure the provision of the right item to the right place at the right time. It is generally found that for every rupee spent to purchase an item, another rupee or a major part of it is spent on storing and delivering it. This makes it imperative for the stores and distribution system to be efficiently organized.

There are at least two methods of distribution in hospitals: (a) requisition system, and (b) par-level system. A third method, cart exchange system, which is rated highly, particularly from the point of view of management control, is not practised in India.

The requisition system is the most widely used method in the Indian hospitals. Each user department maintains and keeps track of its inventories. Periodically or when inventory levels are low, a requisition is prepared and sent to stores which delivers the requisitioned items to the user department.

In the par-level method, each user department stores a certain amount of supplies in the department. The levels of stock for each department are pre-determined based on the usage rate and on how frequently the stock was being replenished. At pre-determined intervals, the supply personnel check the stock and bring up stock to the par level.

In the requisition system, space utilization and management control are almost negligible, whereas in the par-level system, they are good. We, therefore, recommend the par-level system.

The type of distribution system should be decided early in the planning stage. This decision is crucial to the materials handling system. It should be based on cost consideration, but attention should also be paid to reduced employee travel time, reduced steps in material flow, and the number of employees needed for different activities of the distribution system.

Elevators are essential for the movement of goods. Large hospitals designate a separate elevator, called a "service elevator" for distribution of goods among other purposes.

➤ LOCATION

Ideally, the major divisions of materials management should be located in one area of the building. But this may not be possible or practicable. The department generates moderate traffic of hospital personnel and vendors to and from outside the hospital.

For vendor traffic, the purchase office should have a subwaiting area with a receptionist and seating and an outside entrance to avoid routing it through the main lobby and the administrative corridor (Fig. 3.1). The department should have easy access to a multi-purpose room that can be used for product display and demonstration.

The central stores should be on the ground level and as far away as possible from other traffic areas. It should be close or conveniently accessible to an unloading dock which should have a covered area big enough to handle trucks bringing in the goods and supplies. The receiving office, equipped with a platform scale, and the unpacking area should be adjacent to the unloading dock.

➤ ORGANIZATION

Purchase, stores, receiving and distribution may all be integrated is one department, called materials management, under a materials manager or an assistant administrator. Or, these major functions, especially purchasing and central stores, my be informally linked but function independently under separate managers. Consolidation of all functions may produce savings.

There are professionally trained materials managers. Many universities offer graduate and post-graduate courses in materials management. Others acquire skill in the department by beginning at entry level in various posts and advancing as they gain experience.

➤ FACILITIES AND SPACE REQUIREMENTS

The following offices and facilities are required for the materials management department.

- Materials director/manager's office
- Secretarial and clerical area
- Office for assistant director (if there is one)
- Materials management clerical work area
- Office(s) for purchase officer(s)
- Vendors/visitors' waiting area
- Receiving and unpacking area
- Central stores
- Unloading dock (common area)
- Multipurpose room for product display and demonstration. Can be a common area shared with other departments.
- Catalogue library. Space for filing product data, catalogues, specification sheets, brochures and samples.

➤PROBLEM SITUATIONS

Theft, Fraud and Kickback

Large scale theft and fraud can take place in the purchasing process, stores and at the receiving point, resulting in fictitious bills, overpriced purchases, inflated invoices, fraudulent payments and downright theft. Following are some examples of these fraudulent practices.

- Frequently, rules require merely the countersignature of purchasing department on the purchase order. This allows departments to carry on corrupt practices while the purchasing officer exercises little or no control over the purchase.
- Ancillary departments often stipulate specific manufacturer or supplier along with the price while submitting purchase requisitions. This leaves the purchase officer with no option to negotiate prices. Quite often, the purchase officer will be too happy if someone else does the preparatory work for him. Such practices provide ample scope for receiving a commission or kickback.
- Sometimes a staff member in departments like x-ray or dietary which have the privilege of making autonomous purchases places purchase orders, sometimes over the phone, and certifies that the goods have been received regardless of their quantity or whether they have actually been received or not.
- A single person performs the functions of purchasing, receiving and passing of bills for payment making it easy for him to accept a commission or kickback.
- Purchase of large scale stores items is contracted on a yearly basis by one individual without committee sanction or approval of an administrative officer.
- There is lack of attention to and control over the receiving function which is one of the most

vulnerable areas. Often packing slip is not scrutinized against the purchase order and documents, quantity and quality are not checked.

♦ General stores and the receiving area may be adjacent to each other, and stores personnel may double as receiving personnel and vice versa making manipulation of records easy.

♦ The stores supervisor may be too busy to supervise the work of stores clerks who are left to carry on the work of issuing and distribution as they please.

It is difficult to comprehend in how many inscrutable and uncanny ways people embezzle money and commit fraud. In every institution, some employees hold strategic positions where they can easily embezzle money. The accounts clerks may keep two sets of books, write cheques to fictitious suppliers, give refunds to materials that are not returned and collude with suppliers in obtaining quotations and supplies. Examples can be multiplied.

Wastage

Most managers recognize the enormous loss they suffer on account of fraud, defalcation and theft, and the imperative need for internal control to prevent such losses. But what they fail to realize is that losses due to wastage are considerably greater.

Wastage may occur in almost any area. In the area of human resources, it may be due to loss of time on the job, tardiness and absenteeism, unnecessary overtime, superfluous personnel, lack of skill, inefficiency, poor morale and turnover, etc. Material wastage occurs when materials drawn from the stores lie unused till they become unfit for use, or a bigger quantity than is necessary is indented for and the unused materials are thrown away. Other items most prone to wastage and theft are the stationery and office supplies. Efforts to control office supplies often cause resentment. Hospitals may do well to study this problem.

FOOD SERVICE DEPARTMENT

►OVERVIEW

Good food is important in the treatment of the patient and is a part of his total care. The food service department in today's modern hospitals ranks as one of the major departments. It is headed by a specialist who is either a professional manager or a chief dietitian.

Most people readily accept the professional service of their doctors with minimum criticism. They do, however, tend to pass judgements on the cleanliness of the hospital, the personal care and attention given to them as patients and visitors and on the quality of food. The coffee shop is one of the places where a visitor often stops by on entering the hospital, and it sets the overall impression of the hospital for the first time visitor. The decor and the service here, as in other parts of the hospital, should be consistent with the image the hospital wishes to project. An irritated customer here may give vent to his feelings at the patient's bedside, and look for faults in patient care. Hospitals have long recognized the public relations value of the food service department. Unfortunately, criticism of the food is one of the most frequently heard complaints in any hospital. The major share of this criticism can be avoided by a properly planned, designed and administered food department.

➤FUNCTIONS

The functions of the food service department are as follows.
* Provide the best possible food at a cost consistent with the policy of the hospital.
* Buy to specifications, receive supplies, check their quantity and quality, and store, produce, portion, assemble and distribute food.
* Establish standards for planning menus, preparing and serving food, and controlling meals. Standards should be established before setting up food purchase specifications.
* Establish policies, plan layouts and equipment requirements.
* Plan and implement patient therapy, education and counselling, advise patients and their families on special dietetic problems prior to their discharge from the hospital or as referred from the outpatient clinics.
* Train dietetic interns.
* Impart instructions to nurses, medical and dental students, interns and residents about principles of nutrition and diet therapy.
* Cooperate with medical staff in planning, preparing and serving experimental or metabolic research diets.

➤LOCATION

In the earlier times, hospital kitchens were generally allocated space unusable for any other purpose. Food service department located below the ground level is certain to have a deleterious effect on the quality of food and efficiency of the department. A kitchen in the basement, for example, is likely to be dingy, dark and poorly ventilated. A ground floor location is preferable, and is also convenient for delivery of supplies.

The department should be close to the materials management department. The storage area should be in close proximity to the unloading dock. Easy access to vertical transportation system serving patient care units is important to facilitate delivery of patient meals and return of used trays and utensils. The cafeteria and dining room(s) should be close to food preparation and production area, and within convenient access to the hospital staff.

➤DESIGN

The design and physical facilities of the food service department have an important bearing on the standard of food service, labour costs and the morale of the employees. For example, storage rooms far removed from the work area, poor arrangement of the preparation and production area for work flow, and long travelling distance for prepared food lower the level of efficiency and increase unnecessary steps for employees resulting in increased costs.

In the general layout, the most important factor to be borne in mind is the logical work flow, that is, receiving supplies, storing and refrigerating them, preparing and serving food, returning trays and washing dishes (Fig. 6.5). There should be adequate space and facilities for performing the work in each of these functional areas.

➤FUNCTIONAL AREAS

The major functional areas (Fig. 6.5, 6.6) of the food service department, following the sequence in the workflow, are described below.

Receiving Area and Control Station

The food service department requires a substantial amount of supplies and materials. The receiving area which may be common to all other hospital supplies should be large enough for handling bulk supplies. The receiving clerk inspects and checks all the supplies both for quantity and quality. In the case of dietary supplies, the dietitian or a staff member of the food service department personally checks the supplies. The receiving area should be equipped with scales for weighing materials and supplies. All internal control measures described under materials management apply to this area too.

Storage and Refrigeration Room(s)

The storage area which comprises dry and refrigerated storage should be adjacent or close to the receiving area. Dry storage is for staples and refrigerated storage, for perishables. Hospitals generally store several days' supplies to meet any eventuality. Some dry foods are bought and stored in bulk. Wooden or steel racks and platforms are used for storage. Large hospitals have walk-in coolers and refrigerators with varying degrees of temperature for meat, meat products and poultry, dairy products

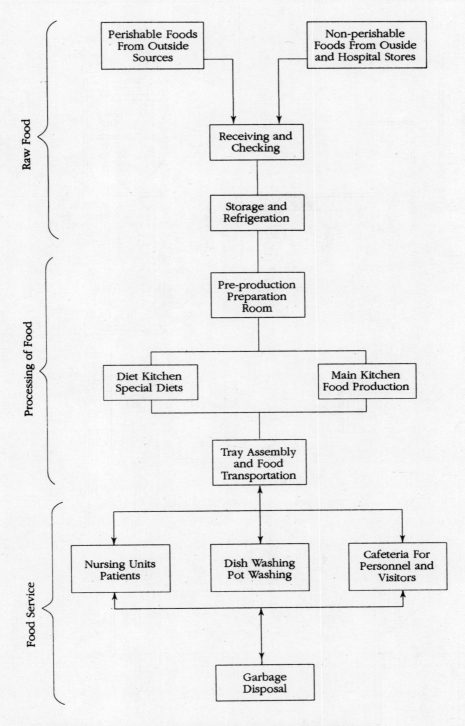

➤ **FIG. 6.5** Food Service Department: Activity Flow Chart

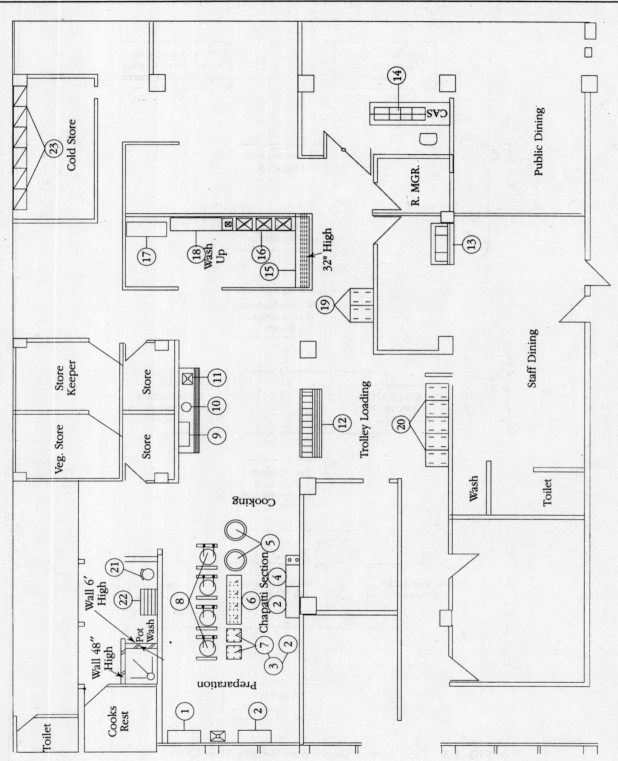

➢ **FIG. 6.6** Layout of a Hospital Kitchen—An actual Kitchen in one of the hospitals

and eggs, and fruits and vegetables. As in restaurants it is common practice in such hospitals to freeze all leftover food for later use. The refrigerators should have a thermometer in each unit for daily temperature check. The walk-in refrigerator should also have an alarm connected to a place with a 24-hour personnel coverage in case someone accidentally gets locked up inside.

Preparation and Production Areas

Some hospitals prefer to have a separate pre-production preparation area where sorting, peeling, slicing, chopping and washing of materials can be done prior to cooking. A double sink with draining boards, work tops, peelers, and grinders are the necessary equipment. There should be efficient arrangements in the production area so as to permit best possible work flow and minimum cross traffic (Picture 61).

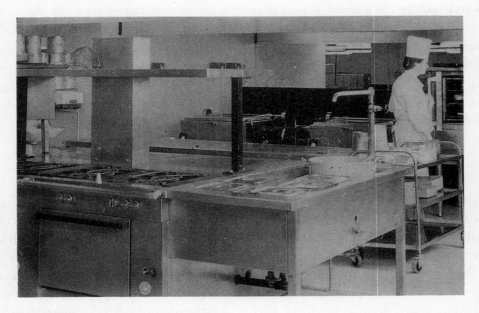

➤ **PICTURE 61** Central Kitchen

Special attention should be paid to the size of the production area. Early in the planning stage, it should be decided whether the hospital will serve only vegetarian food or non-vegetarian food as well, and if the latter also, whether there should be separate kitchens for these. Some raw foods when cooked may produce disagreeable odours and also taint other food. It may thus be necessary to handle them separately.

Food in hospitals is prepared in batches using the progressive approach. In progressive cooking, food is prepared in small batches at regular intervals during the serving time. This provides freshness and palatability, and the food remains hot.

The essentials of good production are:
* Good physical layout which ensures easy flow of work.
* Use of standardized recipes.
* Correct techniques of preparing each kind of food which preserve natural flavour and nutritional value.
* Progressive cooking and preparation in the shortest possible time.
* Good management and supervisions.

Serving Room

The serving room is a place where patient food trays are assembled. It receives prepared food in bulk from the kitchen and the refrigerators. After the trays are assembled, they are loaded on to tray carts or food trolleys and sent to the patient floors. It is imperative that the serving area be close to the elevators.

The equipment and facilities in the serving room include refrigerators, table tops, cupboards for storing trays, dishes, cutlery, and other articles necessary for assembling trays.

The dietitian has overall responsibility for the inpatient food service. She has the last immediate duty of checking the trays for proper identification, accuracy and temperature of foods, and ensuring that the food is palatable and served attractively.

Food Delivery

Food trolleys that can be plugged into an electrical outlet to keep the food hot are now available. An airline truck is a tray truck with separate heated and refrigerated sections for hot and cold foods, and bulk thermal containers for liquids. The hot bulk cart contains hot food in bulk which is dished on to the patient trays on the patient floors (Pictures 62, 63). Some hospitals distribute foods in individual hot food containers carried in open food carts. Smaller hospitals may serve them in ordinary tiffin carriers. Beverages like coffee and tea are poured into cups in the patient rooms. Whichever method of distribution is used, the patient serving should not take more than forty five minutes; if it does, the system should be evaluated.

Special Diet Kitchen

The special diet kitchen is an integral part of the hospital kitchen. The special diets should be prepared under the supervision of a qualified dietitian, the actual preparation being done by student dietitians or interns as part of their training. Since special diets are usually modifications of the basic menu and since the special diet kitchen derives its supplies from the main kitchen and transports the trays through the same tray carts, it should be located in the main kitchen or in close proximity

> **PICTURE 62**   Mobile Hot Food Carrier

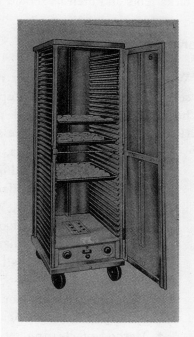

> **PICTURE 63**   Mobile Food Cabinet

to it. The diet kitchen also requires pots, pans, vessels, etc. like the main kitchen but on a much smaller scale. In addition, it requires scales for weighed diets.

Dishwashing Area

Dishwashing, an otherwise noisy job, is made easy with large modern dishwashing machines. In these, a continuous stream of soiled dishes are loaded at one end and clean dishes unloaded at the other. Wire baskets may be used to place glasses and cups in individual compartments.

An abundant supply of hot and cold water should be piped to the dishwashers and sinks. Drainage and plumbing should be well engineered.

Soiled dishes are brought to the dishwashing area and scraped and waste collected in a garbage receptacle. Dishes are then checked and placed in dishwashing trays, and loaded for washing. After this, they are stacked in appropriate places for reuse.

Potwashing Area

Washing of pots, pans and utensils is normally done by hand. It is a dirty and noisy work best done in a separate room. The place must have deep sinks, abundant supply of hot and cold water, and drying racks. Pots and utensils should be identifiable so that they can be returned to their respective user units.

Cafeteria

While accepting the proper nutritional care of patients as the primary responsibility of the food service department, most hospitals also provide food to non-patients and non-patient areas, such as hospital staff, visitors and patient bystanders, functions and meetings through the cafeteria (Plate 10: Picture 64), coffee shop and the snack bar.

In planning the cafeteria, the following factors should be considered.

- The number and kind of groups to be served — day staff, resident medical and nursing staff, visitors, patient attendants and bystanders; whether there should be separate dining rooms for medical staff, officers, VIPs, and other staff.
- Types and extent of food selection — vegetarian or non-vegetarian, number of food items, a complete meal for a fixed price or items by selection a la carte.
- Kind of service — self service at the counter or table service; whether separate counter for doctors, etc.
- Size of the dining room and number of shifts — whether all persons can be accommodated in two or three sittings during one or one-and-a-half-hour meal period.
- Method of clearing table. If self-service, whether personnel will be required to return their trays to a designated area, e.g. a trolley or a cart, and if they will also be required to dump garbage in the garbage bin before depositing the trays.
- The hospital cafeteria works like a fast food business operation — cash and carry. The customers buy coupons at the counter, pick up food items in exchange for them, carry their trays to the tables, and eat. The hospital cafeteria should be designed for this kind of operation.

A customer-oriented menu is the key to successful management of hospital cafeteria. The chief of food service must recognize certain fundamental principles which ensure an efficient and profitable running of the cafeteria. They are:

- Satisfaction of the customers who enjoy good food. In the case of hospitals, they are more of a semi-captive customers.
- Variety in food. Patients may or may not be accustomed to luxury but most of them are used to variety in their diets at home. If it is not provided, they may quickly develop distaste for the food.
- Purchase of high quality food at economical prices.
- Receiving and storing food supplies properly.
- Exercising effective control on supplies at the points of receiving, storing and issuing.
- Preparing foods according to standard recipes and standard quality and serving them attractively in standard portions.
- Accounting for the sale of food.

Coffee Shop and Snack Bar

The coffee shop-cum-snack bar should preferably be away from the main kitchen and dining rooms to cater largely in-between-the-meals coffee, tea and snacks to mostly outpatients, visitors, and personnel. This way, the main cafeteria can remain closed except for breakfast, lunch and dinner as keeping the whole cafeteria open over two shifts is costly. The coffee shop should be easily accessible to outpatients, particularly the emergency patients. This is important in the night when the cafeteria is closed and the patients need refreshments. It should be designed like a fast food restaurant for a quick turnover of patrons and not as a lounge where people settle down for an informal chat.

➤ORGANIZATION

Traditionally, a dietitian has been the chief of the food service department, also called the dietary or nutrition department. But in larger hospitals, professional managers with degrees in management and specialty degrees in food service or hotel management are now becoming more common with the dietitian as the dietetic supervisor. In smaller hospitals, she may serve a dual role as both dietetic supervisor and department manager. The manager usually reports to one of the associate administrators.

The question is often raised as to whether the food service department should be placed under the management or the clinical services, and to whom should it be responsible — to the administrator, or the medical director. While mostly it is responsible to the administrator, the department also has a close relationship with the medical staff in professional matters.

The department has two main functional divisions: one relating to the administration of the department and food production, and the other relating to therapeutic food service and instructions to patients, and their counselling.

Administrative duties ranging from purchases to planning of menus occupy most of the manager's time. The therapeutic and educational duties include diet therapy, planning patient menus and special diets, supplying special diet list to patients and counselling, educational activities and teaching students and training dietitian trainees.

Unskilled workers constitute the bulk of workers in the department. The trend in hospitals is to employ workers at the lowest salary level. This results in instability, lack of responsibility, and poor quality of work. The department is often a breeding place for unions. Many hospitals require those involved with food service to undergo physical examination to make sure that they are free of communicable diseases.

Dietary aides, if properly trained, can perform a variety of functions such as checking supplies, writing requisitions, checking and reporting census, making out time schedules, checking routine tray line, and making out charge slips.

Every hospital must develop a diet manual prepared by the food service department and approved by the medical staff, and make it available for use of medical and nursing staff.

Early in the planning and design stage, hospitals should decide as a matter of policy whether the hospital food is to be compulsory for all the patients or they have the option to bring food from home, perhaps with the exception of special diets. The size of the department and the concomitant facilities are dependent on this decision. Both these systems have their merits and demerits. In some hospitals, food is included in the charges for the room.

Meal planning is one of the primary functions of the department. It is the determination of meals that are to be served to the patients and the non-patients. Cycle menus that are commonly used consist of a series of skeleton menus to be served over the length of the cycle — weekly, biweekly, or monthly. Variations are sometimes made to take advantage of the seasonal foods.

Some progressive hospitals allow the patients to select their own meals using menu cards as in restaurants. Dietitians help patients in giving their orders.

Therapeutic nutrition requires a qualified dietitian to assist in patient therapy. In most cases, nutrition therapy, as ordered by a physician, requires modification of the normal diet in its content, consistency and preparation. It is necessary that a specially prepared diet is written for every individual patient although customarily a master cycle menu may have been developed. Therapeutic and special diets and meals should be clearly marked, preferably by colour coded labels.

►FACILITIES AND SPACE REQUIREMENTS

The following facilities and space are required. Some of these have already been discussed in detail.

- ◆ Food service manager's office. It should offer an unobstructed view of all the parts of the department, and be well-ventilated and preferably soundproofed.
- ◆ Secretarial, clerical office with space for file cabinets and other equipment, seating for visitors, vendors, etc.
- ◆ Office space for chief dietitian and staff dietitians. Some hospitals locate the office of therapeutic dietitians on the patient floors for making them quickly available to the medical staff and patients.
- ◆ Receiving area.
- ◆ Storage and refrigeration area with walk-in refrigerators, coolers and dry storage.
- ◆ Pre-production preparation area.
- ◆ Cooking or food production areas, separate for vegetarian and non-vegetarian foods.
- ◆ Special diet kitchen.
- ◆ Serving or tray assembly area.
- ◆ Dishwashing area.
- ◆ Pot washing area.
- ◆ Trolley, cart washing and clean cart storage area.
- ◆ Deep sinks and handwashing facilities in various places.
- ◆ Garbage disposal facilities.
- ◆ Storage with racks and cabinets for clean trays, dishes, cutlery, etc.

- Storage with racks for clean pot, pans, vessels, etc.
- Employee facilities like lockers, staff toilet, etc.
- Janitor's closet.
- Dining hall with self-service counter, cashier's booth, clean tray storage area, seating for adequate number of people, used tray depositing area, handwashing facilities, drinking water fountains, etc.
- Special (private) dining rooms for officers, medical staff, special guests, meetings, etc.
- Coffee shop/snack bar, preferably off site.

➤PROBLEM SITUATIONS

Conflicts

Conflicts often arise between the food service staff and the nursing and admitting staff when patient admission, discharge and transfer result in last minute requests, cancellations or changes in preparation and delivery of scheduled meals. Some times food also gets wasted. A degree of tolerance, understanding and effective communication will help reduce such conflicts. Another point of conflict between the food service and nursing department is as to who should pass and pick up the patient trays. This is an administrative decision.

It is hard to provide a menu that pleases everyone. Complaints against the food service department are common and frequent. The work of the department is rendered more difficult because of the need to contain costs. Dietitians can play an effective role in this regard both in the preparation of menu and in talking to patients, especially in the matter of special diets which may not always be palatable or pleasing to the eye.

Many hospitals provide subsidized food to personnel and charge a much lower rate to them than to visitors and patients. Some hospitals provide free food to employees of the food service department while on duty. Most hospitals like to continue this tradition, but if because of the rising cost they have to reduce or abolish the subsidy, it may breed resentment among employees.

Theft

Petty theft and pilferage are common in the food service department. These mostly involve food dishonestly consumed on the premises, stealing patient food, eating food left in patient trays, and pilfering food from the store room and pantries on the patient floors. The biggest offenders are the employees of the department, housekeeping, maintenance personnel and guards. An effective method to curtail this is to lock the place where food is stored. Good supervision is necessary.

Bigger frauds can take place at the materials management level, particularly in the purchasing process. All safeguards discussed in the section on materials management should also be employed here.

Commonly Used Kitchen Equipment

A list of commonly used kitchen equipment is given below.
- Cooking vessels
- Bulk cooker
- Idli plant
- Bain marie
- Meat mincer
- Wet & dry grinder
- Chapathi plate cum puffer
- Dosa plate
- Toaster
- Juicer
- Potato peeler
- Dough kneader
- Water boiler
- Milk warmer
- Dish washing machine
- Cooking ranges
- Bulk cooking battery
- Walk-in cooler
- Refrigerators
- Water cooler
- Food trollies
- Shallow fryer
- Deep flat fryer
- Tea urns, Coffee dispenser
- Multipurpose ovens
- Baking ovens

LAUNDRY AND LINEN SERVICE

►OVERVIEW

When most people think of hospitals, they usually think in terms of doctors and nurses because of their high visibility, the kind of services they render and the close contact and relationship they have with the patients and their relatives. But those who are familiar with the day-to-day operations of hospitals realize that no hospital can operate without the less glamorous and not-so-conspicuous services such as housekeeping, laundry and kitchen that go by the name of supportive services. A great deal of space and money is allocated to them which is naturally a major consideration in the planning, designing and construction of a hospital. Laundry and linen services is one such vital department of the hospital.

Criticism of linen service is one of the most frequently heard complaints in the hospital. Attention to patient's personal needs and comfort is as important as the physician's medication, the care rendered by the nurse and the appetizing food served promptly and attractively. An adequate supply of clean linen sufficient for the comfort and safety of the patient thus becomes imperative. Besides helping in maintaining a clean environment which is aesthetically significant to patients, clean linen is a vital element in providing high quality medical care. The other aspect of this is the personal appearance of the staff who attend on patients. Pleasant, neatly dressed employees in fresh, neat uniforms go a long way in creating a positive image of the hospital.

A reliable laundry service is of the utmost importance to the hospital. In today's medical care facilities, patients expect daily linen changes. In some areas, linen has to be changed even more frequently. This rigorous schedule can be very exacting both on the laundry and the capacity of linen to withstand the repeated cycles of use and wash. To enable the laundry to meet such a demand, the hospital should have a sufficient quantity of linen for circulation and for providing a rest period in storage. This will not only minimize wear and tear of linen but also add to its life.

➤FUNCTIONS

The functions of the laundry and linen services are as follows.
- ◆ Collection of or receiving soiled and infected linen
- ◆ Processing soiled linen through laundry equipment. This includes sorting, sluicing and disinfecting, washing, extracting, conditioning, ironing, pressing and folding
- ◆ Inspection and repair of damaged articles, their condemnation and replacement
- ◆ Assembling and packing speciality items and linen packs for sterilization.
- ◆ Distributing finished linen to the respective user departments.
- ◆ Maintenance and control of active and back-up inventories and processed linen.

➤LOCATION

The laundry should be so located as to have ample daylight and natural ventilation. Ideally, it should be on the ground floor of an isolated building connected or adjacent to the power plant. This is because laundry is one of the largest users of power, steam and water. A location that allows movement of linen by the shortest route saves steps and time. The department should also be close to the service elevators. Some hospitals have linen chutes through which linen bags are dropped to a designated place from where they are picked up by laundry personnel. However, these are now becoming obsolete.

Every time a load of linen is handled, the cost of laundry services goes up. The location and physical plan layout are important in keeping the cost down. One way of doing this is to keep the traffic flow line as short as possible on vertical and horizontal transportation between the laundry and the user departments. This can be more easily accomplished in a vertical multi-storeyed building where the services are in the basement.

➤SOME PLANNING ELEMENTS

Size of Active Inventory

In planning and maintaining linen stock, a stratified inventory system is generally used. This means that for every piece of linen in use, there are four others either being processed or held in store. Therefore, the active inventory consists of items used daily multiplied by five. For example, for each hospital bed in use, one sheet or pillow case will be found in the following places:

- ♦ A soiled one in use on the patient's bed.
- ♦ A clean one in the linen closet in the nursing unit.
- ♦ A soiled one in the hamper or dirty linen collection area.
- ♦ One piece being processed in the laundry.
- ♦ A clean one in the linen store or back-up store for replacing active store.

Laundry Capacity and Load

A final assessment of the plant and machinery required for a new laundry can be made only by compiling a list of types and quantity of articles to be laundered weekly. At the planning stage, however, the information required can be projected by using the following guidelines.

American Standard An average of 15 pounds per bed per day plus 25 pounds for each operation or delivery.

British Standard Sixty articles per bed per week at 0.39 kg per article.

Indian Standard The rule of thumb is five kgs per bed per day.

All soiled linen in hospitals can be classified into two categories: (a) ordinarily or normally soiled linen and (b) fouled or infected linen. The latter category comprises an estimated 10 per cent of the total load of work. All babies' soiled napkins should be treated as infected.

For arriving at the actual daily workload, the total load of seven days' soiled linen should be washed on six working days of the week. The laundry should have the capacity to process at least seven days' collection within the regular six-day work week.

Soiled and infected linen comprises large flats (sheets, etc.), small flats (pillow cases, etc.), tumble work (bath towel, bed spreads, blankets, etc.), press work (garments, etc.), operating room and obstetrical linen, nursing and paediatric linen, and isolation linen.

➤DESIGN

The laundry functions effectively only when it is planned in strict accordance with the work sequence, namely, receiving, processing and dispatching (Figs 6.7 and 6.8).

The activities of the hospital laundry are in many ways similar to those in hotels and other institutions. However, the hospital laundry also handles specialty items and tasks with the most important being disinfection and infection control. It should be designed for asepsis and for removal of bacterial contamination from linen. Many hospitals fail to see that the layout and system of processing in a hospital laundry should follow the principles involved in the central sterilization and supply department. There should be a strict barrier separation between the normally soiled linen and fouled or infected linen on the one hand, and between the soiled area and the clean processing area on the other. The latter can be accomplished by installing double-door, pass-through washing machines in the wall separating the soiled area and the clean processing area. Linen is loaded on the soiled side and unloaded on the clean side (Plate 10: Picture 65).

This physical separation of soiled and clean areas has an important bearing on the design of the laundry and infection control. Traditionally, the various steps involved in the processing of linen are carried out in the same room as, say, in a hotel laundry. This is contrary to the now well-established concept of complete separation of clean and soiled functions throughout the hospital. An enormous quantity of bacteria is released in the air of the processing area during sorting of linen before wash. This airborne contamination pervades the whole area and eventually settles down on the clean processed linen which is delivered to the patient care areas. This can be avoided by the separation of clean and soiled areas.

Ideally, the process of separating normally soiled linen from the infected linen should start from the time they are soiled or infected in the user departments. Infected linen should be bagged in linen bags that are distinguished by colour. There should be sufficient space left to make for complete closure of the bags. The bags should be temporarily deposited in a well ventilated holding area to await collection by the laundry personnel.

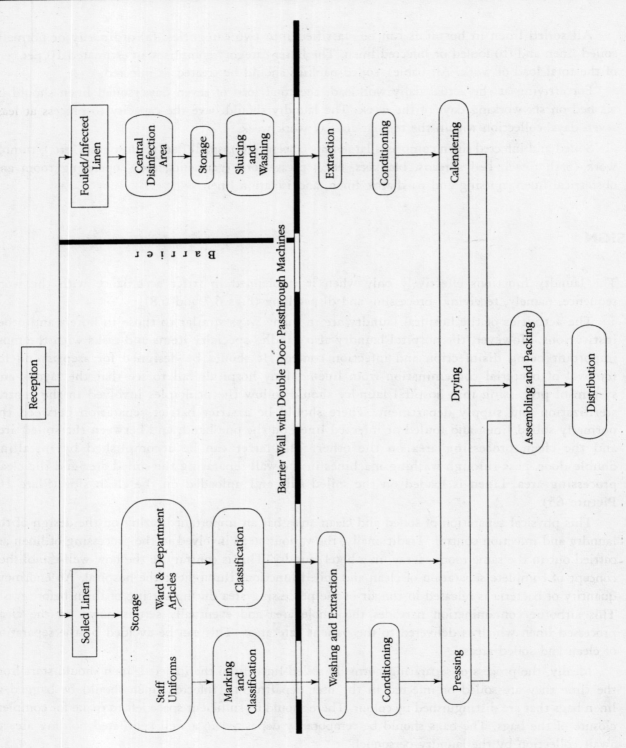

▶ FIG. 6.7 Flow Chart of Laundry & Linen Services

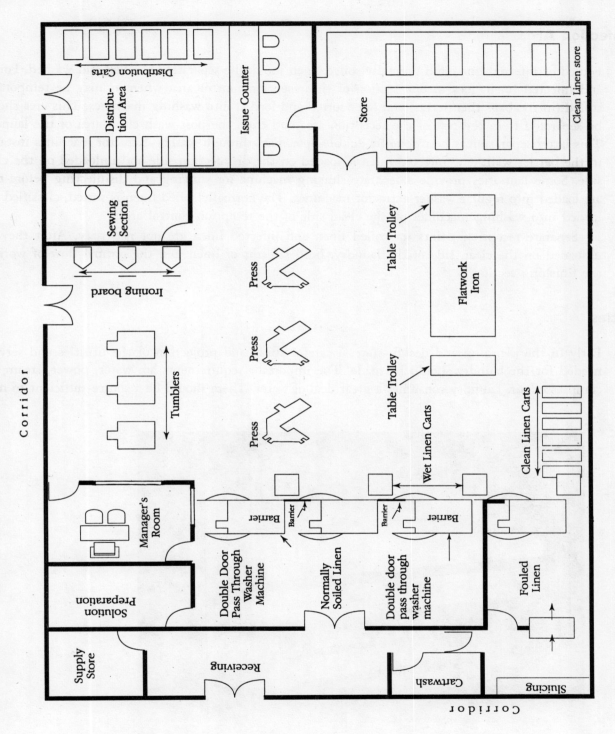

▷ **FIG. 6.8** Layout of a Typical Laundry

Disinfection Area

Fouled or infected linen and normally soiled linen should be separately handled and washed. Fouled and infected linen goes to one section of the reception-control area where it may be temporarily stored (in the bags themselves) and later sorted and loaded into washing machines. This area should be separated from the rest of the reception area and from the post-wash clean area of the laundry. This latter separation is best done by double-door, pass-through washer-extractor machines installed in the barrier wall. The infected linen is loaded on the disinfection side and unloaded on the clean side. Some laundries provide a separate sluicing machine for sluicing and disinfecting before they are loaded into regular washer-extractor machines. The normally soiled linen is sorted, classified and loaded into washing machines on the clean side of the reception-control areas.

Separate reception points for soiled linen and infected linen are not necessary. After they are unloaded on the clean side of the laundry, both streams of linen join the normal flow of work in the finishing section.

Utilities

Early in the planning and design state, a careful study and projection of the utilities and services needed for the laundry should be made. The important requirements are water, power, steam and compressed air. Laundry consumes a great deal of water. There should be a source sufficient to meet

➤ **PICTURE 66** Rotary Ironer, also Called Flat Work Ironer or Calender

the entire need. Discharge of effluent should also be dealt with at the earliest stage. Adequate electric power must be available. Hundred per cent of the normal power should be provided as standby power. Adequate quantity of steam and correct temperature are important. Steam should be delivered by the shortest route to minimize line losses and at the same time provide ample heat to flat work ironers (Picture 66) and presses. The laundry also needs compressed air for operating these flat work ironers and presses.

The maximum demand for all utilities must be projected. The correct way to compute it is by the simultaneous use of various types and sizes of equipment keeping in mind the future expansion and increase in workload.

➤ORGANIZATION

The operational chief of the laundry is a laundry manager who may have been trained in laundry operation or has adequate experience in the field. He reports to one of the associate administrators. Many laundry managers come up through the ranks. However, with increased automation and better opportunities to train people in the technical schools, more and more hospitals are recruiting ITI-trained personnel to head their in-house laundries.

No formal training is required for the other personnel and most of them learn their responsibilities on the job. Hospitals will do well to recruit personnel who are able to read and understand simple instructions. This is necessary in a hospital laundry.

➤FACILITIES AND SPACE REQUIREMENTS

The following facilities and space are required for the laundry aid linen services.
- Reception-control area with facilities for receiving, storing, sorting and washer-loading of soiled linen.
- Sluicing and central disinfection area.
- Clean linen processing room.
- Laundry manager's office with provision for an unobstructed view of the laundry operation.
- Sewing, inspection and mending area. A light table is necessary for inspection.
- Staff facilities.
- Supply storage room.
- A lockable store to accommodate materials for reclothing calenders and presses.
- Solution preparation and storage room.
- Handwashing facilities in each room where clean and soiled linen is handled or processed.
- Provision for supply of water, power, steam and compressed air.
- Cartwashing and cart storage area.

- Clean linen storage room.
- Clean linen issuing counter.
- Electrical distribution switchgear room.
- Water recovery and recycling plant, if necessary.
- Water softening plant, if necessary.

The following facilities are required off-site.

- A central clean linen storage and issuing room.
- Clean linen (lockable) storage in every nursing unit and user department.
- Separate room(s) for receiving and holding soiled linen from the wards and departments until ready for pick up by the laundry personnel.

➤SELECTION OF EQUIPMENT

The last few years have witnessed a revolution in laundry machinery and processing system. Automatic machines and labour saving devices have resulted in economies in the number of personnel and operational time, increased productivity, better utilization of water, heat, power, steam and washing materials, and the maximum utilization of men and machines. Some of the features commonly found are automatic formula dispensers, automatic operation controls, sorting and counting devices, machines combining washing, rinsing and extraction, and flat work folding machines for automatic folding.

The selection of equipment of a proper size is of utmost importance for balanced and economical production. The laundry equipment should be carefully selected. The following factors should be kept in mind.

- Reasonable capital cost.
- Reliability of design and compliance with the Bureau of Indian Standards.
- Availability of spare parts and ease of maintenance.
- Efficiency in working under normal conditions.
- Economy in consumption of utilities like water, power, steam, etc. and in washing materials and other consumables.
- Continuity of workflow and reduction of manual effort.

List of Equipment

The following is a list of commonly used equipment in a typical laundry.

- Washer-extractor sluicing machine
- Double-door washing machine
- Hydroextractor (Machines combining washing, rinsing and extraction are also available.)
- Flat work ironer, also called rotary iron or calender
- Tumble dryer

- ◆ Utility press
- ◆ Mushroom press
- ◆ Table trolley
- ◆ Ironing table
- ◆ Hand iron
- ◆ Dry linen trolley
- ◆ Wet linen trolley
- ◆ Linen hamper
- ◆ Hanger trolley
- ◆ Distribution trolley
- ◆ Motorized sewing machine
- ◆ Platform scale
- ◆ Air compressor

➤ PROBLEM SITUATIONS

Theft of Linen

Theft of linen in hospitals is common. Linen in good condition is a very marketable commodity. Besides, people use sheets and pillow cases in their homes and pilfered linen items become handy. Theft of linen takes place usually at night on the patient floors and departments. Interestingly, soiled linen is not a significant target of theft.

All linen should be kept under lock and key, and linen in stock should be made accessible only to those who need it as part of their duty.

The linen closet in the nursing unit should be located directly facing the nurses' station to deter pilferage. The supply level of linen in the wards should be kept low to correspond with the bed occupancy. Thefts are proportionately higher when a large quantity of linen is accessible to the employees, visitors and patients.

$$\boxed{\text{HOUSEKEEPING}}$$

➤ OVERVIEW

Housekeeping services, also called environmental services, are of paramount importance in providing a safe, clean, pleasant, orderly and functional environment for both patients and hospital personnel. A clean and hygienic environment has a tremendous psychological impact on the patients and visitors, which immediately sets for them the overall impression of the hospital. Since it is difficult for the lay people to judge the practice of medicine in any hospital because of their lack of medical knowledge, they often form their opinions about the hospital on the basis of its appearance and cleanliness. Good housekeeping is an asset and a powerful public and patient relations tool which has a direct bearing on the prestige and reputation of the hospital.

Consider two scenarios. In the first, the patient or the visitor finds the floor and walls of the hospital refreshingly clean. He will naturally be pre-disposed to speak well of the hospital, thus adding to its good reputation. In the other scenario, a visitor finds the lobby dirty and untidy, and the wards smelling of offensive odours. He will immediately entertain doubts about the quality of care his loved ones may be receiving in the hospital. This is similar to the experience of a guest staying overnight in a dirty and untidy hotel. He will not return to it again, and may even tell others about the appalling conditions, thus discouraging them from going there.

Good housekeeping is far more important to patients than many of us are inclined to think, and for two good reasons. Firstly, the hospital is their temporary home for the duration of their stay. Secondly, knowingly or unknowingly, they are exposed to the risk of cross-contamination or hospital-acquired infections. Every patient has a right to be protected from the hospital-acquired infections and from germs brought into the environment by other patients, visitors and the hospital staff. The hospital may have the best doctors on its staff and the most modern equipment, but if its housekeeping is of a poor quality, it will overshadow the effect of all other things.

Maintaining a clean, orderly and sanitary hospital is important from the point of view of economy as well. Properly maintained buildings have potentially longer and less-expensive life while poorly maintained ones deteriorate fast and consequently prove more expensive in the long run.

Although the housekeeping department constitutes, in a manner of speaking, an insignificant and the least glamorous department, its work sets the tone and contributes greatly to the overall efficiency of all other departments. A clean, attractive and orderly work environment enhances employee productivity, efficiency and morale. On the other hand, much work time may be lost in a store room, for example, which is disorganized, cluttered and chaotic.

➤FUNCTIONS

Specialization of medical and paramedical services on the one hand and changes in building materials and designs on the other have brought about revolutionary changes in housekeeping with newer methods and gadgets, increased and more specialized service and changes in the overall functions, operation and design. While many hospitals still continue to provide in-house services, some also contract the housekeeping services either totally or in part.

The department is responsible for performing a variety of tasks. The following are some of the common functions.

- Daily cleaning: This includes sweeping and mopping floors, dusting furniture, cleaning fixtures, walls, ceilings, windows and bathrooms, emptying trash cans, and defrosting refrigerators in nursing stations.
- Periodic cleaning: This includes washing windows, waxing floors, cleaning carpets, dusting high ceilings and changing draperies.
- Trash and garbage removal: This includes collecting trash and garbage from various points within the hospital and moving them to incinerator or dumpster.
- Discharge cleaning: This includes cleaning patient room after discharge or transfer of a patient and readying it for another patient.
- Watering indoor plants, if required.
- Exterminating bugs and pests.
- Preventing spread of infection and ensuring conditions for good patient care by using proven infection control procedures and techniques.

In addition to the above, the housekeeping department has certain incidental responsibilities, as follows.

- Saving electricity by turning off lights, fans, etc. when not in use.
- Ensuring an economical use of supplies (which is one area where there is much waste).
- Developing goodwill by a courteous, helpful and caring attitude toward patients and visitors.
- Promoting safety rules and measures by observing them and reporting dangerous conditions.
- Maintaining a harmonious working relationship with the employees of other departments.
- Being actively involved in the plans and activities of the disaster committee, firefighting, simulated disaster and fire drills, etc.

The workload of the housekeeping department fluctuates depending on the patient census. When the bed occupancy is high the workload increases. When it is low, empty rooms are generally kept locked making daily cleaning unnecessary. Discharges and transfers call for extra cleaning. They require extra staffing on certain days of the week and time of the day when most discharges occur, as for example, on Saturdays and around noon.

In some hospitals regular housekeeping personnel do not carry out housekeeping functions in certain specialty areas such as surgical and recovery rooms, labour and delivery suites, laundry, kitchen and maintenance department.

►LOCATION

Housekeeping serves all areas and departments of the hospital. Although it can be situated in a non-prime area, it should as far as possible be centrally located and close to the vertical transport system to facilitate easy movement of housekeeping materials and equipment.

►ORGANIZATION

The head of the housekeeping department is called the executive housekeeper who is assisted in the administration of the department by an assistant executive housekeeper and floor supervisors. The training and qualification of the executive housekeeper are very important. She must have a degree in science and basic knowledge of health care sanitation including principles of bacteriology applicable to prevention and control of infection and communicable diseases. She must also have a working knowledge of medical terminology as applied to sanitation practices, and the ability to plan, administer and develop all phases of housekeeping. Good interpersonal skills and leadership qualities are essential.

In smaller hospitals, the director of nursing or nursing superintendent may be in charge of housekeeping operation as she is of the CSSD and laundry, but in larger institutions, the executive housekeeper is responsible to an associate administrator.

The importance of the position of executive housekeeper is evident by her placement as a member of the hospital infection control committee.

The assistant executive housekeeper and the floor supervisors must be well trained and fully conversant with the housekeeping procedures. They should be able to effectively manage the housekeeping employees who are a rather recalcitrant group difficult to handle. Good interpersonal relations are thus important.

Housekeeping employees are largely unskilled workers employed at the lowest salary level. They should receive a good orientation, in-service education and on-the-job training.

The importance of a sound selection programme in the hiring of housekeeping personnel cannot be overemphasized. In most hospitals the department is a hotbed of trade union activities and indiscipline. In the selection and development of personnel, the following principles should be kept in mind.

Good Selection Check applicant as a person, his antecedents, ability to work with others and potential to do the required job.

Good Training Give the employee the help and training he needs to perform the job.

Good Supervision Provide guidance in performing work, correct poor work, and support him on the job.

Recognition Recognize and praise good performance. Make the employee feel wanted and appreciated.

The key to efficient and effective functioning of the housekeeping department is good communications, particularly with the admitting department and nursing units — the two potential areas with which conflicts may arise. There should be cooperation and good working relationship between housekeeping and these two departments.

The admitting department should promptly notify the housekeeping of all admissions, transfers, and discharges, and the housekeeping in turn should get the rooms cleaned and ready without delay and notify that they are ready. This is particularly important when the occupancy rate is high. Otherwise the newly admitted patients will have to wait for their rooms. Computerized admission operations also speed up work.

Infection control is an important aspect of housekeeping activities. Many hospitals have an infection control committee of which the executive housekeeper is a member. Housekeeping staff should be aware of hospital-acquired infection and principles of infection control.

Standards are very important for the efficient and successful functioning of housekeeping. They equalize the work load of employees, ensure that each employee knows what is expected of him and in what amount of time, and maintain high quality service. Standards of quality, for example, define the required cleanliness of each object. Standards relating to frequency determine how often an employee is required to clean an area. Time standards allot the time required to complete each cleaning task. Performance standards provide a description of what is expected of an employee in various areas of work including attendance, punctuality, cooperation, etc.

➤FACILITIES AND SPACE REQUIREMENTS

The following facilities and space are required for housekeeping.
- ◆ Office for executive housekeeper.
- ◆ Clerical work area.
- ◆ Office for assistant executive housekeeper and desk or office space for supervisors. It is desirable that the floor supervisors are physically located in their assigned areas of supervision where the personnel under them work.
- ◆ Storage room for housekeeping equipment.
- ◆ Storage for housekeeping supplies.
- ◆ Housekeeping (janitor's) closets on all floors throughout the hospital, equipped with floor sinks, and space and shelves for housekeeping equipment, carts, buckets and supplies. As a matter of daily routine, supplies for daily use on the floors and departments should be delivered to their respective closets directly.

➤PROBLEM SITUATIONS

Conflicts and Interruptions

The housekeeping work in certain areas of the hospital such as the patient rooms, operating rooms, ICUs, labour-delivery suites, emergency department and nurseries is constantly interrupted and its time severely restricted so much so the housekeeping personnel's time is often wasted in waiting. The timing of work of certain other departments precludes the economical and efficient functioning of the housekeeping department. Frequently, these departments consider housekeeping personnel as a nuisance and an interruption in their busy schedules. Conflicts are bound to arise in these areas.

A more serious problem may arise in the patient care areas. Sick people who are generally sensitive to noise, odours and irritating bustle of activity around them may not cooperate with the housekeeping personnel in carrying out their work. The following are some other areas of conflict which raise certain issues that should be resolved by administrative decisions.

- Who cleans up after a maintenance man has done his work?
- Who moves furniture and heavy equipment?
- Is it the responsibility of housekeeping personnel to notify the maintenance department of the need for repairs that they observe while cleaning?
- Whose duty is it to make the patient's bed when he is discharged?
- Who washes surgery and delivery room equipment?
- Should housekeeping personnel help in moving patients from one room to another and a dead body to the mortuary?

One frequent complaint against housekeeping personnel is their attitude towards the nursing staff, that is, their lack of cooperation with and responsiveness to the nursing personnel's requests. They either walk away when asked to do some chores, or do them grudgingly or raise their voice in open defiance. Suitable action should be taken to nip such cantankerous attitude in the bud.

The same problem arises when the floor supervisor fails to gain the cooperation of the cleaning staff. In larger hospitals where housekeeping personnel are widely spread out, supervision is a problem and employees tend to become lax in their work. Inadequate supervision results in poor performance, inefficiency and low employee morale so evident in many of our hospitals.

Theft and Waste

Theft in housekeeping is not a major cause for worry. Since most of the housekeeping supplies are centrally stored and supplies like cleaning fluids are issued on a daily basis, there can be only petty thefts. However, what should be a cause for concern is the waste of housekeeping supplies which is often a far greater source of loss to the hospital. This problem of waste is compounded by the difficulty of supervising individual employees who are scattered throughout the hospital.

Bigger thefts of supplies and equipment are committed by housekeeping employees from the departments to which they are assigned to work. Theft can be of any material or equipment and from any department; every place affords an ample opportunity to the unscrupulous employee. Some are assigned to work in remote and deserted areas where there is not much supervision. Trash and garbage disposal provides an excellent means to conceal and take away stolen articles. If, however, trash is taken straight to the incinerator, chances of pilferage may be reduced.

VOLUNTEER DEPARTMENT

➤OVERVIEW

One of the beautiful things that one sees in hospitals in the developed countries like the USA is the volunteer programme that permeates their whole social fabric and affects the economic health and social progress of the people. In America, for example, hospitals of all sizes and scopes have well established volunteer programmes so much so the well being of the nation's economy is inextricably linked to that of the volunteer programme. Nearly forty million volunteers render several billion dollars' worth of free service to the hospitals.

Volunteers provide many extra services that supplement the essential functions of the professional staff — services that add to the quality of care as well as comfort and happiness of the patients. They assist in promoting understanding of the hospitals in the community. As established channels for community participation in hospital affairs, they provide a base on which to build an extended hospital-public relationship.

Volunteers fall into two general groups: hospital volunteers and auxiliaries or women's auxiliaries. Hospital volunteers work primarily in the functional areas of the hospital while the auxiliaries work largely outside the hospital in activities not directly related to the functions of the hospital. The hospital volunteers have entered all aspects of the functioning of the hospitals and provide free service to them.

Volunteers are public-spirited and service-minded people who are disciplined and have a high sense of commitment to the hospital. They derive satisfaction and joy that come from helping others. They comprise people of all ages, backgrounds and abilities: students, housewives, working people and retired people. To qualify, a volunteer must possess interest, dependability and the physical fitness necessary to perform the volunteer assignment. Each volunteer is required to participate in an orientation session to learn all about the volunteer programme as well as the hospital before starting his(her) assignments. Some volunteer assignments require special training courses while others receive on-the-job training. There are special programmes for teenage volunteers tailor-made for summer holidays.

Volunteers subscribe to the philosophy of person-centred holistic health care and concern for wellness. They believe that an individual is more than a body that is hurt or broken. Each patient is an integrated whole with the physical, emotional, intellectual, social and spiritual aspects bound together. They, therefore, work to promote the wellness of the whole person.

Although auxiliaries are found in some of our hospitals, any organized hospital volunteer programme rarely exists in India. Hospitals should be encouraged to tap these valuable resources.

➤ AVENUES OF SERVICE AND WHAT VOLUNTEERS DO

The services of volunteers can be utilized in a variety of ways and in almost every area of the hospital. Some of the more common areas where they can contribute their services are as follows.
- Gift shop
- Coffee shop/snack bar
- Information desk
- Admitting
- Sorting and delivering all incoming mail to staff and patients, and forwarding mail to the discharged patients.
- Recreational therapy
- Occupational therapy
- Book and magazine cart and patient library
- Delivering surgery schedules
- Clerical and secretarial work
- Visitor control and issuing passes
- Flower and newspaper delivery to patients
- Assisting as hostesses during hospital functions, conferences, and workshops.
- Conducting hospital tour to visitors, student groups, etc.
- Serving as interpreters for patients who do not know English or the regional language.
- Conducting free blood pressure screening clinic.
- Assisting in public relations programmes in writing human interest stories, hospital publications, newsletters, etc.
- Organizing and assisting in hospital fund-raising programmes and events like entertainment, sale, fair, etc.

➤ LOCATION

The volunteers may work in almost every department. A central location is therefore desirable. The volunteer department should be easily accessible to the administrative services, particularly the public or community relations department. Because of the volunteer department's numerous contacts with the public and social and civic groups, it must be located to provide easy access to the representatives of these groups visiting the department.

➤DESIGN

The enquiry desk, gift shop, book shop and coffee shop are areas where volunteers often work, and where the visitors stop by before entering other areas of the hospital. The overall impression of the hospital is generally set for the visitors immediately upon entering the hospital facility. The service and decor in these areas should be consistent with the image the hospital wishes to project. If the coffee shop or book shop is designed poorly, the volunteers may not be able to render proper service. It is the service and decor in these areas that often set the pace for family members who later visit the patient on the floor. The irritated or dissatisfied customer may complain about the unsatisfactory condition at the patient's bedside. They are likely to find fault with the care provided to the patient.

Coffee shop should not only be cheerfully decorated, but also designed to provide tasty food and for a rapid turnover of the guests. It should not be permitted to be used as a lounge area for hospital personnel who sometimes tend to take a long coffee break. The design should be counter-type with self-service and the seating on the pattern of the fast food restaurants. The low, wide, non-swivel type seating arrangement is convenient for the sick and the handicapped.

The design elements should extend to other areas where volunteers work, including the director of volunteers' office where she meets members of the public and also interviews applicants to join the volunteers. There should be a place near the reception-secretary for logging hours of volunteers. A place for shelves for books and magazines and for the patient book truck(s) is also necessary.

The gift and book shop should have large display windows and adequate space to keep a large number of selling items.

➤ORGANIZATION

The volunteer department in the hospital is usually headed by a chief or director of volunteers who may be a full-time, salaried person. This person, generally a lady, reports either directly to the chief of the hospital or to one of his associates. She recruits, interviews and selects volunteers, organizes orientation and training programmes, and assigns them to various functional departments in the hospital. She maintains strong liaison with departments where volunteers are placed. It is her responsibility to keep the volunteers informed of hospital policies and procedures.

Volunteer department, like all other departments in the hospital, is organized by the authority of the governing board. Volunteers have their own bylaws governing their organization, work and relationship with the board, administration and other areas of the hospital. In its internal government, volunteer department is autonomous and self-governing. Outside its internal affairs, the department has no authority over other departments or personnel of the hospital, and should not attempt to exercise any.

CHAPTER 7

Planning and Designing
Public Areas and Staff Facilities

CONTENTS

7 CHAPTER

- Entrance and Lobby Area
- Gift Shop, Book Shop, Florist's Shop
- Coffee Shop/Snack Bar
- Meditation/Prayer/Quiet Room
- Staff and Employee Facilities
- Bank Extension Counter

The public areas of a hospital serve as an important reference point in the context of space and traffic in the facility. In planning and designing a new facility, attention should be paid to the following.

ENTRANCE AND LOBBY AREA

The main lobby (Plate 10: Picture 67) of the hospital is used primarily to accommodate patients and their families and friends. Unless a separate entrance is designated for them, the main lobby also serves as a convenient access route to the medical and other staff proceeding to administrative and other areas of the hospital. Patients usually enter the hospital through the main lobby and are also discharged there. Many patients may be tense and anxious. Some may be infirm and some in wheelchairs. It is important that they are courteously received and favourably impressed at this point. Staff should be caring and friendly. On the other hand, discharge is a happy occasion; nothing in the discharge process and in the activities that go on in the lobby at the time of discharge should mar this happiness.

The composite parts and adjacent areas of the lobby (Fig. 7.1) to which attention should be paid are the following.

- Parking area. It should be adequate with a separate space for medical staff and officers.
- Main public entrance. It should be at ground level, sheltered from weather and accessible to wheelchairs.
- Reception and information desk or counter. It should be conveniently located for people as they enter the lobby. In small hospitals, telephone switchboard and information are often combined in one area.
- Public waiting area.
- Public toilet facilities. Most visitors have difficulty in locating public toilets. These should not be hidden from view but should be easily accessible to people in the lobby.
- Water coolers or drinking water fountains.
- Circulation. Circulation area which is an integral part of the net area of the main lobby should be generous.
- Direct access to horizontal and vertical circulation area. Elevators should be conveniently accessible from the lobby. They should not discharge people into the main lobby but into a recess.
- Alcoved space for wheelchairs and stretchers, out of traffic but easily accessible.
- Doorman's station and security post, if necessary.
- Easy to follow signage system (Plate 11: Pictures 68, 69). One of the most frustrating experiences for patients, particularly in large hospitals, is to find their way through the complex maze of the hospital structure. A simple and effective programme of directional graphics and signage is therefore essential.

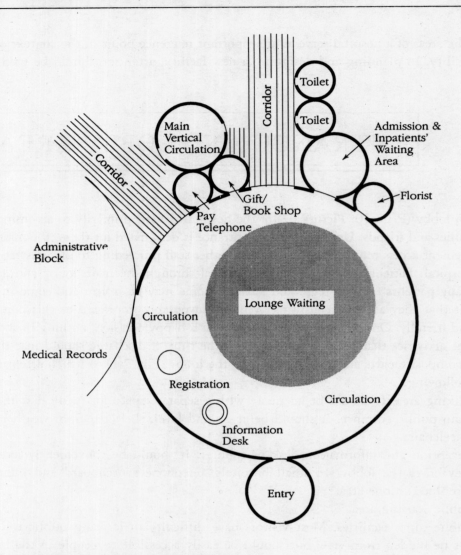

> **FIG. 7.1** Main Entrance and Lobby Area of Hospital

- Programme board announcing seminars, conferences and other special events with proper location. It should be visible to the guests as they enter the lobby.
- Directory of floor plan and the building.
- Senior doctors' name board(s).
- Cashier's counter. It should be conveniently accessible to the main flow of lobby traffic.
- Coffee shop-cum-snack bar.
- Gift and book shop.
- Florist's shop.
- Retiring facility for anxious or bereaved relatives is desirable.

PLATE 11

> **PICTURE 68** Signage System

> **PICTURE 69** Signage System

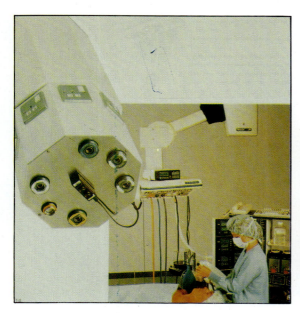

> **PICTURE 71** Centralised Medical Gas System: Ceiling services

> **PICTURE 72** Centralised Medical Gas System: Outlets and other fixtures in patient room

➤ MAIN WAITING AREA

The main waiting area should provide seating for the largest estimated number of people who may occupy it at a given time. The fact that all patients are usually accompanied by one or two relatives or attendants should be taken into account while planning. The main waiting area looks better organized, efficient and presentable if most of the patients and their relatives are dispersed to the appropriate sub-waiting areas. Some of the special sub-waiting areas are: room for expectant fathers on the obstetrical floor, waiting area for relatives of patients undergoing surgery, area adjacent to admitting for patients waiting to be admitted, sub-waiting areas in front of clinics and ICUs, and in administrative offices, for example, vendors in front of purchase office and job seekers in front of personnel office.

GIFT, BOOK AND FLORIST'S SHOPS

Gift shop, book shop and florist's shop should be prominently located off the main lobby but visible from it. They should have plenty of display space with full glass walls in front and on sides. Gift shops are becoming increasingly popular and are a good source of income to the hospital. Often, they are run by volunteers. These shops occupy a prime area and serve a useful purpose. Nevertheless, they should not be larger than warranted for the simple reason that they are entirely secondary to patient care.

COFFEE SHOP-CUM-SNACK BAR

It is an essential facility for the outpatients and emergency patients, especially in the night, and so should be close to the outpatient and emergency services. It should also be in close proximity or adjacent to the food service department from where it should receive food items as it is not economical to run two separate food establishments.

The coffee shop is another place which visitors and outpatients often have difficulty in locating and so it should be easily accessible. The design and organization of the coffee shop have been described elsewhere.

MEDITATION ROOM

Most Christian institutions have a chapel for worship for the staff and patients. They also have a chaplaincy department to cater specifically to the religious needs of the large number of nursing staff and nursing students who are resident on the campus. Other hospitals also have what may be variously called a meditation room, quiet room or prayer room where staff and patients can spend time in prayer and quiet.

STAFF FACILITIES

The staff facilities usually include locker rooms, staff toilets, staff lounges and library. Some hospitals also provide recreational facilities for the staff.

Many hospitals establish bank extension counters in their premises for the benefit of staff and the institution itself. Hospital collections are conveniently deposited in and salaries of staff disbursed through the bank where every staff member operates an account. One consolidated cheque for all staff is given to the bank which credits salaries to individual accounts of the staff. Arrangements may be made for outstation patients carrying a large amount of cash to temporarily deposit it in the bank.

Planning and Designing Engineering Services

CHAPTER 8

CONTENTS

☐ **Engineering Department**
☐ **Maintenance Management**
☐ **Clinical (Biomedical) Engineering**
☐ **Electrical System**
☐ **Air-Conditioning System**
☐ **Water Supply and Sanitary System**
☐ **Centralized Medical Gas System**
☐ **Telecommunication System**
  ▪ **Telephone System**
  ▪ **Nurse Call System**
  ▪ **Dictation and Central Transcription System**
  ▪ **Paging System**
  ▪ **Public Address System and Piped Music**
  ▪ **Television, Closed Circuit Television (CCTV)**
☐ **Environmental Control**
☐ **Solid Waste Management**
☐ **Safety and Security**
  ▪ **Safety in Hospitals**
☐ **Security and Loss Prevention Programme**
☐ **Fire Safety**
☐ **Bomb Threat**
☐ **Alarm System**
☐ **Disaster Preparedness**
☐ **Code Blue Procedure**
☐ **Transportation**
  ▪ **Elevator, Lift, Dumbwaiter**
  ▪ **Stairways and Ramps**

ENGINEERING DEPARTMENT

➤ **OVERVIEW**

In the last two decades or so, hospitals around the world have become a dynamic industry which is expanding and becoming more complex with every passing year. Technological advances have permeated virtually through every hospital department. The unprecedented proliferation and sophistication of medical equipment pose a big challenge to those responsible for the installation and maintenance of the hospital plant and its equipment. This challenge, thrown by the specific needs of the hospital's central plant, includes all aspects of design and space requirements for the generation, storage and distribution of steam, cold and hot water and for refrigeration, ventilation, air-conditioning and electricity. Besides adequate capacity, reliability is a major concern, primarily because of the many life saving services provided by the hospital. The activities of the engineering department include highly specialized services which should be determined by specialist engineers, consultants and architects of the planning and design team.

Frequently, there is a tendency to minimize and underestimate the overall scope and requirements, particularly space needs, of the mechanical plant and other engineering services of the hospital during the planning and design stage. As a result, engineers are often compelled to fit equipment into inadequate space. This will seriously hamper the operation and maintenance of equipment throughout its operational life.

It is not difficult to see the importance of engineering department for the safety and smooth operation of the hospital if we consider that mechanical and electrical components represent 50 to 60 per cent of the total cost of construction of a hospital barring the cost of medical equipment.

The space requirements of engineering and maintenance department have increased greatly in recent years because of the development of more complex and sophisticated equipment which calls for more space and facilities and a greater number and variety of trained personnel. This includes facility for a complete, 24-hour monitoring system so essential to run an efficient modern hospital.

The engineering and maintenance department is charged with the responsibility for ensuring safe and economical operation and maintenance of hospital facilities and expensive equipment.

The department should be capable of providing technical and management support to the hospital. Organizing it into a professional, well-disciplined department responsive to the challenges of a rapidly changing modern hospital is a formidable task. The central figure charged with this task and putting together a capable and skilled team is the chief engineer. The hospital engineer of yesteryear who was more of a skilled mechanic has now given place to a highly educated professional possessing impressive academic qualifications and vast experience. In today's hospitals, he is a prominent member of the top management team.

➤ **PICTURE 70**   Engineering Department Machinery

➤FUNCTIONS

The engineering department performs a wide range of functions which may be assigned to various units of the department. It is responsible for the operation of all equipment, machinery and distribution lines, and preventive maintenance and repair. Specifically, the department performs the following functions.

- ◆ Plant operations (Picture 70).
- ◆ Building operations and maintenance.
- ◆ Mechanical and electrical maintenance.
- ◆ Preventive maintenance.
- ◆ Clinical engineering and biomedical equipment and electronics maintenance.
- ◆ Landscaping and grounds maintenance.
- ◆ Vehicle operation and maintenance.
- ◆ Elevator, lift and dumbwaiter maintenance.
- ◆ Plumbing, water supply and sanitary system.
- ◆ Contracted services.
- ◆ Carpentry, painting and sign shop.

- Solid waste disposal and incinerator.
- Electrical system including equipment, machinery, power, lighting, emergency generators, UPS and refrigerator maintenance.
- Communications system.
- Fire prevention, fire detection, firefighting methods and devices.
- Minor plant alterations, renovations and repairs.
- Equipment and instrumentation evaluation.
- Equipment control and pre-acceptance check.
- Condemnation and disposal.

►LOCATION

The ideal location for the engineering and maintenance department is on the ground floor in a non-prime area. Convenient access to elevators, unloading dock, mechanical areas, and the boiler plant is essential. The main shop area should preferably have an outside wall on one side for ventilation as well as future expansion. The storage area for grounds maintenance equipment should have an outside entrance.

►DESIGN

Engineering and maintenance department should be planned keeping in mind three important functional areas: the administrative area, the shop area and the mechanical equipment area. The administrative area should be similar to the other administrative areas of the hospital in decor, lighting and finishes. The shop area should be distinctly separate from the administrative area. It should have adequate space to accommodate various shops and the mechanical, electrical, plumbing, carpentry, painting and equipment repair activities. Large open spaces should be allocated to different shops. Movable partitions may be used, if necessary. Adequate lighting is essential, and it is desirable that the place is soundproofed.

The chief engineer's office should be located between the entrance to the department and the main shop area. There should be a small waiting area and a clerical area so located as to control access to the offices. The assistant engineer's office should be adjacent to the clerical area.

The control monitor room should be adjacent to the chief engineer's room. This room which houses equipment and control panels and is manned round the clock monitors mechanical and electrical systems. It locates and prevents potential breakdowns. However, at no time should this be considered a substitute for preventive maintenance programme and scheduled inspections and services.

The main shops are large work areas. Their locations with reference to one another are important. For example, mechanical and plumbing shops should be placed together or close together. Carpentry which produces much dust should be somewhat isolated from the other repair areas, and so also the paint booth which should be in an enclosed space but adjacent to the main shops. The paint shop should be large enough to accommodate equipment and furniture, largely hospital beds and lockers, sent for repainting from all parts of the hospital. The electric shop should be adjacent to equipment repair. A lockable room should be provided for the storage of supplies and materials required by the various work areas. A separate and distinct area should be provided for electronic and clinical engineering. This should be well ventilated and isolated in order to eliminate problems caused by dust. This is discussed in detail later.

The shops and other sections of the engineering department like the boiler plant come under the purview of various statutory and regulatory bodies.

➤ORGANIZATION

The chief engineer is in-charge of the department. He is responsible to the chief executive officer or one of his associates. In large hospitals, he may be assisted by one or more assistant engineers who may be put in charge of major sections such as clinical engineering, electrical engineering, maintenance management and plant operations. A clinical (biomedical) engineer is usually placed in charge of clinical engineering unit. Other personnel in the department are: engineering supervisor, mechanic, skilled artisan, electrician, carpenter, HVAC mechanic, boiler operator, plumber, painter, biomedical engineering technician, medical equipment repairman, helper, incinerator attender, secretary, clerical worker, stockroom clerk and storekeeper.

For an efficient organization and administration of the department, the chief engineer will do well to establish certain specific objectives to help him fulfil his responsibilities meaningfully.

His first objective should be to create an environment in the hospital that is conducive to the well-being of the patient and safe and attractive to the visitors and personnel. His second objective should be to ensure that the utilities like steam, electricity, air-conditioning, etc. are in a constant supply except for scheduled shutdowns for repairs and maintenance. His third objective should be to ensure economical operation and maintenance of the physical plant. Keeping things operating is important but even more important is keeping them operating economically.

The construction contract should include:
* Colour coding of all pipes and ducts so that supply and return conduits are easily identified.
* Provision of copies of as-built drawings to the hospital authorities and engineering department.

Colour coding of the piping system is more easily done and is relatively inexpensive at the time of installation. The drawings are also invaluable to the engineering department staff.

➤FACILITIES AND SPACE REQUIREMENTS

The following facilities and space are required for the engineering department.

- Chief engineer's office with space for storage of protected drawings, records, manuals, etc.
- Clerical and waiting area.
- Office(s) for assistant engineer(s).
- Office or desk space for supervisors.
- Separate building(s) or room(s) for boilers and for mechanical and electrical equipment.
- General maintenance shops for repair and maintenance.
- Central control room.
- Storage room for equipment and building repair and maintenance supplies and materials.
- Storage for inflammable liquids, diesel, oil, etc. A license may be required for this.
- Clinical engineering suite for storage, repair, and testing of electronic and other biomedical equipment with office room for the clinical engineer. It should be isolated, enclosed, dust-free and preferably air-conditioned. Supply of compressed air and nitrogen is necessary.
- Storage area for yard equipment and supplies. It should be off-site so that the equipment and supplies may be moved directly outside without interference to the other areas.
- Sanitary storage and disposal of solid waste, or its removal.
- Incinerator. This is discussed later in detail.

➤ PROBLEM SITUATIONS

Theft and Fraud in Engineering Department

Theft of tools and maintenance materials is very common in hospitals. Service personnel working in the evenings, nights, weekends and on holidays when fewer staff is on duty with little or no supervision easily manage to steal equipment and materials unnoticed. One effective way of preventing or reducing such thefts is by instituting a rigid system of documentation of all materials requisitioned and used, particularly those requisitioned and charged to individual work orders. There should also be an inventory of all equipment and instruments. Every worker should be responsible and accountable for the tools given to him.

Many large hospitals permit their engineering department to do its own materials management — purchasing, receiving and storing — which provides the greatest opportunity for employees to steal. More than what is needed for a work order may be requisitioned and the balance kept for personal use. Thefts also take place when materials are transferred from one building to another. Although thefts of this nature cannot be eliminated completely, they can be reduced drastically by effective supervision and instituting control measures and procedures for storage, inventory and issuing.

The engineering department, particularly its maintenance section, by the very nature of its work provides ample opportunities for theft and fraud. In this context, special attention should be paid to the following areas and activities.

- Maintenance workers often bring outside jobs to the hospital and engage themselves in their private repair work during the hospital working time. Electrical and electronic repair work easily lends itself to this practice which may often go unnoticed. Weekends and holidays are the ideal time for staff to do private work.

♦ Maintenance workers are often assigned to work for prolonged periods of time in remote or outlying areas of the hospital where there is no supervision. As a result there is a constant risk of their idling away their time or disappearing from the place of work and engaging themselves in private work. Employees have been found moonlighting during their regular working hours.

♦ When workers do not finish their work because they have not put in a full day's work, overtime becomes necessary. Overtime is a financial burden which can often be avoided.

♦ Work record should be maintained for all jobs. This will show how a worker spends his time. The concerned supervisor should carefully scrutinize the hours an employee spends on individual jobs.

♦ Another subtle loss to the hospital stems from the abuse of privileges and positions in the hospital by its staff and employees. Maintenance workers often oblige their supervisors by doing the work that they should not be doing. This may be some work in the department which has not been sanctioned or approved. This practice does not stop there. Some of them do small repair jobs in the homes of their superiors, first in their free time and on holidays, and then during the regular hospital time using hospital tools and equipment. In large hospitals, these activities go unnoticed. Integrity of the staff apart, there should be internal control measures to check such malpractices.

MAINTENANCE MANAGEMENT

➤OVERVIEW

Planned or scheduled maintenance is simply planning and scheduling the maintenance of equipment and facilities in order to extend their life, reduce costly breakdowns or failures, and attain maximum operational efficiency. It includes such functions as preventive maintenance, functional testing, performance verification, calibration and safety testing. The goal of planned maintenance is to provide for a safe and functional environment by ensuring the proper maintenance of all equipment and facilities. Planned maintenance usually provides for one major procedure which may include inspection, lubrication, calibration, safety testing and testing for wear for each piece of equipment at regular pre-determined intervals. This may be supplemented by as many minor procedures as necessary to keep the equipment performing at the desired level or standards.

The reasons for planned maintenance are many. The following are more important ones.

- The cost of equipment replacement is skyrocketing and hospitals find it difficult to bear this cost.
- There has been a proliferation of sophisticated hospital equipment and no clinical engineering department, however well equipped and staffed, can cope with the heavy demand of repairing it on an emergency basis.
- The indirect cost to the hospital and patient care of the non-availability of essential equipment and of lost opportunities due to equipment failure is high.
- Poorly maintained equipment is hazardous and expensive. It wastes electricity and power.
- The only way to handle a maintenance job is to do it in a systematic and planned manner.

➤ CLASSIFICATION OF HOSPITAL AREAS

For the efficient scheduling of a maintenance programme and securing the best results, the various areas of the hospital are classified as follows.

Non-Flammable Anaesthetizing Locations   Usually, these are areas designated for the administration of an inhalation anaesthetic agent. For example, operating rooms.

Critical Care Areas   These are areas where patients are subjected to invasive procedures or are directly connected to the line-operated medical devices. For example, operating rooms, ICUs and catheterization lab.

General Care Areas   These are patient care areas where patients come in contact with ordinary electrical appliances or are connected to medical devices. For example, patient rooms and wards.

Wet Location   It comprises patient care areas normally subject to wet conditions. For example, hydrotherapy room of the physical therapy department. It does not include areas rendered wet because of routine housekeeping procedures and incidental spillage.

Non-Patient Care Areas   These comprise administrative offices, laboratory, storage, etc.
Incorrect classification may result in inadequate or too much and unnecessary maintenance as in the case when a non-patient care area is classified as critical care area.

➤ ESSENTIAL EQUIPMENT

The type of equipment also has a bearing on the maintenance programme. An equipment is classified as "essential equipment" for the purposes of maintenance if it falls under one or more of the following groups.

- Equipment considered essential for life support. For example, monitors, emergency generators, etc.
- Equipment that is potentially risky and involved in incidents. For example, boiler which has a high incident risk.
- Equipment needing a more intense maintenance schedule. The more mechanized a piece of equipment or the more often it is used, the more intense the maintenance.
- Equipment maintained by an external agency on contract service and not by the in-house personnel. It does not form a part of the scheduled maintenance.

➤HOSPITAL EQUIPMENT CONTROL SYSTEM

One effective way of keeping the costs down is to manage the maintenance of hospital equipment through the establishment of what is called the equipment control system which has been widely used with a high degree of success. This system provides for detailed information regarding every piece of equipment in the hospital: its location, date of purchase, description of equipment, date on which the equipment was last serviced, due date for next service, etc. This information also helps in preparing the annual report for the maintenance programme in terms of total maintenance cost, total cost of spare parts, etc. during the year on every piece of equipment.

This topic is dealt with in greater detail in the section on clinical engineering.

➤TYPES OF MAINTENANCE WORK

Typically, the hospital maintenance work falls into one of the following classes.

Preventive Maintenance Maintenance work on a piece of equipment on a planned schedule to prevent breakdown. The work consists of inspection, adjustment, calibration, cleaning, repair, etc.

Emergency Maintenance Immediate repair of a vital piece of equipment. Requests are usually made by phone demanding immediate attention.

Routine Maintenance Maintenance on a routine basis. These are non-urgent repairs such as painting, replacing tiles, etc. which can wait until scheduling is possible.

Contract Maintenance Repairs to equipment made through contracted labour and materials under annual maintenance agreements. This can be done either on a routine basis or when a piece of equipment under service contract breaks down.

Project Work One-time-only work such as building renovation and tiling a roof. The department generally calls for quotations for this kind of work. Requests for projects are generally required to be submitted prior to the budget year so that approval may be accorded and budget allocation made.

While establishing a preventive maintenance programme, the engineer should be clear in his mind of his priorities. Three significant factors should be borne in mind: (i) the importance of the piece of equipment to the functioning of the hospital or patient care; (ii) the direct cost of the failure of the equipment; and (iii) the indirect cost of failure and non-availability of an essential piece of equipment.

➤PRIORITIES AND GUIDELINES

Some maintenance departments establish guidelines and a time frame for different types of maintenance work. They classify all types of work into the following categories on the basis of their priorities and criteria.

Priorities	*Criteria*
Emergency	Life is threatened or endangered; hospital services will come to a standstill or curtailed, and an emergency situation may arise if not carried out.
Within 4 working hours	Normal repair in the patient areas.
Within 8 working hours	Normal repair in the support areas.
Routine	Minor tasks; some delay is of little consequence.
Long term	Requested and registered but not yet scheduled.

➤GUIDELINES FOR SELECTION AND MAINTENANCE OF EQUIPMENT

Hospitals can prolong the life of hospital equipment, reduce costly maintenance and attain greater operational efficiency by following certain rules and guidelines in the selection and maintenance of equipment. The following are some such rules.

- Avoid buying equipment that is expensive to repair and maintain and one that is not reliable.
- Involve technical personnel in the purchase of equipment and use their expertise.
- Make sure that the equipment, machinery and instruments that are planned to be purchased are not obsolete.

- Appoint qualified and experienced maintenance personnel. Unqualified and inexperienced persons may cause irreparable damage to expensive equipment and loss to the hospital.
- Provide necessary tools and test instruments to the maintenance personnel to work with. Although this may mean some initial investments, it will pay dividends later and offset the initial cost.
- Check if contract service for maintenance of certain key equipment works out cheaper. If it does, do not invest on costly tools and test instruments.
- Provide a thorough training to the maintenance personnel and operators of equipment and encourage them to update their knowledge through special programmes and field visits.
- Establish policies and procedures and ensure adequate records for all work and the materials used. Establish internal control measures for accounting, purchase, stores and issue.

➤POLICIES AND PROCEDURES

Written policies and procedures should be formulated. The following are some of the procedures that should be established and enforced.

- All maintenance requests should be approved and signed by the head of the department or unit.
- Maintenance request form should be in duplicate and should give details regarding the name and title of the person making the request, location, date, requested action, etc.
- The repair or work order should be in duplicate giving instructions to the repair man regarding time, nature of the job, materials to be used, etc.
- Instructions should be issued to the maintenance personnel not to undertake any job without specific work order signed by the maintenance chief or his assistant.
- A procedure for emergency repair should be established with a clear definition of what constitutes an emergency.
- There should be an effective system to sanction and control overtime.
- There should be a comprehensive preventive maintenance programme for the entire hospital.

CLINICAL (BIOMEDICAL) ENGINEERING

➤OVERVIEW

Clinical or biomedical engineering is one of the latest and most dynamic programmes in hospitals.

In today's high technology environment with a proliferation of advanced and complex medical equipment, clinical engineering has assumed great significance. This brief section has been included in this chapter to provide information about how to set-up and operate a clinical engineering programme as no hospital which has invested a small fortune on costly equipment can afford to remain without such a set-up.

The aim of clinical engineering programme is to provide technical expertise and management support to hospital administration, engineering department and the medical staff.

➤FUNCTIONS

The following are some of the important functions of the unit.

- ◆ Writing specifications for all the new equipment and machinery.
- ◆ Evaluating equipment and machinery. Evaluation must include, among other things, not only the initial cost of the equipment but also its operating cost. Quite often, the high maintenance and operating cost of the equipment turns out to be many times the initial cost.
- ◆ Inspection of incoming equipment and machinery and performing pre-acceptance checks before official acceptance and payment.
- ◆ Setting standards and ensuring their compliance.
- ◆ Organizing in-service training programmes and training for personnel in the clinical engineering department as well as all the user departments to use the equipment properly.
- ◆ Evaluating the need for new or replacement equipment and for major repairs.
- ◆ Advising and providing expertise to medical staff and administration.
- ◆ Organizing a planned maintenance programme for all equipment and attending to emergency breakdowns and repairs.
- ◆ Instituting an effective equipment control system.
- ◆ Establishing equipment inventory of all existing and incoming equipment.
- ◆ Maintaining work record and maintenance history record.
- ◆ Active involvement in the activities of the hospital's safety committee and checking safety hazards.

➤DESIGN, SPACE, FACILITIES AND UTILITIES

The productivity of the clinical engineering unit as a whole has a direct relationship with the quality of the facilities. When facilities are inadequate or poorly designed, much time is wasted in extra steps and in making makeshift arrangements. A congenial working environment influences the efficiency of employees. A pleasant and comfortable work area creates a favourable attitude and enhances productivity of personnel.

It is difficult to establish a standard space layout for a clinical engineering laboratory for all the hospitals. Many factors influence the space programme. Some of them are: the size of the hospital, the extent of sophistication of medical and other equipment, the size and training of personnel and the extent of contract service. Each hospital should tailor its structure and design to meet its individual needs. However, layouts are important and must meet the functional needs.

Generally, in a small hospital where there is only one biomedical technician, one room of 150 sq. ft. (15 ft. × 10 ft.) is adequate to perform the basic functions.

In a typically large layout, on one side of the room, there is a desk with wall-mounted cabinets or bookcases above it for reference books and catalogues, and adjacent to the desk, a file cabinet for manuals and records. Next to the cabinet is a lockable cabinet, shelves and drawers for storage of test instruments, spare parts and smaller instruments awaiting repair. Large instruments awaiting repair are stored at the end of the room on one side.

On the other side of the room, there are workbenches with storage drawers underneath for tools and more spare parts. This is the main work area. The bench is 12 ft. long so that the technician can work on more than one item at a time. A large laboratory type sink is provided at the end of the row for cleaning the instruments. The repair area is divided into mechanical repair area and electronics repair area. These two are separated and provided with more cabinets for storage. There is a small secured store room as well.

Irrespective of their size, all laboratories require the same facilities. The lab should be air-conditioned. Temperature control is necessary because electronic instruments and spare parts are temperature-sensitive and if they are not kept in an air-conditioned room, their operation may be affected. The lab also requires good lighting.

There should be adequate power supply for several equipment and instruments to be tested and operated at the same time. Both single and three-phase outlets are required for testing various types of instruments.

Provision must be made for both hot and cold water for cleaning the equipment. Some equipment on the mechanical side may require water for operation. Besides these facilities, the lab requires compressed air and vacuum system — either piped or furnished by a compressor and a vacuum pump — for cleaning equipment. Some equipment require compressed air and others vacuum for operation. Certain instruments and equipment require moisture-free cleaning which is done by using nitrogen.

►ORGANIZATION

The head of the clinical engineering unit is a clinical or biomedical engineer with a degree in clinical engineering technology, electronic technology, or electromechanical technology along with many years of hands-on experience. He should also be competent in handling all the work of the unit. A professional engineer who is registered in his engineering speciality is not suitable for this position unless he possesses an additional degree or diploma in biomedical engineering. Various colleges offer specialized courses. The Coimbatore Institute of Technology, for example, offers a one-year postgraduate diploma course in medical instrumentation technology with a B.E. or A.M.I.E. degree being a prerequisite for admission. If clinical engineering is one of the units of the engineering services, its chief reports to the chief engineer.

Other technical personnel in the unit are:

- ◆ Clinical or biomedical engineering technician who has a certificate or diploma in biomedical engineering or instrumentation. He is a highly skilled technician competent in doing the actual maintenance and repairs of most medical instrumentation.
- ◆ Medical equipment repairman who may be an electrician or a radio-TV repairman with some in-service training or hospital experience, or someone who has undergone some training specially designed for medical equipment. Although not an electronics technician, he has acquired skill and is capable of repairing many of the instruments or mechanical devices in the hospital.

➤INSTRUMENTATION EVALUATION

One of the primary responsibilities of the clinical engineering unit is instrumentation evaluation which involves documentation review as well as actual hardware inspection and testing. The unit plays a key role in assisting the administration and medical staff in selecting equipment and in determining the need for new or replacement equipment. When a new piece of equipment is received, the unit carries out the incoming inspection.

Incidence of operational failure or malfunctioning of newly received medical instruments is high in hospitals even in western countries. It ranges from 25 per cent to an alarming 50 per cent. It is not difficult to see the risk arising from placing expensive instruments in service without their proper inspection.

➤INSTRUMENT CONTROL

Every hospital must develop an instrument control system, making the clinical engineering unit responsible for developing and implementing procedures. The steps involved are:

- ◆ The equipment control system starts when a new or a replacement equipment is to be selected. A review must be made concerning the age, condition and utilization of existing equipment, and whether there is any maintenance problem and also how many units exist similar to the one requested.
- ◆ When a new piece of equipment is received, the following steps should be gone through.
- ◆ The equipment is received by the clinical engineering unit for a pre-acceptance check. The

unit should check that the operator and maintenance manuals have been received, the equipment is of acceptable standards, and it meets the manufacturer's and purchase specifications.

* Calibration is checked for accuracy, and a calibration procedure and records are initiated.
* Tags and labels are affixed to the equipment.
* Necessary entries in the inventory (assets register) are made, and maintenance records started.
* The warranty card is filled and sent to the manufacturer. In some cases, warranty benefits may be forfeited if the warranty card is not signed and returned. The pre-acceptance inspection is the best time for the maintenance personnel to study the warranty conditions and to be aware of maintenance procedures and time, and to ensure that the equipment performs properly.
* The equipment is then taken to the user department and operators are shown how to use it properly. It is important to remember that most malfunctions with hospital instrumentation result from operator error; therefore, this initial training is very important.
* The warranty period is an important part of the instrument control system. During this time, hospital personnel should learn about the maintenance of the equipment. One month prior to the expiry of the warranty period, the equipment should be checked to make sure that it functions properly. If any repairs and adjustments are to be made, they should be done under the warranty.
* The equipment should then undergo safety checks, preventive maintenance, calibration and routine maintenance. This can be done either under a service contract or by hospital personnel. All checks and work should be entered in the proper records.
* Replacement of equipment is the final step in the instrument control system. The system evaluates and projects the need for equipment replacement so that a budget provision can be made for it. However, technology is changing so rapidly that quite often a new equipment becomes obsolete even before its normal life is over.

Tagging and Labelling When a new piece of equipment is received, it should be identified with a tag. If this was not being done earlier, a programme should be started to tag every item in the inventory. The tag preferably with a metal plate affixed to it, should have the name of the hospital and a unique identification number. The number is important because in a hospital there may be several items of the same model of an equipment.

Besides this tag, other labels concerning calibration, preventive maintenance and safety checks should be affixed to the equipment. There should be a separate label for each one of these functions because different checks may be due at different times.

Instruction or operation manual should be available for each piece of equipment.

Equipment Inventory One of the first tasks of the clinical engineering unit is to establish an equipment inventory listing all the existing equipment and their location. The inventory should record all details regarding every piece of equipment such as nomenclature, manufacturer, model number, serial number, date of acquisition, cost, location, and a unique number.

Work Record In every hospital, especially in larger ones where employees are scattered all over and supervision is remote, workers waste time, either doing nothing or going slow at work. They spend a lot of time walking between jobs and the shop for reasons like fetching tools or parts. As a result, their productivity comes down to as low as 50 per cent. In order to maximize the efficiency of the personnel time and to see how they spend their time, a work record system should be established. The work record has other uses too, like determining to what department and equipment and to what extent the personnel time on a particular work should be charged, and also for studying the extent of instrument failure and the cost of down time.

Maintenance History Record The maintenance history record is the historical record of all the work done on a given equipment, and is an important tool of the instrument control system. It is recommended that the record is not computerized but is kept on a 5-inch by 8-inch card.

A review of the maintenance history record show what has gone wrong with an instrument and what part of it has failed frequently so that timely repairs or trouble shooting can be done, and necessary stock of spare parts maintained. All these measures help in reducing down time.

Preventive Maintenance Programme Preventive maintenance is the most effective method of ensuring that the instrumentation in a hospital functions reliably, accurately and safely. It is also the most economical way of maintaining the equipment. Pre-planned scheduling of maintenance work and of the technician's time is far better than neglecting the equipment till it breaks down and becomes inoperative — often when it is most needed — and then arrange panic repair.

Under preventive maintenance, equipment is cleaned, lubricated, adjusted and checked for wear and tear at predetermined intervals, and components that might cause breakdown or serious impairment are replaced. This is done on a scheduled rather than a user-demand basis. The result is a clearly improved performance along with a major reduction in economic costs which are certainly lower than the costs arising from demand repair work besides loss of revenue while the equipment is not functioning.

Calibration The establishment of a calibration system is one important component of the preventive maintenance programme. The function of calibration is to control the accuracy of all equipment used in a hospital. As a rule, all instruments should be calibrated at intervals established on the basis of the extent of use of the equipment, accuracy, expected life span of trouble-free service and wear. Whenever an instrument malfunctions or is damaged, or does not meet the manufacturers's specifications for accuracy, it should be serviced, repaired and/or calibrated. Most of the mechanical equipment requires calibration once in every six months.

Written calibration procedures should be prepared and used for calibrating all equipment in the hospital. Calibration of an instrument should be initiated during the week prior to its expiry date.

Safety It is the responsibility of the hospital to provide a safe environment to the patients and personnel. There is no guarantee that they will not be injured even when all precautions and safety measures are taken including compliance with statutory regulations. However, such precautions and compliance ensure that everything possible has been done to provide a safe environment in the hospital. It should be remembered that a contract for equipment maintenance by an outside agency does not absolve the hospital of any kind of responsibility or liability.

A hospital has three major kinds of responsibility as far as handling of equipment is concerned. These are as follows.

Warn It is the responsibility of the hospital and the manufacturer to identify equipment which, if improperly used or maintained, may cause injury to the patients or the operator of the equipment. For example, an equipment operated in excess of safe limits.

Educate The manufacturer should provide information on the proper use of equipment. Operators should then be properly trained in the correct use or operation of the equipment. This responsibility lies with the hospital.

Record It is not enough simply to check or inspect an equipment, or make a check mark. A record should be kept of all the work. From the legal point of view also, documentation comes in handy during investigations or trial proceedings.

ELECTRICAL SYSTEM

►OVERVIEW

Electrical energy is an essential source of power, the pivot around which almost every function of the hospital revolves, and the system is increasingly becoming more demanding, complex and crucial. This is partly because of the specialized medical and electronic equipment used for diagnosis, treatment and rehabilitation of the patients and partly because of the larger load of power needed in today's hospitals. In the electrical system, the main concern of the design team and the hospital engineer is the power distribution system which, it is rightly said, is the electrical lifeline of the hospital. There are also other concerns like an adequate and dependable supply. No less important, which many owners do not take seriously, are the electrical equipment and fixtures which should be of the best quality. They should conform to safety codes and regulations.

The highest dependability of electrical service, so essential for high quality patient care and functioning of the hospital on a 24-hour basis, is made possible by using high quality equipment, careful design, good construction and efficient operation in the hands of top quality engineers.

Emphasis should be on the design of an electrical system that will operate economically and provide for easy maintenance rather than the least possible installation cost.

Electrical system is one of the major costs of operation and many hospitals find themselves saddled with a heavy financial burden due to an inefficient system.

There should be an emergency generator to supply power to the essential and critical areas immediately if normal electrical service is interrupted. Under certain circumstances, even a ten-second power interruption may not be permissible. In such cases, an uninterruptible power supply (UPS) is the only answer.

➤ DESIGN

In the design of a hospital electrical power system, the major elements that should receive serious consideration are: safety, reliability, cost, voltage quality and ease of maintenance.

Safety Safety encompasses the protection of life and property, and continuity of hospital services. Protection of human life, both of patients and personnel, is of paramount importance. But safety of equipment is also essential. A faulty electrical system devoid of adequate safeguards may cause extensive damage to essential equipment and machinery which in turn may cause loss of service and a delayed return to normal operation because of repairs.

Designers often ignore the electrical residual current generation due to static electricity or earth leakage current which may endanger the lives of patients without even the doctors knowing about it.

Economics It is necessary to consider the cost of the total system and not just of its components. Cost of installation and cost of operation must be balanced. So also, cost and reliability. Cost of equipment is a major percentage of the initial cost of installation. Cost of operation is frequently not given adequate consideration with all attention focused on equipment and installation.

Voltage Stability of voltage is very important in the hospital power supply. It reflects the quality of electric power. With the increasing use of automated and electronic equipment in the hospital, voltage regulation under normal operation and abnormal changes in load merit special attention. Hospitals will do well to consider additional investment in the use of special devices in certain critical areas vulnerable to voltage fluctuations. This may prove more economical in the long run.

Maintenance A proper maintenance of electrical system is necessary for its safety and reliability. The system should be designed and streamlined in such a way as to make maintenance work easy and safe, and to enable routine maintenance and inspection without shutting down the essential hospital supply. For this, the use of a circuit arrangement providing an alternate source of power should be arranged.

Some Design Elements

- Nature and magnitude of load.
- Source of power.
- Cost of electric power system.
- Voltage levels.
- Circuit arrangement, whether radial, ring main, etc.
- Most economical size of substation.
- Secondary distribution.
- Combined light and power systems. This is usually most economical, but flicker problems must be watched.
- Means of voltage regulation, if required.
- Short circuit protection.
- Grounding.
- Overcurrent protection.
- Lightning protection.
- Proper metering of all circuits.
- Power factor correction.
- Antistatic electricity precautions.
- Isolation transformers for operation room complex.

➤PLANNING

The hospital electrical system calls for careful planning. Consideration should be given not only to immediate requirements but also to future needs both in terms of expansion and increased workload. Early in the planning stage, the engineer, the consultant or whoever is in charge of planning should closely work with the architect on the one hand and the hospital administrator, the medical staff and the other hospital staff on the other.

The views of the hospital staff should be given due consideration because they are the ones who operate and maintain equipment and facilities of the system. The medical staff should be consulted on crucial, specialized medical areas and the administrative staff, on the other areas such as vertical transportation and computer network. A third group whose views should be considered is the engineering and maintenance staff charged with keeping the system going.

The power distribution system should be adequate to meet the service reliability requirements of the hospital and yet it should be made as economical as possible. This requires that the power system engineer should plan the power distribution system on an all-inclusive basis. To do this effectively, he must constantly search for facts on which to base his decisions.

The right decisions in setting up the substation, electrical H.T. and L.T. panels, diesel generator room, load centres, etc. save not only installation but also the operational costs.

Finally, the engineer should be conversant with the various statutory regulations, codes and standards applicable to hospitals.

►DESIGN PROCEDURES

In planning and designing an electrical system, the engineer should observe the following basic steps and procedures.

◆ Work out the actual connected load and the demand load for the present as well as the future.
◆ Develop a site plan of the hospital plant showing the size and location of the present and future loads.
◆ Work out the essential loads and then determine the capacity of diesel generator (D.G.) sets.
◆ Establish voltage levels throughout the hospital plant.
◆ Determine the size, number and location of power centres.
◆ Determine the service reliability, select circuit arrangements required in each hospital area, and design the circuits to provide the reliability.
◆ Provide adequate power supply points to the various pieces of equipment.
◆ Provide protection against lightning, earth leakage current, short circuit current, undervoltage and overvoltage.
◆ Observe special precautions required for the hospital safety.
◆ Consider special lighting design for patient rooms and other areas.
◆ Provide necessary specifications. Specifications supplement the working drawings and furnish the information not shown in them by describing equipment and its functions. They also prescribe qualities of materials and workmanship required under the contract. They tie the entire job together.

►EMERGENCY GENERATORS

The public utility electricity supply in our country being inadequate, hospitals must expect to go without normal source of power frequently. However, hospitals are especially vulnerable to even a short term loss of electrical current because patient care depends on an uninterrupted power supply. Recognizing this, hospitals do provide for an alternate (emergency) source of power to serve essential portions of the hospital's distribution system. This emergency power should be reliable and is generally used for lighting and operating essential equipment.

The usual source of power used in hospitals for emergency purposes is the generator which is driven by an internal combustion engine operating on diesel oil.

The components of the alternate electrical system are the same as those in the normal distribution system except for the alternate source of supply and transfer switches. The engineer must determine the load to be placed on the system.

The load can be transferred from the normal source of power to the emergency source either manually or automatically. The automatic transfer switch should be capable of transferring the load within ten seconds of the power failure. However, in normal practice it may take longer than that. Manual operation may take a few minutes depending on where the operator is and how long he takes to reach the generator site.

At the planning and design stage, the engineer should specify which loads have to be transferred automatically and which loads manually, depending on the urgency or how critical is the function of the areas. For example, all lights may be connected to the automatic transfer load because patients and visitors often panic in the dark.

►UNINTERRUPTIBLE POWER SUPPLY (UPS)

Under certain circumstances, even a 10-second interruption of power supply in hospitals may prove life threatening. In the operating rooms where an open heart surgery or kidney transplant is in progress, or in the ICUs, cardiac catheterization lab or stress test laboratory, such interruptions may be fatal. In the non-medical areas, all computers depend on an uninterrupted supply of power. The engineer with the assistance of administrative and medical staff should identify critical areas that should be hooked on to the UPS.

Large hospitals often use advanced clinical and diagnostic equipment, sophisticated medical instruments and voltage-sensitive computers. These demand clean, computer grade power which is totally free from the momentary interruptions, transients, sags, surges and brownouts that often occur in the utility power lines. With the actual and projected rapid deterioration in commercial power, power conditioning is becoming a standard requirement for reliable operation of EDP and clinical equipment and is no longer viewed as a luxury in the same way as air-conditioning has become more of a necessity for them now.

Several power conditioning alternatives are available ranging from isolation transformers to the UPS. One of the choices is the motor-generator set which can protect the computer and other medical equipment from many power problems. However, it suffers from an inability to ride through longer commercial power outages.

UPS sets have over a period of time, established themselves as highly dependable pieces of equipment capable of supplying an output voltage of high quality (that expresses a degree of stability) and of harmonic neutralization (that implies total independence from the commercial supply source) and having a high rate of reliability.

Voltage sensitive medical equipment, computers and perhaps a small percentage of lighting are connected to UPS which typically has a battery back-up of 30 minutes. Advanced models of some medical equipment have a built-in battery back-up which provides uninterrupted power supply to keep the equipment operational for some time.

Technical consideration pertaining to electrical system are beyond the scope of this book. But as stated earlier, hospitals should utilize the expertise of design and specialist engineers and consultants in the early stages of planning itself so as to provide optimum electrical system facilities in their hospitals.

AIR-CONDITIONING SYSTEM

➤OVERVIEW

There is a need for conditioning air in particular areas of the hospital to achieve a certain level of temperature, humidity, filtration and circulation.

Operating rooms, labour-delivery suites, ICUs, nurseries, morgue and autopsy rooms are some of the areas which require air-conditioning. Sophisticated medical equipment is sensitive to temperature and humidity which affect its readings and performance. It should, therefore, be housed in air-conditioned rooms. The same applies to the central processing unit of the computer system.

Air-conditioning systems range from a battery of small self-contained window units to complex central air-conditioning systems. The type and selection of air-conditioning systems are a major consideration. The self-contained window units used in earlier times have now become obsolete and no longer serve the purpose. It has also become prohibitively expensive to use a battery of units, and thus a central air-conditioning system is now preferred for reasons of flexibility, application and maintenance.

The basic elements of most air-conditioning systems are air and water. Cooling can be achieved by utilizing chilled water, direct expansion refrigerant, or outside air when conditions are correct. The system of cooling using chilled water is discussed here.

The system components of controlled chilled water system are as follows.

- ◆ Compressor
- ◆ Condenser
- ◆ Chiller
- ◆ Chilled water pumps
- ◆ Condenser water pumps
- ◆ Cooling towers

 ◆ Expansion valves
 ◆ Chiller water pipes
 ◆ Air handling unit, cooling coil and its accessories such as ducting, prefillers, terminal filters, etc.
 ◆ Centrifugal fans for exhaust.

➤DESIGN CRITERIA FOR CERTAIN SPECIALTY AREAS

Planning and designing an air-conditioning system is best left to the AC engineers who are specialists in that field. But what is of particular significance to the facilities planning in hospitals and to the design team are the special air-conditioning needs of certain specialty areas of the hospital which require special attention. The following are some such areas.

Operating Rooms The main objectives of air-conditioning the operating rooms are to:
 ◆ introduce fresh, uncontaminated, dehumidified and cool air into the operating rooms;
 ◆ exhaust the air contaminated during surgery;
 ◆ provide working comfort for the surgical team; and
 ◆ prevent contamination from the adjacent areas.
Some of the design parameters considered for the operating rooms are:
 ◆ Temperature range should be between 23–24°C.
 ◆ Positive pressure to be maintained with respect to adjacent areas.
Special care should be taken to reduce air turbulence in the room by supplying air at velocity of no more than 200–250 fpm near the operating table. Low level exhaust air pick-up points should be placed at peripheral walls to have a scavenging effect.

It is recommended that individual air handling systems are provided for each operating room along with its changing rooms. This avoids cross contamination and has more flexibility. The relative humidity shall be maintained at 55 +/– 5%.

All items of equipment should be selected for low noise criterion. The noise levels should not exceed 52 dB at 125 Hz in occupied conditioned space.

Labour-Delivery Suites Unlike surgery, the labour-delivery suites operate on a 24-hour schedule. These suites are air-conditioned at the same temperature as the operating rooms, usually 23°C. Even in hospitals which are not centrally air-conditioned, labour delivery suites are one of the areas that are generally air-conditioned.

In the maternity department, nurseries are another area that requires air-conditioning. Full term nurseries are usually designed for 25°C temperature and 55–60 per cent relative humidity, and the premature nurseries for 28°C temperature and 65 per cent relative humidity.

Morgue and Autopsy Rooms As the morgue is generally designed to hold bodies in refrigerated body boxes, it poses no particular problem. However, the autopsy rooms have odour problems. A large air supply and an exhaust system should be provided. The exhaust should be carried above the roof, if possible. The objective of air-conditioning the autopsy rooms is to provide comfortable working conditions for doctors and other personnel. The air pressure in the autopsy rooms should be negative compared to the adjacent areas to overcome odour problems.

Central Sterilization and Supply Department The CSSD should be air-conditioned. In a modern CSSD, the decontamination (cleaning) area, the sterilizing area and the sterile storage room are physically separated. The sterilizing room generates a great amount of heat, and working in that condition is difficult. Special precautions should be taken regarding air-conditioning and ventilation if the sterilization work uses ethylene oxide.

In many hospitals, CSSD is either a part of the operating room (OR) complex or located adjacent to it, and if the OR complex is air-conditioned, in all probability, the CSSD too is. With washers, sterilizers and other equipment, the place always stays warm, and heat continues to be generated no matter how much air is removed from the area.

Equipment Rooms Places such a radiology, catheterization lab, etc. which house sophisticated and sensitive equipment like CT Scan and MRI, should be air-conditioned. This should be done after evaluating the accurate data requirements of all medical equipment located in these areas. These pieces of medical equipment are very sensitive to large deflections in temperature and humidity. Thus, special care should be taken to control these two factors.

➤SUPPLY TO OPERATING ROOMS, ETC. AND RETURN AIR/EXHAUST AIR

Air supply to operating rooms and labour-delivery suites should be from ceiling outlets near the centre of the work area. This will effectively control air movement. Exhaust or return air should be from near the floor level. Great care should be exercised in the design to ensure that turbulence and other factors of air movement do not cause particulates to fall into the wound site. Special procedures such as organ transplants may require "laminar air flow" or such other special designs.

Air supply to nurseries, birthing rooms, LDRP suites and rooms used for invasive procedures should be at or near the ceiling. Exhaust or return air inlets should be near the floor level.

The place used for administering inhalation anaesthesia should be provided with a scavenging system to vent waste gases. The gas collecting system should not interfere with the patient's respiratory system. Gases from scavenging should be exhausted directly to the outside atmosphere.

General Area

The corridors should have adequate ventilation. Individual private rooms should have independent fan coil units and thermostatic control for flexibility. All toilets should be exhausted for odour with a central exhaust system.

WATER SUPPLY AND SANITARY SYSTEM

➤WATER SUPPLY

Water is one of the critically important utilities in a hospital, yet its supply is often taken for granted. Much of the hospital's engineering service is concerned with installing, repairing, and maintaining the systems that deliver utilities and services — water being one of them — in a functional, continuing and safe manner.

A hospital requires a copious supply of water. While the nation's population and the demand for water are increasing, its water supply is diminishing. This puts a tremendous pressure on the ability of administrators and hospital engineers to supply pure water to hospitals. Every effort should be made to conserve this precious but fast diminishing resource.

Hospitals should as far as possible rely on public water supply system for the necessary quantity and quality of their water supply. This may be supplemented by their own water supply. Some hospitals have on their premises private wells or bore wells which they can use to augment their water supply.

Designing a water supply and sanitary system for a hospital is a complex task which calls for the expertise of competent, experienced and specialist engineers. Since most of the service lines are concealed, faulty design and installation or any compromise in the quality of materials will lead to disastrous results difficult to rectify later. Therefore, planning and installation of this system are of the utmost importance. Defective installation would affect not only the functioning of the hospital but also its hygienic conditions.

Two important components of the system are:

* water supply distribution network, and
* sewage disposal.

The design should allow flexibility to recycle waste water, if need be, for reuse as A.C. cooling tower make-up water or in gardening and toilet flushing.

➤ SOURCES OF WATER SUPPLY

The primary source of water supply to hospitals is generally the public utility supply system. Invariably water is in short supply in most cities. To meet the high and ever growing demand for water, the following alternative sources of water supply may be considered:

- Borewells.
- Tanker supply.
- Recycled water. The treated waste water from wash basin, shower, laundry, etc. can be used in W.C. flushing, landscaping, etc.

To ensure a continuous supply of water, adequate storage capacity of underground sumps and overhead tanks should be provided.

➤ DESIGN ELEMENTS

In designing the hospital water supply system, the major elements that merit attention are as follows.

- Continuous and reliable supply of water
- Quality of water
- Proper distribution network
- Cost
- Ease of maintenance and operation

Continuous and Reliable Supply of Water

The designer should work out accurately the water requirements of the hospital for various purposes. Armed with this and the information regarding the sources and the quantum of water available from each of them, he should proceed with the design of underground and overhead water tanks of adequate capacities.

Reliability of water supply depends largely on: (a) incoming water supply, (b) equipment and system design, and (c) maintenance and operation.

Quality of Water

The quality of water supplied to the hospital affects virtually every aspect of hospital operations. It is, therefore, essential that suitable water of microbiological quality is provided for drinking, laboratory procedures and solutions used in medical and surgical treatment. Chemically acceptable water is essential for the operation of equipment, for laboratory tests and dietary purposes. A regular surveillance programme should be instituted. This normally includes evaluation of the source of supply, equipment and distribution system, and routine microbial and chemical analyses of water.

The engineer should have samples of water collected routinely and have them sent for analysis. In addition to potability, water should be tested for hardness and iron. Hard water is detrimental to equipment and increases operating costs.

Routine water processing programmes carried out in hospitals include chemical treatment of water, deionization, distillation, filtration and sterilization. The engineer should be familiar with hospital operations and know where specially treated, deionized or distilled water is used and when it is necessary to provide it. He should realize that deionizing process removes only the ionizable contaminants and not the bacteria or other organics and inorganics, and that the deionized water often becomes heavily contaminated with bacteria that grow on the resins. Distillation provides water of the highest purity. However, distilled water is not sterile. Sterile water is produced by processing water in sterilizers or autoclaves. Hemodialysis requires specially treated water.

The quality of water depends largely on two factors:

1. the quality of raw water supplied to the hospital, and
2. level of quality required for various purposes.

Since there are varied sources of raw water, it is generally not possible to control its quality. It is thus essential to first analyse the quality of water obtained from various sources and then recommend the treatment process before use.

Distribution Network

For the proper utilization of water, the distribution system should be designed taking into consideration the pressure and quantum of water required at various outlets. Due care must be taken while selecting distribution equipment and pipes because of the risk of contamination and corrosion. The water supply distribution system is often designed using the hydropneumatic system for achieving a uniform pressure at all the outlet points.

Great care should be exercised in protecting hospital's potable water against contamination which can result from a poorly designed and installed plumbing system.

The potable water supply system should not be connected with other piping systems, nor should it be connected with fixtures having submerged inlets. This could cause contamination.

➤PLANNING

The hospital water supply and sanitary system calls for a high degree of careful planning. The design should take into consideration not only the present needs but also the future requirements.

Adequate provision should be made for future expansion. The location and size of plant rooms, service ducts, etc. play a very important role in designing an economical and convenient system. Attention should be paid to the maintenance and operation of the system. In large hospitals, the service floor concept may be considered for running all sanitary and water supply lines horizontally below the toilets and terminate them in a common vertical duct. Careful interaction with the architect and other service engineers is necessary before finalizing the layout for water supply and sanitary system.

➤HOT WATER SYSTEM

Hot water supply is one of the prime requirements in any hospital. It is required in patient bathrooms, kitchen, laboratory, laundry, CSSD, etc. If the hospital is located in a cold climatic area, it is essential to provide hot water in all the toilets and wash basins. Hot water is supplied through the central distribution system and is usually generated using the oil-fired hot water (HSD) generators. The temperature of hot water ranges between 55–60°C, and stored in an insulated, closed pressure hot water mixing tank.

The ideal location for boilers, mixing tank and associated pumps and equipment is either the pump room or a separate room adjoining it. The exhaust flue gas from the hot water boiler should be taken above the building as per statutory regulations and discharged into the atmosphere. The complete hot water system should be distributed to various utility outlets using insulated G.I. pipes. To eliminate wastage and get immediate hot water when the tap is opened, a hot water circulation pump may be necessary. The hot water pressure in the pipe line should be maintained in the same way as in the cold water supply by connecting the hot water supply to the hydropneumatic system.

➤STEAM

Steam is required in the kitchen, laundry, CSSD and other sterilization areas. It is generated at 8–10 kg/sq. cm pressure using the oil-fired steam boilers. The requirement of steam in the various areas will be at different pressure. Therefore, pressure reducing stations with headers should be provided to tap the required steam. Provision has to be made to recover the condensate and conserve waste heat. The steam boilers should be located next to the hot water boilers. If the steam requirement is not much, it is possible to have only one combined steam boiler supplying both steam and hot water.

➤DRAINAGE SYSTEM

The drainage system of the hospital should be simple, effective, economical and serviceable. It should be designed keeping in view the kind of septic and toxic waste envisaged which in turn needs to be effectively disposed off. It is advisable to adopt a double stack system in which separate stacks are provided for the collection of waste and soil from the toilets and other areas. Pipe lines should run with sufficient slopes so that the sewage could be conveyed to inspection chambers by gravity.

Provision should also be made to terminate the collected sewage in the municipal sewer line via a battery of inspection chambers or manholes. Where it is not possible to do so, alternate arrangements like sewage treatment plants should be made. It should be ensured that the effluents from the treatment plants are further treated to acceptable standards before their utilization in landscaping or recirculation.

For sewage disposal, it is preferable to use PVC pipes as they are easy to install and repair, and also allow for smooth flow of sewage.

➤SEWAGE TREATMENT PLANT

The objective of the sewage treatment plant is to stabilize the decomposable organic matter present in sewage to produce effluents and sludge. These can then be disposed of in the environment without causing any health hazards or nuisance. The treated sewage water can also be reused for various purposes like gardening, landscaping, flushing of W.C.s, A.C. cooling towers, etc.

CENTRALIZED MEDICAL GAS SYSTEM

➤OVERVIEW

Gases administered to patients are called medical gases. Centralized medical gas system is increasingly becoming an essential requirement in hospitals in the same way that other essential services and utilities such as electricity, water and air-conditioning are.

The centralized medical gas system provides an efficient, economic and dependable medical life support network that supplies medical gases (oxygen and nitrous oxide), vacuum (suction) and compressed air to the operating and special procedure rooms, ICUs and patient floors. The system makes for better patient care in all the areas of the hospital.

The following are the important components of the centralized medical gas system.

* Source of supply. It comprises the central supply room with control equipment or panel.
* Distribution system. It is a system of piping that extends to the points in the hospital where medical gases are required and used.

* Point-of-use delivery connections. These are suitable station outlet valves and pendants at the use points (Plate 11: Pictures 71,72).
* Monitoring and control equipment and alarms.

➤CENTRAL SUPPLY ROOM: MANIFOLD ROOM

The central supply room usually consists of a cylinder manifold and a control panel. The manifold may be as small as two banks of two cylinders each, or as large as two banks of twenty cylinders each. The control panel consists of primary and secondary pressure regulators to ensure delivery of gas to the pipeline at the required pressure provided there is gas in the cylinders. When the cylinders become empty or gas is not delivered to pipeline due to any malfunctioning of the regulators, a warning lamp on the control panel illuminates. The control panel also has pressure gauges to indicate pressure in the cylinders.

The vacuum unit comprises of a vacuum pump with an electric motor. A cylindrical reservoir tank stabilizes the pressure of the pipeline system at all the outlet points. The motor has a negative pressure switch for automatic start and stop. Vacuum pumps are duplexed.

The compressed air unit consists of a compressor with electric motor, after cooler, air receiver and air dryer. The centralized compressed air system instantly provides compressed air through the pipeline wherever it is required. The compressors are also duplexed.

The primary supply is that part of the equipment which actually supplies the system. A secondary source automatically supplies the system when the primary supply becomes exhausted. The primary and secondary equipment provide the normal operating supply. When the operating supply fails or is exhausted, a reserve supply automatically takes over.

In the event of the control panels of oxygen and nitrous oxide breaking down, an emergency kit ensures gas supply through the pipes. The emergency kit comprising a regulator and high pressure tubing is connected to a bulk cylinder and the gas is fed directly to the pipeline through a service outlet. In case vacuum and air supply systems fail, a standby motor-cum-pump is used to provide uninterrupted supply.

Usually, half of the total daily consumption of oxygen and nitrous oxide is kept in the manifold room as reserve supply. In the case of vacuum and air, there is always a reserve capacity in the system itself which lasts for a while.

Oxygen and nitrous oxide should be stored separately from flammable gases and liquids. The storage location should be free of combustible materials such as paper, plastic materials, cardboard, etc. When the quantity of gas stored exceeds 2,000 cubic feet, the storage area should ideally be outside the building. However, if it is situated inside, the room should have at least one-hour fire resistance construction. Additionally, there should be an automatic fire extinguishing system.

The size of the manifold room depends upon the systems it houses. The dimensions should be decided keeping in mind space for all equipment and for stocking full and empty bulk cylinders, Adequate space should be left around the equipment for servicing.

➤ADVANTAGES OF CENTRALIZED MEDICAL GAS SYSTEM

The following are the advantages of centralized medical gas distribution system.[*]

For the Patients
- No distressing sight of oxygen cylinders at the bed side.
- Elimination of the irritating noise from the movement of cylinders in and around the hospital.
- Protection of sterile areas from contamination caused by the use and movement of cylinders.
- Uninterrupted and clean supply at desired locations.

For Hospital Staff
- Instant availability of gas on taps.
- Clean, safe and reliable delivery of gases
- Continuous flow of gases whenever and wherever required.
- Minimal accident hazards due to mishandling of cylinders.

For the Hospital Administrator
- Easy purchase of gases in bulk quantities at favourable terms.
- Economy on purchases of cylinders.
- Fewer breakages.
- Rationalization in ordering, storing, and transporting a wide variety and sizes of gas cylinders.
- Minimum damage to building due to handling of cylinders.

➤OXYGEN AND NITROUS OXIDE

In view of the increased demand for oxygen in hospitals, centralized distribution has become necessary. Oxygen for hospital use may be supplied from a bulk tank or from cylinders, It may be stored as a gas or a liquid; the oxygen containers may be stationary or movable. The bulk oxygen system is defined as an assemblage of equipment such as oxygen storage containers, pressure regulators, safety devices, vaporizers, manifolds and interconnecting piping with a storage capacity of more than 20,000 cubic feet of oxygen including reserves. The system terminates at the point where oxygen first enters the supply line at service pressure.

---

[*]The authors are indebted to Indian Oxygen Limited for the following material which is adapted from one of their brochures.

Nitrous oxide is not as widely used as oxygen. The largest users are the operating and special procedure rooms.

►VACUUM (SUCTION)

Piped medical-surgical vacuum system is used for patient draining, aspiration and suction. Vacuum is also used in the hospital laboratory. Like oxygen and nitrous oxide, the medical-surgical vacuum is distributed through a network of pipes. There is a central vacuum supply system with control equipment, an alarm signalling system and a network of piping extending to areas in the hospital where patient suction is required. The piping terminates with outlet valves at each user point.

The central vacuum system (Picture 74) consists of two (or more) vacuum pumps which operate either simultaneously or alternately, depending on demand. Each pump should be capable of maintaining 75 per cent of the peak calculated demand. The pumps should in normal course alternate automatically. This may be done by an additional circuit board. Also, if the first pump is incapable of maintaining the required minimum vacuum, the alternate pump should have a provision to get activated automatically. Normally, the pumps go on and off even if an automatic device is not incorporated.

➢ **PICTURE 73** Medical Compressed Air Plant

The pumps should be equipped with motor starting device and overload protection. As a measure of protection, disconnect switches should be provided in the electrical circuit ahead of the motor starting devices. The failure or shut off of one vacuum pump should not affect the operation of others.

➤ COMPRESSED AIR

The demand for compressed air in hospitals has increased greatly. Air is used for both medical and non-medical purposes. The medical uses include laboratory work, inhalation therapy equipment and powering of surgical tools, for example, powering of pneumatic drills in orthopaedic surgery and in the dental department. A word of caution. Drills may require 90–120 psi pressure whereas the pressure in the piped system may be only 60 psi.

In the non-medical areas, compressed air is used for maintenance of tools and equipment, and also in the clinical engineering department. Compressed air used for medical-surgical purposes should be free of dust and moisture. This is done by means of oil filters, dust filters, moisture separater and the 'dryer' at the equipment level. A separate dryer equipment is provided to the air compressor system. This absorbs moisture in the air before delivering it to the pipeline. If the compressed air is not used properly, it can damage equipment and contaminate chemicals, foods and drugs.

In the compressed air system (Picture 73), the standard items of equipment and accessories are:

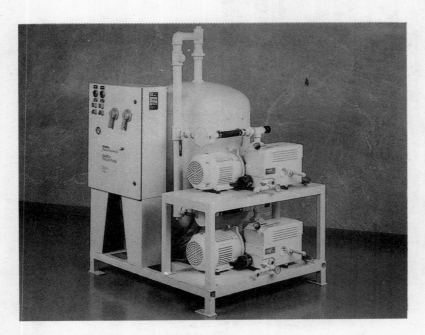

➤ **PICTURE 74** Medical Vacuum Plant

compressors, shutoff valves on tank outlets, valves on drains, pressure gauges, safety valves, check valves, pressure switches, inlet filter-mufflers, and automatic electrical controls.

All personnel should be warned that air under pressure, as from a hose line, should not be allowed to touch any part of the human body. Minute particles propelled by the airstream can cause painful puncture wounds. Lives have been lost when compressed air was used in practical jokes like when air under pressure is introduced to any body cavity like the anus.

➤PIPES

Pipeline should be seamless type, of high quality copper tubing for medical use with Lloyd's approval. Embedded piping should be protected against physical damage and corrosion. Exposed oxygen pipelines should not be installed in such areas as kitchen, laundry and rooms where combustible materials are stored. The gas content of pipeline should be readily identifiable by colour of the tubing and by labels that bear the name of the gas.

Before they are erected, all pipes, tubes, valves, and fittings should be cleaned thoroughly and washed with tetra chloride, or a hot solution of sodium carbonate or trisodium phosphate mixed with water, and rinsed thoroughly with warm water to free them from oil, grease, dust or other combustible materials. After installation of the pipes but before the air line is connected to the compressor on the one hand and the outlets on the other, the pipes should be blown clear using oil-free air or nitrogen. When the whole system is in place, it should be subjected to a test pressure of 150 psi or one and a half times the working pressure for 24 hours by means of oil-free air or nitrogen to check if it could withstand the required pressure, and also for any leakage. This test called "pressure lock" is done section by section.

➤PRECAUTIONS AND CONTROLS

The following precautions and controls are necessary in various areas of the hospital served by centralized medical gas system.

* High quality regulators and other gas flow control devices approved by the Bureau of Indian Standards should be used to reduce the pressure of every cylinder used for medical purposes. All these devices should have connections designed for attachment only to the appropriate gas cylinders.
* The main supply line should be provided with a shutoff valve for use in an emergency. This should be located for easy access.
* Each riser from the main line should be provided with a shutoff valve adjacent to the riser connection.
* There should be a shutoff valve at each anaesthetizing location on each oxygen and nitrous

oxide line. Valves should be readily accessible for use in an emergency. The valves should be safeguarded against physical damage and should bear a sign such as "Oxygen: Do Not Close." This will prevent tampering or inadvertent closing.

- Shutoff valves accessible to unauthorized personnel should be kept in boxes with view windows in a manner that can be operated manually. The box should be labelled: "Caution. Oxygen Valves. Do Not Close Except In Emergency. This Valve Controls Oxygen To...."

- For continuous surveillance and to monitor the pressure of various gases at different areas of medical pipeline distribution system, gauges and alarms should be installed.

- A written procedure manual on prevention of fire and fire fighting should be developed and circulated among staff.

- Suitable flow meters which can provide flowrate as per the desired user-settings should be installed for direct use on the pipeline pressure. For suction, pressure controls should be used at each terminal outlet to adjust the suction pressure at the required level.

- Regulators of proper size should be installed in front of each piece of gas-using equipment to ensure proper control of pressure and flow.

- The manifold room should have at least one-hour fire resistant construction and door and an automatic fire extinguishing system.

- For routine maintenance and repair, each vacuum pump should be provided with a shutoff valve to isolate it from the vacuum system.

- Wall outlets for vacuum, supplied by the manufacturers, should be legibly marked "Vacuum" or "Suction" so that they can be easily identified when they are disassembled for hooking up to the piping system.

- Compressed air should be compatible to be connected to ventilators. Oxygen and compressed air in the newborn nursery are to be attached to the ventilators.

- All piping should be colour coded as per the international code for each gas. Additionally, the pipes should be labelled with metal tags or stencils.

- Colour and valve schedules should be given to the hospital authorities for permanent record and reference.

- A procedure manual must be provided to the hospital authorities both for the operation and maintenance of medical gas system and for the prevention of fire hazards.

➤ALARM SYSTEM

Two kinds of alarm are usually incorporated in the centralized medical gas system. The more important one (Plate 12: Picture 75) monitors the pressure of various gases at different areas of the distribution system. If abnormal pressure is sensed, the system sets off an alarm: the normal green signal goes off and the red warning signal glows with audible alarm until line pressure returns to the normal. Alarms should be located in working areas of those who use and maintain the system.

However, it is quite possible that these areas, such as operating rooms, may not be manned all the time. Auxiliary alarm signals should thus be installed in additional places where there is a 24-hour personnel coverage, for example, in telephone operator's room, security office and emergency department.

The other alarm, called the remote signal lamp, should preferably be both visual and audible. This is generally not available in India. What is available is only visual: the lamp lights up when either of the banks of cylinders in the manifold room becomes empty. In this case also, there can be auxiliary signals outside the manifold room in places where there is 24-hour attendance.

The remote signal lamp provides a warning signal. No immediate action is necessary because when one bank is empty, the other bank takes over and delivers gas without interruption.

Table 8.1 shows various locations in the hospital and the number of outlets required for oxygen, suction and air.

$$\boxed{\textbf{TABLE 8.1}}$$

Medical Gas Outlets for Oxygen, Vacuum (Suction) and Air

Location	Oxygen	Vacuum	Air
Patient rooms in medical, surgical, obstetric and paediatric wards	x[a]	x[a]	[d]
Examination/treatment rooms in nursing units	x	x	—
Intensive care, coronary care unit	xx[c]	xxx	x
Nursing, nursery	x	x	x
General operating room, emergency trauma room	xx	xxx	xx
Cystoscopic and special procedure rooms	x	xxx	xx
Recovery room	x[b]	xxx	x
Labour room	x[b]	x	x
Delivery & birthing room	xx	xxx	x
Emergency department	x[b]	x[b]	x[b]
Anaesthesia workroom	x	—	x
Autopsy	—	x	x

[a]One outlet accessible to each bed (one outlet may serve two beds)

[b]Separate outlet for each bed.

[c]Two outlets for each bed.

[d]One outlet for air is required in the paediatric unit only.

Nitrous oxide which is not shown here is required in operating and special procedure rooms.

➤ SAFE HANDLING OF GASES

The following are some of the general practices recommended for the safe handling of gases. More specific rules relating to handling, use and storage should be provided by the supplier.

- Ensure that only experienced and properly instructed persons handle compressed gases.
- Observe all regulations and statutory requirements concerning storage of cylinders.
- Do not remove or deface labels provided by the supplier for the identification of the cylinder contents.
- Ascertain the identity of the gas before using it.
- Know the properties and hazards associated with each gas before using it.
- Establish and be familiar with plans to cover any emergency situation that might arise.
- Never lift a cylinder by the cap or guard; use a cylinder trolley for transporting.

TELECOMMUNICATION SYSTEM

Over the last few years, communication systems in hospitals have made great advances with a variety of new and sophisticated features being added every year leading to the fast obsolescence of existing ones. The communication system in hospitals encompasses intra- and inter-departmental intercom, telephone, paging (overhead and wireless), nurses' call, computerized visual display terminals, television, cable television and closed circuit television (CCTV), alarm system, central dictation, monitoring, and so on.

Instantaneous and reliable communication is crucial to hospitals. A slow response or missed communication can be life threatening. For example, a delay in issuing a cardiac emergency call or failure to reach a specialist on time may endanger the life of a patient. Poor communication can result in overall organizational inefficiency. A tardy response or unfriendly attitude of the telephone operator may establish a negative image in the minds of the public who may lose confidence in the hospital. Since the telephone operator is frequently the first contact of the caller with the hospital, how she responds to his calls sets the overall first impression of the hospital for him.

➤ TELEPHONE SYSTEM

Advanced telecommunication technology of the present day offers a vastly improved and sophisticated telephone equipment with never before features and capabilities. Advanced systems are now available in which a single instrument acts as a multi-button phone. Most telephone systems have flexible circuits that allow telephone calls to be transferred to another areas as, for example, to the admitting office. In smaller hospitals, this eliminates the need for a telephone operator during the night. Some other new features are: touchtone dialling, call pick up, call forwarding, conference capability, transferability of incoming and outgoing calls and direct dialling.

Interconnecting telephones should be provided for all departments and sections including operating rooms, ICUs, nurses' stations, offices, maintenance, housekeeping and elevators. A telephone service outlet should be provided midway in the elevator shaft for connecting the telephone in the elevator. All intercom telephones should be dial type which permit intercommunication without calling the hospital switchboard.

Many hospitals now provide telephones in the patient rooms. Patients can make long distance calls directly with the facility of remote metering or transmission to a computer so that automatic charging of the concerned patient is accomplished. The practice of installing jacks in all patient areas for use of plug-in telephones is now considered obsolete. As far as possible, telephones should be installed in patient rooms, especially in private and special rooms.

Public (pay) telephones should be provided at convenient locations for the outpatients, visitors and staff, particularly in the outpatient area, inpatient areas, emergency department, near the labour-delivery suites and in the fathers' waiting room, if there is one. Pay phones leave hospital switchboard free for patient care and official use. In addition to public telephones there should be a convenient room where visitors, outpatients and hospital personnel can make assisted STD and ISD calls. Conduits should be provided to facilitate installation of telephones, wherever necessary, keeping the future needs in mind.

The Integrated Service Digital Network (ISDN) which is poised to take the business world by storm will revolutionize our lives. Digital switching system which is an advanced computer by itself will be able to handle voice, data, text and image transmission — all on the same telephone line. In other words, telephone, computer, printer, FAX and almost anything else that is electronic can be plugged into a single telephone line to provide an integrated communication system.

➤ NURSE CALL SYSTEM

The nurse call system ranges from the simplest — a mere visual signal system — to the most complex and sophisticated computer controlled system with a visual and audio indicator, two-way voice communication and advanced facilities for management information. It can be linked to the panic button in the patient's bathroom, code blue alarm system and the fire alarm system.

The feature common to all the systems is the switch or button provided at the patient's bedside (Picture 76) which, when activated, registers the call at the nurses' station. In the traditional system, a push button with a flexible cord is provided to each bed. The signal can be switched off only at the bedside. A pilot light is placed over each bed if there are more than one bed in the room. There is a pilot light over the door of the room and a central light panel at the nurses' station.

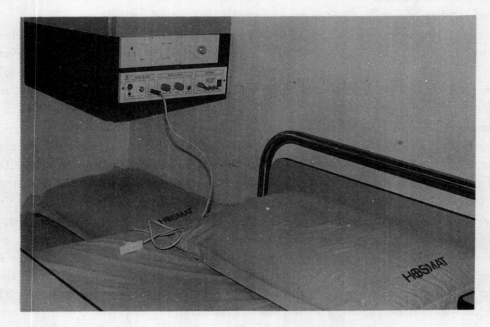

➤ **PICTURE 76** Nurse Call System with detachable bedside cord

A central monitoring panel is provided in the nursing director's office.

The following are the features of the advanced computerized nurse call system.

- Call is registered by the patient.
- Call is acknowledged by the nurse.
- Call is attended to by the nurse. (Nurse switches off the call signal in the room.)
- In the event of delay as programmed by the response time, the signal light flashes.
- When the delay becomes longer, the flash rate increases progressively.
- The signal is both audible and visible.
- There is a provision for two-way voice communication between the patient and the nurse station, This can be programmed in such a way that only the nurse can initiate the voice communication and not the patient.

The system has the following components.

- Panel in the patient room (Picture 77)
- Patient room door panel (Plate 12: Picture 78)
- Main nurse station panel (Plate 12: Picture 79)
- Monitoring panel
- Computer interface
- Software

PLATE 12

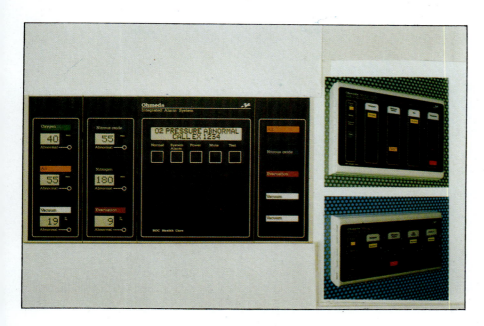

➢ **PICTURE 75**   Centralized Medical Gas: Alarm System

➢ **PICTURE 78**   Display Outside the Patient Room Door

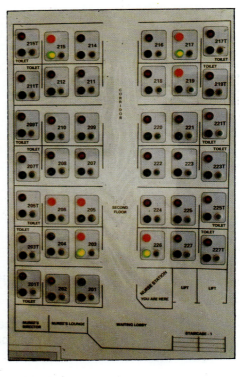

➢ **PICTURE 79** Nurse Station Main Panel

PLATE 13

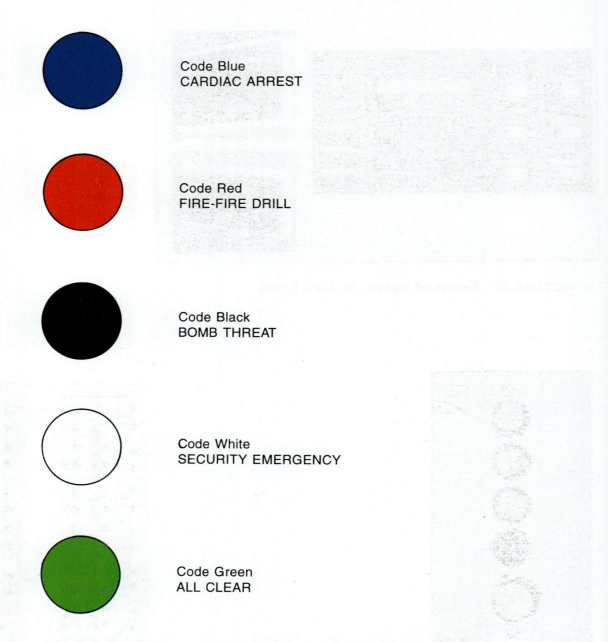

Code Blue
CARDIAC ARREST

Code Red
FIRE-FIRE DRILL

Code Black
BOMB THREAT

Code White
SECURITY EMERGENCY

Code Green
ALL CLEAR

Disaster will be announced as "Code 10-Disaster" over the Public Address System.

"Code Red" is also called "Dr Red" and "Code 10-Disaster" "Dr Major" depending on local preferences.

➢ **PICTURE 80** Emergency Codes

♦ Alert panel
♦ Computerized voice interface

In the patient room panel (Picture 77), there is a red push button to register the call, a red indicator light to show the call is registered, a green indicator light to show that the nurse has acknowledged the patient's call, and a switch to cancel the call.

> **PICTURE 77**   Bedside Integrated Panel

In the panel outside the patient's room, (Plate 12: Picture 78), there is a red indicator to show that the patient requires attention and a green indicator to show that the call has been acknowledged by the nurse. When only the green indicator is lighted up, it shows that the nurse is attending on the patient. This eliminates the need for more than one nurse responding to the call. The panel is provided with a green switch for the nurse to acknowledge the call at the door, if necessary.

At the nurses' station, the panel (Plate 12: Picture 79) has a layout plan of the ward and the beds with bed numbers, red indicator, a green indicator and an acknowledgement switch. The panel clearly shows which bed requires attention. The digital indicator can be added to show the sequence of call registration. An audio beeper is provided to attract the nurse's attention. Delay in attending to the patient results in the signal light flashing and the beeper sounding faster and faster.

The monitoring panel is intended for the supervisory staff to monitor delayed response and improve efficiency. The management information support software monitors the response pattern of nurses throughout the hospital during the day, week and month.

Some hospitals have separate call buttons labelled "Code Blue" or "Cardiac Arrest" on the patient room headwalls. Some others have them at different locations throughout the hospital. When activated, these buttons send out alarm signals to the nurses' station, nursing office and the central telephone room. The signal can be programmed to give a voice communication giving the location of cardiac arrest. The telephone operator simultaneously goes on the public address system alerting the code blue team to respond to the call.

➤ DICTATION AND CENTRAL TRANSCRIPTION SYSTEM

Remote dictation service is set up to allow the doctors to dictate reports from any part of the hospital where a dictating equipment is provided, usually from the operating rooms, ICUs, patient floors, emergency room and the doctors' chart completion room in the medical record department. Modern telephone systems have the capability to be used for dictation as they can now be interconnected with centralized dictation where reports are recorded on tapes and later transcribed by medical secretaries. The system allows any phone in the hospital to dial a code and dictate to the central room. It enables the doctor to start, stop, play back and correct his dictation.

➤ PAGING SYSTEM

Two kinds of paging systems are generally used in hospitals: overhead or loudspeaker and radio paging. These two systems are used largely for different purposes.

The loudspeaker system normally includes a microphone feeding through an amplifier to a number of speakers located hospitalwide. This is used for paging for doctors and other staff, announce code blue emergency, and for issuing instructions in case of fire, bomb threat, disaster or other catastrophes, and for announcing simulated fire, bomb and disaster drills.

Obtaining a license from the Government of India to operate a radio paging system was extremely difficult in the past, but things have changed now. Many leading companies, including international ones like Motorola and Hutchison, are in the field vying with each other for a share in what promises to be booming market.

Radio paging is the system largely used for locating individuals or for transmitting messages to them. The basic components of this system are: a microphone usually located at the telephone switchboard, radio transmitter placed at any suitable location, antenna system which radiates in all directions but usually in a loop around the area of desired coverage, and receivers which are small, lightweight, portable receivers tuned and set to receive messages on only one wave length. These are checked out to and carried by individuals — largely doctors and officers of the hospital who are on the move. Both beep and vibration types are available. Some systems transmit audible signals only, while others transmit voice messages. In the earlier days, all systems in use transmitted messages one way only, that is, no talk-back feature. However, the advanced models now have two-way voice transmission capabilities.

Paging transmissions are normally controlled by the telephone operator. In its simpler form, a code number or a telephone number is transmitted to an individual in a remote place. The individual uses a phone to contact the hospital or any other desired number.

There are now firms which set up and sell paging services in larger cities. Hospitals may utilize their services instead of setting up their own infrastructure and systems which is an investment many of them can ill afford.

As the radio paging system is in a state of flux and many new features and sophistication are in the pipeline, hospitals are advised to tread cautiously before they decide on the system.

➤PUBLIC ADDRESS SYSTEM AND PIPED MUSIC

Public address system or wired or overhead paging is also meant as an emergency backup to the radio paging system in addition to the other functions described earlier. It is invaluable in making announcements to a large number of people in large assembly halls, and other strategic locations. The system should be designed for zone paging so that information can be transmitted to selected places without disturbing patients and hospital staff in other areas. Suitable background music can be piped throughout the hospital during selected hours. Many Christian institutions broadcast devotional songs and worship programmes over the public address system. Individual speakers in patient rooms give patients the option to switch the transmission off if they do not wish to listen.

Where piped music, public address system and television systems are bundled together, a cut-in feature for announcements should be included. Announcements may be made from several places depending on the nature of announcements, the CEO's office being one of them.

➤TELEVISION AND CLOSED CIRCUIT TELEVISION

Television, once considered a luxury, has now become commonplace as a source of news and as a means of entertainment in homes and other places. Many hospitals also provide for patients' entertainment, information and educational and health programmes by way of television, video and closed circuit television.

Star TV and cable TVs provide a variety of entertainment, sports and educational programmes. Many hospitals provide these avenues of entertainment to their patients.

A television system becomes a closed circuit television (CCTV) when the hospital generates its own video programmes and feeds it into the distribution system.

In some hospitals, closed circuit television is used in the operating rooms to transmit information to consulting doctors for advice and to residents and students for teaching purposes. It is also used in cardiac catheterization procedures for displaying x-ray image of the catheter position. In advanced countries, CCTV is used by the nurses to view children in isolation, and for visitor-patient two-way viewing. Inclusion of audio facility provides an opportunity for children to communicate with their parents when the latter are in isolation and children are not permitted to visit them.

When CCTV is used in the operating rooms on a permanent basis, a good quality camera is required, and it should be adapted for use with the surgical lights. Most modern surgical lights are adjustable for positioning and focusing the camera.

The CCTV is widely used in hospitals for security purposes.

ENVIRONMENTAL CONTROL

Hospitals have always considered it their responsibility to control their environment. This involves protection of the patients, staff, visitors and the surrounding community. The concept of environmental control is generally practised through: (i) infection control, (ii) general environmental hygiene and pollution control, (iii) radiological health, (iv) accident and accidental injury protection, and (v) occupational health.

A detailed discussion of this vast subject is outside the scope of this book. We shall, however, highlight some of the important areas.

From the early times when hospital-acquired infections took a heavy toll of human life, efforts have been made to control infection through improved and scientific methods of disinfection and sterilization procedures, better housekeeping system, strict application of basic environmental hygiene procedures, development of laminar airflow, etc. Even then, hospital-acquired infection continues to be a problem in our hospitals.

➤INFECTION CONTROL

Environmental control is based primarily on the science of engineering, microbiology and sanitation. All those who are concerned with environmental control — from the chief executive officer to the hospital engineer down to the housekeeping personnel — should become familiar with the basic facts about microbial growth and death and the transmission of disease.

Every hospital must establish an infection control committee to administer a hospitalwide infection control programme, and formulate and enforce policies and procedures for infection control.

These policies and procedures should cover the following.

* Personnel. This includes employee health, dress code, in-service education, etc.
* Isolation procedures.
* Environment specimens like trash, garbage storage and collection, infectious waste, etc.
* Water supply.
* Maintenance of buildings, etc.
* Ventilation system including air intake and outlet, air filters on air-conditioner units, etc.

- ◆ Preventive maintenance.
- ◆ Rodent controls.
- ◆ Systems control, including evaluation of systems, etc.

➤GENERAL ENVIRONMENTAL CONTROL

General environmental control covers such areas as water supply, plumbing, liquid waste, solid waste, air quality and air pollution control, sterilization (autoclave, use of ethylene oxide), insect and rodent control, interior cleanliness and food service. Some of these subjects have been dealt with in detail elsewhere in this book.

➤RADIOLOGICAL HEALTH

Radiation control and safety practices in hospitals come under the mandatory regulations of the Division of Radiological Protection (DRP) of the Bhabha Atomic Research Centre (BARC). Radiation sources are generally of two categories: ionizing and non-ionizing. Examples of hospital ionizing radiation are x-rays and radioactive isotopes. The most common non-ionizing sources are microwave and laser; the latter is sometimes used in hospitals for diagnostic and treatment procedures.

Radiation control measures are based on the factors of time, distance, shielding, containment, personnel protective and monitoring devices, and ventilation. Hospitals must appreciate the problems concerning radioactive waste disposal, both liquid and solid, and should strictly comply with the regulations.

➤ACCIDENTAL INJURY PREVENTION

Few people appreciate, much less concern themselves with, the problems arising from accidental injuries in hospitals, largely among hospital patients, personnel and visitors, and the resultant financial loss and suffering. Too preoccupied with fulfilling their primary mission of patient care, hospital authorities and staff do not recognize hospital's responsibility in this regard and the fact that most accidental injuries can be prevented.

Causative factors leading to accidents are "environmental" or "human". Control of environmental factors through various means discussed earlier cannot by itself prevent accidents. The human factors are the cause of most accidents. On hospital's part, such seemingly minor considerations as elimination of electrical shock situations, provision of non-slip floor surfaces, handrails and grab bars, wherever necessary, and provision of structurally and mechanically sound equipment and furniture, to mention just a few, go a long way in making the hospital a safe place.

The subject of safety in hospitals is discussed in detail later.

➤OCCUPATIONAL HEALTH

Every employee has the right to work in a safe environment. Frequently, the work place is hazardous in spite of the mandatory safety regulations and requirements under the relevant laws. Hazards and dangers lurk in unsuspected places: in walking, working surfaces, means of egress, hazardous materials, compressed gas and air equipment, materials handling and storage — the list is endless. Hospitals have an obligation to do everything reasonably possible to make the hospital environment a safe place. However, safety is everybody's business. All employees should be made aware of it.

SOLID WASTE MANAGEMENT

Hospitals and civic bodies should share the responsibility for the management of hospital solid waste. However, neither of them seems to appreciate the magnitude and seriousness of the problem and the impact of hospitals' solid waste on the community. Quite often, the municipality is blissfully unaware of the types of waste generated by the hospital and refuses to cooperate with the hospital and be involved in the management of solid waste.

Hospitals need to know the local codes and regulations regarding the management and disposal of waste, for example, discharges to the sewer, discharges to the atmosphere via incinerator exhausts, and discharges on land via disposal trucks. In many respects, hospital's solid waste is not different from any other waste except that part of it has pathogenic organisms, and so the ultimate disposal method will have a different kind of effect on the environment.

A solid waste handling system has five basic components: handling waste at the point of production, transportation within the facility, internal storage, internal processing or treatment, and transportation to the point of final disposal.

Solid waste should be packaged at source, pre ferably in disposable plastic bags, or collected in containers or receptacles with lids. Separate puncture proof containers should be provided for syringes, needles and other sharp objects to prevent accidental injury to those handling them. As far as possible, pathogenic waste, including waste from patients in isolation and waste in specimens from the laboratory, should be sterilized at or near the point of production and prior to removal from the place. Food service department generates an enormous amount of garbage and this may have to be stored temporarily until picked up for final disposal. There should be a sufficient internal storage capacity; otherwise, the waste bags tend to be left in corridors, halls and other public areas.

Many hospitals have on-site incinerators for internal processing or treatment of waste. Careful attention should be given to the selection of the system of internal processing. Frequently, during the planning stage, there is a tendency to reject designs requiring high initial cost without any consideration to the operating costs because the initial cost is very perceptible compared to the future operating costs which may later prove to be very high.

Within the hospital, solid waste is usually transported manually. This in India is the most economical method.

Two systems are available in our country for internal disposal of solid waste: oil or gas-fired incinerators and electric incinerators. These are discussed here briefly.

➤ INCINERATOR

In the past, incinerator was a single-chambered unit, but now improved multiple-chambered versions with auxiliary burners, controlled firing, air washers and scrubbers are available. Technically advanced modern incinerators are designed in such a way that the emission of air and the regulation of chambered temperature are controlled automatically requiring no special operator.

Advantages of the incinerator

- Reduces the volume of solid waste by 85–90 per cent
- Can be installed within the hospital building meeting the statutory regulations. This minimizes transportation costs.
- Pathological and infectious waste can be conveniently disposed of within the premises without much preparation and effort.
- Some incinerators have special heat recovery feature.

Disadvantages

- Maintenance cost is high.
- Specially trained operator may be required to attend to it all the time.
- Fuel consumption is generally high.
- Aerosol cans, though not widely used in Indian hospitals, can create a serious hazard to the incinerator operator.

Incinerator can be a great source of energy. From the enormous amount of waste produced in hospitals which has a high Btu content, hospitals can produce energy while simultaneously disposing of waste. The equipment now available can be used as a standby boiler that generates sufficient steam to operate laundry and kitchen. However, administrators should carefully study and analyse the initial and maintenance costs and the benefits accruing from the system by way of energy generation.

➤PYROLYSIS

Pyrolysis is the thermal decomposition or degradation of a substance or organic compound to other compounds or free elements by the application of heat at high temperature, usually in the absence of oxygen. Waste pyrolysis is defined as the carbonization of organic substances without the air. Thereby, a combustible gas is created, and this is used in further processes as a high quality combustible fuel. Many modern industrial processes are based on pyrolysis. Through pyrolysis, soiled waste can be reduced to sterile, inert and odourless char with a greatly reduced volume.

Pyrolytic incinerators are classified as controlled air incinerators where the heat and air needed for combustion are regulated to first volatalize or gassify the waste in conditions of inadequate air, that is, below stoichiometric air-conditions and heat, and then totally destroy it in adequate heat and excess air.

These incinerators are twin chambered with refractory lined chambers mounted on top of each other. Volatization is achieved in the primary chamber with controlled air supply from one of the twin air ducts. Combustion air is supplied by a forced draught fan. An automatic ON/OFF fuel oil burner supplies the heat source. The volatized or gassified waste burner passes on to the secondary chamber through a specially designed opening to achieve sufficient velocity. There it is subjected to the above stoichiometric air. Again, the heat source is the auto-control fuel oil burner system. The combustion is carefully regulated for minimum turbulence, thus avoiding emission of pollutants and making redundant the use of a dust collector or bag filter.

A specially designed educator system is incorporated at the outlet to bring down the temperature of the exit gases from 950°C to 300°C before releasing them in the atmosphere. This is a safe limit for emission conforming to strict pollution control norms. The mechanism keeps the entire system under a negative pressure, thus totally eliminating the risk of hot gases, flames and volatiles rushing out from the door or any other part.

The optimum balance of time, temperature and turbulence in these incinerators makes them the best technological solution for destroying solid waste.

Advantages of pyrolysis
- No need to segregate waste.
- Needles, syringes and sharps are rendered harmless without any special treatment.
- A volume reduction of up to 90 per cent, and a weight reduction of up to 85 per cent can be achieved.

- With a substantial reduction in emissions, the environmental pollution is drastically reduced.
- The degree of operational comfort is high, particularly in the case of charging furnace.
- The system is mechanically simple and virtually maintenance free.

Disadvantages

- The system may need a large space for installation.
- Pyrolysis has not been time tested in the hospitals. Thus, the pros and cons of the system should be carefully studied before arriving at a decision.

➤ELECTRIC INCINERATOR

Electric incinerators, as the name implies, use electrical energy as the source of heat, and are becoming increasingly popular in hospitals and research labs for the disposal of pathological waste. They come fitted with special nichrome heaters and can be operated on standard voltages. They are energy efficient with low operating costs: reportedly 1/3 of those of oil or gas incinerators. There is no oil or gas pollution and the whole operation is neat and clean. No special igniter is required as in the case of oil or gas incinerators.

It is claimed by the manufacturers of electric incinerators that pathological waste, amputated bones, and highly wet material can be totally destroyed and that the residue by way of sterile ash is less than three per cent. These incinerators can also be designed for manual or semi-automatic charging.

The only disadvantage of these incinerators is that they need stand-by power in the event of a commercial power failure.

Basic Considerations While selecting a solid waste disposal system, hospitals should consider the following factors.

- Cost of equipment.
- Cost of installation.
- Operating and maintenance costs including costs of utilities.
- Labour costs (excluding operating and maintenance costs).
- Cost of replacement parts and accessories like containers, carts and vehicle operation.
- If heat recovery is to be included, the heat recovery and energy-saving potential.

Other factors that should be considered

- Facilities for temporary storage of solid waste at the points of production and at the incinerator site when the incinerator is in operation.
- The method and the cost of final disposal of the residue and its estimated quantity.

• Local codes, the requirements of regulatory and waste collection bodies, and the cost involved in complying with them.

SAFETY AND SECURITY

➤ SAFETY IN HOSPITALS

Overview

The word safety in its purest sense means freedom from injury, risk or harm. The management of any hospital has a twofold responsibility regarding safety: (a) to make the work place and the environment safe by creating safe conditions and (b) establish, communicate and enforce the safety rules. Safety is everybody's business and no safety programme can succeed without the cooperation of the people. Everyone has to work as a team and share the responsibility of safeguarding the patients, visitors and the hospital personnel.

Safety awareness is of paramount importance for the success of the hospital's safety programme. Every task that we perform, whether at the work place or home, entails some risk of personal injury. Our ability to work safely is directly related to our knowledge of the hazards associated with the work. Therefore, a sufficient knowledge of the work-related risks is essential.

Some departments of the hospital are more risk-prone and hazardous than others. The laboratories, nursing floors, laundry and kitchen are areas which call for special instructions and elaborate safety rules. Ignorance about the risks associated with the work place and negligence may endanger the lives of employees and turn them into a liability for the hospital and their families.

It is rightly said that a feeling of safety is, like happiness, a state of mind. It is necessary for the employees to incorporate this feeling into their work and lifestyle. For this they should develop a 'safe' attitude and a 'safe' behaviour.

Accidents do not just happen by themselves; they are caused. These causative factors are more human than environmental, and merely controlling the environmental factors does not by itself prevent accidents. The hidden causes of accidents should also be taken into account. For example, if one slips and falls over a banana peel or an oil spill resulting in an injury or fracture, the cause of that accident is apparent: an unsafe condition created by an unsafe attitude or behaviour on the part of someone.

Some General Safety Rules

The following are some of the basic safety rules and principles which everyone should bear in mind and observe.

- The only correct way to do a job in the hospital is the safe way. Urgency is a poor excuse for neglecting safety.
- Know your job thoroughly. Do not indulge in any guess. It there is any doubt, ask the supervisor.
- Do not handle or operate machinery, tools and equipment without authorization.
- Be alert and observe keenly. Report immediately any faulty equipment, unsafe conditions or acts, and defective or broken equipment. Do not try amateur repairs.
- Stay physically and emotionally fit for your work by maintaining good health and a proper diet. Abstain from alcoholic drinks. Take sufficient rest and practise cleanliness.
- Personal hygiene is important. Wash your hands often. In many areas of the hospital, this is absolutely necessary.
- Prevent the spread of infection and contagious diseases. Cooperate with the hospital infection control committee by observing the established procedures. When you are ill with an infectious disease, report to the doctor immediately and stay at home.
- Wear proper uniform or clothing for your job: neither too tight nor too loose. Tight clothing does not permit freedom of movement, while loose one runs the risk of getting entangled.
  Jewellery and high heel footwear may be hazardous.
- Walk, not run, particularly when you are carrying delicate, breakable articles or instruments. Be extra cautious at the corridor intersections, in front of swinging doors (particularly when they do not have view panels), at blind corners and in congested areas.
- If you see some foreign material, loose wire, oil spill, etc. on the floor which may cause an accident, make sure it is removed at once.
- Never indulge in horseplay or practical jokes involving fire, acid, water, compressed air and other potentially dangerous things.
- Pay attention to all warning boards. These signs caution you about dangers and hazards that may cause injury or harm. For example, smoking in an area where oxygen cylinders are stored.
- Be familiar with your work procedure. All departments have written work procedures which include safety practices at work and for handling equipment.
- Always remember to use handrails on stairways or ramps. They are there to ensure your safety and are meant to be used by all, not just the sick and the old.
- When you want to reach overhead objects, always use a good ladder. Do not climb on chairs or boxes.

Apart from these general safety rules, there are other rules relating to particular areas like fire safety, electric instruments, traffic, patient care, toxic materials, and certain departments. These are discussed in Appendix B at the end of this book.

SECURITY AND LOSS-PREVENTION PROGRAMME

➤ OVERVIEW

Three elements — motive, opportunity and means — are necessary to prompt someone to commit a criminal act. The hospital management can effectively curtail only the element of opportunity. The other two can only be constrained, not countered. For example, the element of motive can be countered to some extent by preaching and practising a code of values, positive morale building, stressing loyalty to the institution and reminding employees of the consequences of theft and fraud. The means may be curtailed by instituting internal control measures like unannounced audits, formulating well-defined policies for the control of materials, cash and other assets, checking and questioning all expense accounts, and so on. Even then, employees are ingenious enough to devise new ways of committing fraud.

The element of opportunity can and should be controlled. The management has a moral obligation to safeguard the assets of the institution by making theft and fraud as difficult as possible, if not downright impossible. Often, the general climate in the organization is such as to provide ample opportunity and temptation to the employees to indulge in fraudulent activities without anybody taking cognizance of those offences or punishing the offenders.

Fraud and theft combine to make a booming business in any society, and a hospital is no exception. It is estimated that one out of every ten hospital employees steals habitually. Another study suggests that 25 per cent of all employees will steal to some extent if they feel that only a small percentage of the offenders is likely to be caught and punished. The study further shows that within that 25 per cent, the management level culprit is responsible for over 60 per cent of the total thefts. Sometimes an employee who would not steal a rupee of hospital funds appropriates valuable articles of supplies for personal use.

Contrary to the popular belief, integrity of a person is not directly proportional to the salary he receives or the responsibility he carries. It is not true that most of the thefts and frauds in hospitals are committed by the lowest category of employees. In fact, the top management personnel engaged in such activities cause most harm to the organization, and are probably the most difficult to detect. What is worse, it is not easy to punish them.

The individuals who get most of the opportunities to steal and commit fraud are: supervisors vested with authority, guards, employees of the housekeeping and food service departments, employees on duty during night shifts, weekends and holidays, people with keys to sensitive areas, storekeepers, receiving clerks, purchase department staff, clerks handling cash, payments, payroll, financial and equipment records, etc. and the maintenance and other service personnel.

Types of Frauds and Thefts

We have discussed the subject of theft and fraud with reference to particular departments. Some of the general factors which cause loss to the hospital are as follows.
- Embezzlement.
- Pilferage (small scale theft).
- Kickbacks and collusion.
- Equipment theft.
- Personal property theft.
- Payroll fraud and theft including fraud in punching time clock.
- Cash theft involving main cashier, subsidiary cashier(s), cafeteria cashier, etc.
- Fraudulent practices in purchasing, receiving and storing.
- Fraud in registers, records and billing.
- Computer fraud.

Some Methods of Internal Control

There are two basic methods of exercising internal control, namely, physical security and procedural security. The following are some examples of these.

Physical Security
- Guarding all means of ingress and egress. Protect the hospital against intrusion from without and illegal movement of goods from within.
- Control of the hospital's perimeter. This is easy if the hospital is housed in a single building, but extremely difficult in a sprawling campus-type layout with several buildings spread across a wide area.
- Control of human traffic like employees, visitors, drivers, contractors, vendors, etc. Conduct body search, if necessary.
- Separate entry and exit points for (a) staff, (b) patients and visitors, and (c) vendors, sales persons, delivery people and contractors. The last category should not be allowed to mix with patients and visitors but instead routed through a permanently guarded entrance.

- Identifying, scrutinizing and properly guiding the non-patient and non-visitor traffic such as vendors at the controlled gates.
- Prohibit pedestrian traffic through receiving dock, receiving area, morgue exits, and truck gates.
- Control of vehicles like delivery trucks, etc. and checking outgoing vehicles.
- Electronic surveillance of strategic and sensitive areas through closed circuit television.
- Controlled or guarded gates at all the patient care areas.
- Install locking devices and alarm systems.
- Issue visitor passes.
- Procedure for and control over the issue of keys and submaster and master keys. Authorization necessary to issue keys and an effective, enforceable procedure for retrieving them.
- Lockers and lockable cabinets for staff against personal property theft.
- Provision of a safe for patients' valuables.
- Secured cabins for cashiers with a panic button or a silent foot or knee operated hold up alarm in their cabins.
- Provide roll-up shutters or grills at strategic places for night time protection.

Procedural Security

- Establish and communicate service rules for all staff members. Each employee should be given a printed copy the receipt of which he has to acknowledge.
- Establish policies and procedure manual for each department.
- Establish committees like general purchase committee, pharmacy and therapeutics committee, etc.
- Establish accountability and control over the flow of hospital supplies and materials, particularly the receiving functions, and regulating the operation of receiving and loading dock.
- Institute inventory control procedures.
- Establish well formulated procedures for requisition, purchase, indenting, supply, and distribution.
- Adopt the policy of not allowing the cashier to have both the keys to operate the cash register. The first key unlocks the mechanism for register operation and gives total readings for money and number of transactions. The second key gives total either cashier-wise or by some other classification, and resets all totals back to zero. If the cashier has both the keys, the prospects of fraud increase.
- Institute perpetual inventory system.
- Conduct surprise checks of all the departmental inventories.

The above list is not exhaustive. The chief executive officer and his team, with sufficient imagination and insight, can devise methods suitable for their own individual system.

FIRE SAFETY

➤OVERVIEW

Fire safety and protection are matters of vital importance concerning everyone in the hospital. The best form of protection from fire is its prevention. Although every possible measure may have been taken to make the hospital buildings as safe as possible, no place can be completely free from fire hazards. A careless employee, a thoughtless visitor, or a confused or disoriented patient can inadvertently set off a fire, and though initially it may appear to be insignificant, it is important to remember that every large fire starts from a small one.

An effective fire safety programme calls for an understanding of the hospital fire plan and the active participation of every employee at all times. There is no better protection against fire than constant vigil to detect fire hazards, prompt action to eliminate unsafe conditions and a high degree of preparedness to fight fire. Panic and confusion are the greatest hazards of fire. They can be countered only by sufficient preparedness.

General Fire Information

Every employee should know how a fire is caused, how it can be prevented, and where the alarm boxes and extinguishers are located. He or she should also have a knowledge of the fire fighting procedure which should be learnt before a fire actually occurs.

For a fire to sustain itself, three elements — heat, fuel and oxygen — should be present. Fire is a chemical reaction which occurs when a material (fuel) rapidly combines itself with oxygen in the presence of heat to produce a flame. If any of these three elements is taken away, the fire will fizzle out. This principle is the basis for fire extinguishment.

Most fires can be classified into three general types. Let us call them class A, class B, and class C.

Class A fire occurs in ordinary combustible materials such as wood, paper, cloth, etc. The best way to put out such a fire is by quenching it with water and thereby reducing the temperature of the burning material below its ignition point.

Class B fire occurs in flammable liquids and greases like oil, petrol, paint, alcohol, etc. It is best handled by the blanketing technique which tends to keep oxygen away from the fire and thereby suppress combustion. Water should never be used. It will only spread this type of fire.

Class C fire occurs in electrical equipment such as motor, wiring, switches, panels, etc. This fire is a combination of the previous two types. Because of the hazards of electrical short circuit, a non-conducting extinguishing agent should be used to put out this type of fire. Again, water should never be used on an electrical fire. The person using water on an electric fire may receive an electric shock.

The fire protection system in hospitals basically consists of a static water supply source within the building. Connected to this are first aid hose reels and landing or hydrant valves with hoses at every floor level, preferably housed in an M.S. hose cabinet with glazed door and strategically placed. If the building is a high rise one, there should be a wet riser serving every 1000 sq, metres of the floor area to which the hose reels and hydrant valves are connected. The required pressure in the line should be provided with suitable capacity pumps. It is necessary to have one working pump and another as standby (diesel engine drive) in case of power failure during fire fighting operation.

In addition to the wet riser system, some unmanned areas require sprinklers. Portable fire fighting extinguishers of the type and capacity suitable for specific areas of application should also be provided in strategic locations.

Fire detection system consists mainly of smoke and heat detectors which sense fire at an early stage and give off an alarm so that the fire could be controlled at the initial stage itself. Smoke and heat detection devices are wired in series and terminated in control panels which are located in areas manned 24 hours of the day. Apart from these detectors, break-glass units and hooters are also provided at strategic areas. When there is a fire, the nearest break-glass unit should be activated by breaking the glass. This automatically sets off the alarm so that precautionary methods such as evacuation of the area can be undertaken.

Basic Responsibilities of Every Employee

Fire safety, fire prevention and, to some extent, fire fighting are everybody's business. Every employee has certain responsibilities in this regard. Specifically, he or she should:

- Be completely familiar with the hospital fire safety programme and the departmental fire plan.
- Be alert and observe the hospital with a critical eye, and report all fire hazards to the concerned authorities.
- Not smoke in the prohibited areas or anywhere if the entire hospital is declared a no smoking area.
- Know the location of fire alarm boxes and be familiar with their operating instructions, use and signals.
- Know the location of fire fighting equipment and be acquainted with its operating instructions and use.

- ◆ Know the location of fire exits and assist the supervisor or head of the department in keeping them clear at all times.
- ◆ Report to the supervisor if he (she) notices any defect in stairway doors which should remain closed and in operational condition at all times.
- ◆ Participate in all fire drills and other training or practice sessions as well as know his (her) assigned duties in the hospital's fire plan and evacuation.

What to Do in Case of Fire

If you discover a FIRE in your area, observe the following points.

Use Code Do not PANIC, RUN, YELL, OR USE THE WORD "FIRE." Use the code DOCTOR RED (Plate 13: Picture 80).

Evacuate Remove persons from immediate danger of smoke and fire. Only patients in immediate danger need be relocated in areas on the same floor but away from the fire. If the fire is in patient room(s), remove the patient(s) and close the door behind you.

Sound Alarm Sound the fire alarm from the nearest fire alarm box. This will notify the telephone operator and fellow hospital employees of the situation. The alarm box will set off a series of sounds or hoots.

Dial Telephone Operator Give the exact location — the floor, wing, area, etc — and the extent of fire. This is important because the telephone operator should be very sure of these details before calling up the fire department. The telephone operator will immediately write the location down.

The telephone operator will announce Doctor Red on the public address system followed by the location of the fire three times. This announcement will be repeated every 30 seconds for a period of two minutes.

To avoid panic among patients and visitors, emergencies in the hospital are announced using codes. For example, Doctor Red for fire. See Plate 13: Picture 81 for other codes.

The operator will also notify important officials like the CEO, nursing director, security chief, engineer and leader of the Doctor Red Alert team (explained later).

If the situation warrants and with the approval of the CEO or the person in charge at that time, the telephone operator will notify the fire department and summon help.

Shut off Ventilating Fans, etc. On notification, the engineering department will shut off all ventilating fans, oxygen (after checking with the area supervisor), gas, electric power to the affected area and if necessary, to any adjoining areas threatened by fire.

Prevent Smoke or Fire Gases from Spreading to Other Floors There is a great danger of people dying of suffocation even on the floors far removed from where the fire has broken out. Smoke and fire gases spread to other floors through air-conditioning ducts, pipe tunnels, etc. This can be avoided by closing all the dampers in the air-conditioning ducts.

Avoid Using the Elevators Walk down the stairs.

Establish Control Centre. The CEO or a senior officer will take charge.

At the Scene of Fire

- Seal off the area of fire. Close windows and all patient room doors. Place wet blankets or towels along the door edges to prevent leakage of smoke. This is an effective fire fighting technique.
- Fight the fire with appropriate fire extinguishers. Use carbon dioxide type extinguishers on electrical and flammable liquid fires. Use fire extinguishers if the fire is small and fire hose if it is large.
  Warning: Do not operate the fire hose if you are not trained to do so. It is risky as you may be swept off your feet. Remember: two people are needed to operate a fire hose.
- Supervisor of the area will take charge.
- Doctor Red Alert team will go to the scene of fire. The team leader will direct operations as they pertain to the actual fire situation.
- When the fire department personnel arrive, they will be in complete charge.
- Personnel on general floor and other patient care areas will remain calm and reassure the patients. They will remain with their patients at all times until properly relieved.
- There should be written procedures for evacuation of patients and on who can make that decision.
- In case you are trapped and are unable to leave your room, do the following.
  - Feel the door. If warm, do not open.
  - Place wet towels, bedding or blankets under the door(s).
  - Stay low on the floor where smoke and heat are the least and air clearer.
  - Go to the window and open it.
  - Attract the attention of fire fighters by hanging a sheet or blanket outside the window.
  - Stay at the window for rescue.
- All Clear signal should be given by a responsible person, and Code Green announced after the fire is controlled (Plate 13: Picture 80).

The Time to Know What to Do is Before a Fire Occurs, and Not After

Regardless of whether it comes under the purview of fire regulations or not, every hospital should be provided with a fire protection system considering the damage fire can cause to life and property. In addition, provision must be made for the following.

- ◆ There should be an effective fire safety programme for the hospital.
- ◆ There should be written policies as well as a procedure manual covering all contingencies arising from fire.
- ◆ Every department should have a departmental fire plan and a fire procedure manual outlining every employee's role in the plan.
- ◆ There should be a pre-appointed standing Doctor Red Alert team to direct all fire fighting operations.
- ◆ There should be written procedures for evacuation of patients in case the fire becomes widespread. The procedures should specify who can decide on evacuation as well as the procedures, methods and the order of precedence to be followed for evacuation.
- ◆ Simulated fire drills which are an essential part of an effective fire prevention programme should be conducted periodically. These drills help ensure that all personnel understand their roles in fire safety programme and perform their assigned tasks well. Practice makes them perfect. Fire drills should be conducted in a realistic manner.

Summary

If the fire is in your area

- ◆ Remove persons from immediate danger.
- ◆ Activate fire alarm.
- ◆ Alert personnel calmly. Never use the word Fire. Use the code Doctor Red or code Red.
- ◆ Dial the telephone operator. Give exact location and extent of fire.
- ◆ Seal off the affected area. Close all windows and room doors in the area. Use wet blankets to confine smoke.
- ◆ Unless lives are at stake, do not attempt to re-enter a room if the fire has gone out of control. Wait for help to arrive.
- ◆ Shut off all equipment, gas, etc, which may compound the risk.
- ◆ Fight the fire. Use a proper extinguisher.
- ◆ Follow your department's specific fire plan and procedures.
- ◆ Set up a fire control area.
- ◆ Take head count of patients and staff.
- ◆ Post staff at the elevator.
- ◆ Prepare for evacuation of patients or other duties as prescribed in the department fire rules.

◆ Establish contact with engineering, security, etc.
◆ Establish and maintain communication with the control centre, and inform it about staffing needs.
◆ Relinquish control when fire department personnel arrive at the scene.
◆ When the fire is completely put off, send an All Clear message to the control centre. This should be agreed to by the fire department personnel if they are present.

If the fire is not in your area
◆ Stop what you are doing.
◆ Report to your department head or supervisor.
◆ Continue your duties within your department if instructed by your supervisor.
◆ Take head count of patients and staff.
◆ Shut off equipment, gas, etc. which might aggravate the risk. Check with the supervisor before shutting off the oxygen.
◆ If you are in a patient care area, communicate with the patients and reassure them.
◆ Send staff to control centre or the assignment area, if required.
◆ Be prepared to assist in evacuation of patients, if necessary.
◆ Post staff at the elevator.
◆ Maintain a stand-by alert for any eventuality

Do not
◆ Panic.
◆ Run or shout in the corridors.
◆ Use the word Fire. Refer to it as Doctor Red.
◆ Use elevators (unless you are already on your way down).
◆ Leave your department unless permitted or directed by your supervisor.

Within a reasonable time after the fire is extinguished, head(s) of department(s) where fire had broken out should write a fire incident report and send it to the administration. The engineer should assess the damage caused by the fire, make an estimate of the loss suffered by the hospital and send a report to the CEO.

BOMB THREAT

►OVERVIEW

About 95–99 per cent of all bomb threats are found to be false. However, the implications of this kind of threat make it imperative that immediate action is taken to avoid personal injury or any loss of life and property as well as to arrest the spread of panic. The hospital, where such a large number of helpless people are concentrated and are utterly dependent on other people for their safety, should be alert to the dangers of a bomb threat and take all measures for its prevention. A bomb threat thus should not be dismissed lightly.

In most cases, telephone is used to communicate a bomb threat. The telephone operator is, therefore, the most likely person to receive the threat call. However, every employee should be familiar as to what to do after a bomb threat is received.

Upon Receiving a Bomb Threat

- Do not take the threat lightly.
- Be calm, do not panic. Do not display emotion or fear.
- Be courteous to the caller.
- Attempt to prolong the conversation with the caller and try to get as much information as possible.
  - Ask the exact location of the bomb, how it looks like and at what time it is set to explode.
  - Be alert to the distinguishing background noises like street noises, aircraft, crockery, machinery, bells, etc., or anything that would help determine the location of the caller.
  - Note the distinguishing voice characteristics as well as the age, sex, mental condition of the caller which may lead to his identification. (See the Bomb Threat Procedure Report Form at the end of the section.)
  - Note if the caller appears to have knowledge of the hospital by description of the location, departments, personnel, etc.
- Keep the caller on the lines as long as possible in an attempt to trace the call. Ask more questions to prolong the conversation. Some specimen questions to engage the caller are given below.
  - Where is the bomb right now?
  - When was it placed?
  - What does it look like?
  - Did you place the bomb? If so, why?
  - Do you represent any organization?
- Inform the caller that the hospital has a lot of helpless and sick patients and a bomb explosion could result in the death of or serious injury to many innocent people. He should help to avert this crisis.
- Immediately after the caller hangs up, the operator should notify the CEO, director of nursing,

medical director, security officer and others concerned. She (the operator) will tell them, "There is a bomb threat situation at the hospital."

- Outside the regular working hours, she will notify these officials at their residences and notify simultaneously the person on duty in a supervisory position.
- Upon getting clearance from the CEO or his deputy or the person in charge of the hospital at that time, the telephone operator will notify the police and the fire department.
- In the night, the supervisor or the person on duty will take full charge of the situation until one of the senior executive officers arrives.
- After the top management has been informed, the operator will make announcements on the public address system informing hospital personnel of the bomb threat.
- Using a code, she will announce Code Black (Plate 13: Picture 80) three times. This announcement will be repeated every 30 seconds for a period of two minutes.
- If the bomb threat is received by anyone other than the telephone operator, details of the threatening phone call should immediately be conveyed to the telephone operator who, in turn, will initiate the steps described above.
- If the caller has indicated the location of the bomb, the telephone operator will contact the CEO, director of nursing and others and inform them of the exact location. The security supervisor will seal off the area.
- If the caller has not indicated the exact location, a general search will be conducted in all work areas. There is no better or more qualified person to search a particular area of the hospital than the employee who works there. He or she can readily recognize any foreign or unusual object placed in the employee's work area.

Warning

It is important that the personnel involved in the search clearly understand that their only mission is to search for and report suspicious object(s). Under no circumstances should they touch, move or jar (give a sudden shock or jolt to) the object or anything attached to it.

Disposal of suspicious objects will be the specific responsibility of the bomb squad which comprises specially trained police or fire department personnel.

- On hearing the Code Black announcement, each ward sister or head nurse will prepare an inventory of patients in her area and send copies to the nursing service and control centre, retaining the third copy in the nursing station.
- The office of the chief of the hospital will become the control centre for telephones and other forms of communication. Any report regarding the search or any unidentified or suspicious objects should be relayed to this office. The other senior officers will assist the chief of the hospital. Secretaries will be in attendance. Only authorized personnel will be allowed entry in this area.

- All personnel should remain calm and reassure the patients. They should notify the control centre of any significant development in their areas.
- If an unexploded or suspected bomb is found, the following procedure should be followed.
  - Close all doors in the immediate area. This will reduce the impact of the bomb.
  - Prevent the suspected object from being moved, touched or jarred in any way.
  - Report all details to the control centre which, in turn, will inform the police and the fire department.
    When they arrive, police and fire department personnel should be met and escorted to the scene.
  - Isolate the area.
  - Evacuate patients, visitors and other personnel from the danger area in an orderly fashion.
- The decision to evacuate patients will be made by the chief of the hospital or the person in charge at that time, in consultation with other senior officers.
- Upon arrival of police and/or fire department personnel, complete authority will be given to them for search and investigation. If a suspected bomb is found, police will either defuse it or remove it from the hospital.
- As soon as possible after informing the people concerned, the person who received the bomb threat call will complete a pre-printed Bomb Threat Report form (see specimen form at the end) and send it over to the control centre.
- Upon a signal from the control centre that everything is clear, the telephone operator will announce the All Clear message. She will announce code Green three times, and repeat it in the next two minutes.
- The hospital personnel will then return to their respective places of work and resume their duties.
- Immediately after the All Clear message has been given, a critique will be held in the control centre. All senior officers will attend the meeting. The meeting will discuss the established procedure, the difficulties encountered during the procedure, and the possibilities of using the experience gained to further improve the system.

The time to know what to do is before a bomb threat is received and not after. For this purpose:

- Hospitals must develop a procedure manual setting forth in detail the procedure to be followed in the event of a bomb threat.
- There should be periodical drills for all staff of the hospital so that they become familiar with the procedure.

BOMB THREAT PROCEDURE

REPORT OF PERSON WHO RECEIVES THE BOMB THREAT CALL

Name of the person who received the call

_____

Title   _____

Time   _____   Date   _____

EXACT WORDING OF THE THREAT

_____

Sex of the Caller _____   Age _____

Length of Call   _____

CALLER'S VOICE

_____	Calm	_____	Slurred
_____	Angry	_____	Stutter/stammer
_____	Excited	_____	Deep
_____	Slow	_____	Harsh
_____	Rapid	_____	Clearing throat
_____	Soft	_____	Deep breathing
_____	Loud	_____	Cracking voice
_____	Laughter	_____	Disguised
_____	Crying	_____	Accent
_____	Normal	_____	Whispered
_____	Distinct	_____	Familiar

If voice is familiar, who did it sound like?

_____

_____

BACKGROUND SOUNDS

_____ Street noises	_____ Factory machinery
_____ Crockery	_____ Animal noises
_____ Voices	_____ Clear
_____ Music	_____ Long distance
_____ House noises	Other
_____ Motor	_____
_____ Office machinery	_____

THREAT LANGUAGE

_____ Well spoken (educated)	_____ Incoherent/not clear and logical
_____ Foul	_____ Taped
_____ Irrational	_____ Message read by threat maker

REMARKS

_____

_____

Reported call immediately to:

_____

_____

_____

Signature

ALARM SYSTEM

A hospital, more than any other institution, is exposed to emergencies and life threatening situations: from medical emergencies like cardiac arrest, accidents, casualties and disasters to dangers arising from fire and bomb threat. It has to be all the more alert to these situations because nowhere else are such a large number of helpless people concentrated in one place and are so utterly dependent on other people for their safety and health.

Built-in safeguards and preparedness are the essence of all safety programmes. The alarm system is one such programme. We discuss here some of the alarms that hospitals should have.

Fire Alarm

Every hospital must have a fire alarm system which should be a part of the hospital's electrical system. Wherever possible, it should be designed to transmit an alarm signal directly to the telephone operator so that she can contact the fire department and notify the hospital personnel without any loss of time. The fire alarm system can be automatic or it can be operated manually.

Smoke and fire detection devices are installed in the patient rooms and other high risk areas in the heating and ventilating ducts between the floors. These actuate the fire alarm system. On activation, the system sounds audible alarms throughout the premises or zones, including distinctive visual and audible alarm signals at the respective nurses' station. To indicate the location of fire, there is an indicator light outside every patient room. This is activated when there is a fire in the room.

In the automatic system, smoke detectors not only actuate the fire alarm signals, but also close smoke doors and simultaneously shut off fans in the central air handling system. If the fire alarm system is not automatic, then anyone noticing or hearing the fire signal should immediately inform the telephone operator who, in turn, will call the fire department and notify the hospital personnel.

Medical Gas Alarm

In the centralized medical gas system, oxygen and nitrous oxide which are stored in bulk in the manifold room are distributed to other areas of the hospital such as the operating rooms, ICUs and patient rooms through pipelines. Compressed air and vacuum (suction) are also supplied through pipes to certain areas.

Two kinds of alarm are incorporated into the medical gas system. One monitors the pressure of various gases at different areas of the distribution system. If abnormal pressure is sensed, the system sets of an alarm — the normal green signal goes off and the red warning signal glows with audible alarm until the line pressure returns to normal. The second alarm is called the remote signal lamp which is generally only visible. The lamp lights up when either of the banks of cylinders becomes empty.

The remote signal lamp is only a warning signal. No immediate action is necessary because when one bank is empty, the other takes over and supplies the gas without interruption.

The alarm should be located in the medical gas user areas such as the operating rooms and patient floors as well as the main working area where medical gas system is maintained. However, these areas, especially the maintenance area, may not be manned all the time. Secondary signals should therefore be installed in places like the telephone operators' room, security office and the like where a 24-hour attendance is assured.

Blood Bank Alarm

Most hospitals use specially crafted refrigerators — a cold room or walk-in cooler is ideal — to store whole blood in the blood bank. These refrigerators are set to a particular temperature to maintain blood in good condition and are provided with an alarm. The alarm which is both audible and visual goes off whenever it senses high temperature or a drop in voltage. If the blood bank or the laboratory of which it is a part is not manned round the clock, the alarm signals should be located both in the blood bank and in a place having 24-hour attendance like the telephone operators' room or the security office.

Narcotics Alarm

Narcotics are stored in locked cabinets in the nurses' stations as well as the pharmacy. These are restricted drugs which are constantly stolen by persons addicted to them. Some hospitals install a signal system that illuminates a light bulb which is visible from the nurses' station and the corridors whenever the narcotics cabinet door is opened.

Cold Room and Walk-in Cooler Alarm

Many hospitals have walk-in cold rooms or coolers in their food service department and laboratory.

There have been instances of the staff of the food service department getting accidentally (or even deliberately) locked up overnight inside the walk-in coolers. There should be an alarm button that can be used in such an emergency with a distinguishable audible and visual alarm indicator in a prominent area where there is a 24-hour personnel coverage.

Voltage Fluctuation Alarm

In any hospital where crores of rupees worth of sensitive and expensive equipment is used, stabilized voltage is essential. Motors are usually designed to withstand only a 10 per cent fluctuation in voltage supply. Beyond this limit, the motor will get damaged unless it is disconnected.

Low voltage poses the biggest threat to electrical system and equipment. Diagnostic equipment often gives erroneous readings in low voltage conditions. There are certain areas and sensitive equipment that do not tolerate excessive low or high voltage. Such areas or equipment may be fitted with a simple voltage-sensitive alarm along with a voltmeter. The alarm can be set at any desired point.

Elevator Alarm

Many hospitals have more than one passenger and bed-cum-passenger elevators which are in continuous operation. Whenever there is an electric power failure, elevators with their passengers get stranded, often in between the floors. In order to rescue the stranded passengers, a panic or emergency push button is provided in each elevator. When it is pressed, a battery operated alarm installed in the electric room or the security room which is manned round the clock is actuated to alert people about the rescue operation. Elevator operators or maintenance crew then manually winch down the elevator car from the machine room to the next lower floor to rescue the stranded passengers.

Modern elevators have an optional levelling feature which automatically takes the elevator car to the next floor level in case of power failure.

Security Alarm

Certain sensitive ares of the hospital like the cashier's office, the psychiatric ward, bank extension counter and pharmacy which are prone to theft and burglary or where patients suddenly become violent need to summon immediate help from the security personnel. Some hospitals provide alarm systems in these areas. The alarm may be of two kinds. One is an automatic alarm like the one used in strong rooms of banks or jewellery shops, which goes off when someone tries to break in. The other is similar to the one used by bank tellers. The device is activated by the employee to summon security or police help.

Patient Emergency Alarm

Various new features are now available that can be incorporated into the conventional nurse call system to meet emergency situations in the patient rooms. If the nurse does not respond to the patient's call immediately, the system makes the light outside the patient's room and on the nurse call panel in the nurses' station go blinking. If there is still no response, the blinking of lights and the bleeping signals from the beeper on the panel gradually keep on increasing in frequency.

An additional feature that can also be fitted into the nurse call system is the panic button in the patient toilet which the patient can activate by a pull cord in case of an emergency.

Code Blue Alarm

Code blue is a term used in hospitals to announce or signal an emergency of a serious nature such as a cardiac arrest. In some hospitals, in all patient rooms and other strategic locations, there are independent buttons — not a part of the nurse call system — named Code Blue which when activated emit distinguishable emergency alarm signals both at the nurses' station and at the telephone operator's room. While the nurse attends to the patient instantly, the telephone operator goes on the public address system announcing code blue three times giving the location of the emergency. In such hospitals there is a written procedure to deal with such situations and a pre-appointed code blue team which responds to the call instantly. The members of the team are trained to deal with medical emergencies including cardiac arrest.

To avoid panic among patients and visitors, emergencies in hospitals are announced using codes: code blue for cardiac arrest, code (doctor) red for fire, code black for bomb threat, code white for security emergency, doctor major for disaster and code green for all clear (Plate 13: Picture 80).

DISASTER PREPAREDNESS

➤ OVERVIEW

Disasters are of two types: natural and man-made. Natural disasters include storms, heavy rains and floods, landslides, tidal waves, earthquakes, tornadoes and hurricanes. Examples of man-made disasters are riots, plane crashes, complex motor vehicle accidents, train accidents, building collapses, dam breaks and blackouts. In most of these accidents, fire may also break out.

Any hospital could be involved in a disaster. It may be an internal disaster such as a fire, bomb threat or a medical gas leak, or it could be an external disaster such as a complex automobile accident, train or airplane accident, explosion, tornado or other man-made disorders like riots.

Hospitals are concerned with the conservation of human life as well as the restoration of individuals to society. They are cognizant of their responsibility towards the hapless victims of a disaster. However, merely being cognizant or feeling a sense of responsibility is not enough. Hospitals should be properly prepared to effectively meet any eventuality. Too often, their preparedness to deal with disasters is grossly inadequate.

For a hospital to be properly prepared, there should be a well documented and well established disaster preparedness plan which is the result of intelligent and cooperative thinking and planning by all concerned. However this is only one step towards being prepared. No matter how well a plan has been conceived and written, it is of little value if filed away without implementation. To be effective, the plan must be clearly understood by every employee of the hospital. Personnel should know what they have to do and how to do it. Rehearsal drills with conditions simulating the actual scene can provide basic training and practice that are so necessary for its successful implementation.

Drills should be scheduled for different shifts to get as many personnel trained as possible. To make them more realistic, some drills should be held without warning while others may be practice drills announced ahead of time. The importance of these drills which ensure preparedness cannot be overestimated. The disaster preparedness plan should be implemented as a "dry run" at least twice a year at approximately six-month intervals.

Inasmuch as the disasters cannot be forecast, it is necessary that the plan makes provision for all possible types of them. It should also cover detailed assignments and specific tasks to be performed by each department. It is important to involve as many of the personnel as possible.

In dealing with an external disaster, especially of a large scale, it is not only necessary to involve other hospitals in the neighbourhood but also the entire community in addition to the police and fire departments and other civic bodies. For this, it is desirable that occasionally disaster drills are conducted involving other hospitals and the community including the police and fire departments.

A detailed and comprehensive disaster plan is beyond the scope of this volume. However, the essential elements of such a plan relating to the steps that should be taken prior to and during the time when it is in operation are discussed here.

Some Definitions

Disaster   Any serious or unusual situation which cannot be resolved by the personnel on duty in the hospital at the time.

Doctor Major   A code used to indicate a disaster of undetermined magnitude which is likely to tax the hospital facilities.

Triage   The process of classifying the sick and the injured according to the urgency and type of conditions in order that each casualty receives treatment according to his or her immediate need.

Information Team   A team consisting of a medical person and others who reconnoiter the disaster site and report its estimate of the casualty situation to the disaster medical director or the on-site medical director who may then decide if some of the cases have to be sent to other hospitals.

All Clear   The time when the acute phase of disaster terminates. This determination is made by the disaster chief, medical director, CEO or the nursing director of the hospital.

Initial Steps Before the Disaster

- Appointment of a disaster committee with the medical director as the permanent chairman and consisting of a representative of each department and service and significant others like chief residents of each service.
- Appointment of key personnel such as disaster medical director and/or on-site medical director, information team, etc. and identification of certain areas used during a disaster such as command post, control centre, triage area, medical aid station(s), etc.
- Formulation of a detailed disaster plan.
- All medical and hospital staff should be thoroughly familiar with individual and departmental responsibilities.
- Conduct simulated disaster drills for all employees periodically to establish proficiency in managing a disaster.

When a Disaster Strikes

1. Authorized individuals are called to declare a Doctor Major. Only the CEO, medical director, director of nursing or the senior physician in the emergency room are authorized to activate the disaster plan.
2. Police and fire departments are informed of the disaster.
3. An immediate announcement is made throughout the hospital, if necessary. The telephone operator announces Doctor Major three times and repeats the announcement every 30 seconds for a period of two minutes.
4. Depending on the seriousness and scale of the emergency, physicians and other hospital staff who are off duty are recalled. All departments activate their established procedure to provide immediate and necessary support.
5. Medical staff and house officers report to a designated staff holding room where they are given written instructions for manning important areas or performing crucial tasks. Specific instructions are given to: the disaster chief, chief of emergency room, chief of triage area, chief of emergency waiting room, chief of staff holding room, chief for emotionally disturbed patients' area, chief of waiting room for relatives of casualties, chief of admitting office, chiefs of x-ray department, blood bank and CSSD, etc.
6. "Chiefs" are assigned to key areas. Additional staff is assigned as needed.
7. Security and support personnel control the traffic in and around the hospital.
8. Triage is initiated adjacent to the ambulance entrance. The following instructions should be observed.
   (a) Do not allow casualties to accumulate.
   (b) Sort patients rapidly.

(c) Tag them appropriately.

(d) Route to appropriate treatment or other areas without delay according to the following:
  (i) Hyperacute conditions (life-threatening problems, hemorrhaging, etc.) to treatment area in the emergency room.
  (ii) Serious casualties to serious casualty centre or surgery area.
  (iii) Ambulatory care patients (non-life-threatening and observation cases) to emergency room waiting area.
  (iv) Emotionally disturbed to chapel or meditation room.
  (v) Dead on arrival to the morgue.

9. Definitive care is instituted at the above areas as soon as possible.

10. At each area, data is obtained, recorded and directed to business office, admitting office and public relations office for coordination and release.

11. Patients who may be discharged are directed to the admitting/discharge office.

12. Only the disaster chief, medical director, director of nursing or the CEO may authorize an All Clear announcement indicating the termination of the disaster status. Personnel may then return to their respective places of work.

13. The telephone operator will then make the All Clear announcement over the public address system. She will announce Code Green in the manner as explained earlier.

14. Public relations department
  (a) releases information and messages to media, as necessary,
  (b) acts as liaison between patients and relatives,
  (c) centralizes information about casualties,
  (d) controls all photographs, and
  (e) prepares and posts updated casualty lists.

15. In the emergency department, personnel
  (a) remain in their area unless specifically assigned to other disaster posts;
  (b) defer all routine activities immediately including cast work;
  (c) stand ready to receive and assist disaster casualties requiring urgent life-saving measures;
  (d) see that additional nurses are assigned from the floors, if necessary;
  (e) are responsible that crash carts are brought to and kept in readiness in the triage area, and;
  (f) see that a portable x-ray machine is brought to and kept ready in the emergency department.

16. Following provisions shall be made in the morgue.
  (a) An admitting clerk takes charge,
  (b) Personal effects are not removed from the bodies,
  (c) A morgue register is kept. It lists bodies with their: (i) identification, (ii) date and time of arrival, (iii) time of release and removal by proper person(s), (iv) list of personal effects, (v) signature of person removing each body, witnessed by clerk, and (vi) destination of the body.

17. Similar and appropriate written instructions are given to other areas and personnel, namely:
    ◆ House physician staff
    ◆ Housekeeping
    ◆ Linen services
    ◆ Nursing service
    ◆ Pastoral care, if there is one
    ◆ Pharmacy
    ◆ Hospital security and police department
    ◆ Radiology department
    ◆ Social work
    ◆ Surgery
    ◆ Volunteers
    ◆ Inpatient area
    ◆ Outpatient department
    ◆ Transportation

There is no one standard disaster plan that is tailor made to suit all hospitals. Hospitals vary in size, resources, facilities, type of staff they have and in their ability to render emergency treatment. Some of them may be able to provide only elementary first aid whereas others may have facilities and staff to provide advanced care. The place where hospitals are situated also makes a difference. A hospital located on a busy highway may get more accident cases, or an urban hospital more stab injuries than the ones in a quiet residential area. Each hospital should, therefore, formulate and develop its own disaster plan to suit its own conditions. However, regardless of the type and where it is located, every hospital should have a disaster preparedness plan that can be put into action in any eventuality.

CODE BLUE PROCEDURE

Code Blue procedure is a written procedure to announce, summon help and follow when there is a serious medical emergency such as a cardiac arrest anywhere in the hospital (Plate 13: Picture 80). In some hospitals there are call buttons marked Code Blue in patient rooms or in certain strategic areas of the hospital to summon help immediately. In the advanced computer assisted code blue system, information or announcements can be programmed to be automatic whereas in the manual system, the press of the code blue button will set off alarm signals both at the nurses' station and in the telephone operator's room. While the nurse will attend to the patient, the operator will go on the public address system announcing for the code blue team. It is a team trained to deal with all kinds of medical emergencies. It will respond to the call immediately.

There is a written procedure for code blue. As in the case of fire, bomb threat and disasters, rehearsal drills should be conducted for all personnel. Every staff in the hospital must be given thorough training in cardiopulmonary resuscitation (CPR) procedure so that anyone who is readily available can resuscitate and revive the patient.

See under intensive care units and alarm system elsewhere in the book for more about the code blue procedure.

TRANSPORTATION

➤OVERVIEW

Transportation is an essential function that is performed in every hospital regardless of its size, sophistication and means of transportation. It encompasses a wide range of activities and areas in any hospital. Some of them are as follows.

1. Inpatient escort service upon admission and discharge. Some hospitals make it mandatory for every patient upon discharge to be escorted in a wheelchair to the main entrance of the hospital.
2. Patient transportation to and from ancillary departments like x-ray, physical therapy and pulmonary medicine besides operating rooms, delivery suite, etc. for inpatients, emergency room patients and, on occasion, outpatients.
3. Movement of staff and visitors within the facility through elevator operator service.
4. Movement of supplies, materials and equipment within the hospital.
5. Movement of patient food, generally in trolleys or carts, from food service department to patient floors, and return of used trays, dishes, etc. to washing areas.
6. Ambulance service. Movement of patients from their homes, scene of accident, etc. to the hospital, and discharged patients to their homes.

Some hospitals may have a transportation department for the operation and maintenance of ambulances, vans and other vehicles. Some may have, in addition, what is called the patient transportation or escort service. Where there is no such centralized patient escort service, individual departments may arrange for transportation and escort service for their patients.

The usual means of transportation of people and materials include ambulance, elevators, wheelchairs, lifts , dumbwaiters, stairways, and ramps.

➤ELEVATOR, LIFT AND DUMBWAITER

Elevator

Elevators are a major part of the hospital's transportation system. They handle four types of traffic: patients, visitors, personnel and service. Movement of patient and visitor traffic should be quick. Patients may have to be moved quickly in an emergency, sometimes in their beds and stretchers, to the emergency rooms, operating rooms, ICUs, CCU or the labour-delivery suite. Patients are moved in wheelchairs to the ancillary services and other therapeutic treatment areas. Ideally, patient traffic should be separated from the visitor and service traffic. Some hospitals have separate service elevators for the service personnel and equipment for delivering supplies from outside and for such internal deliveries as food, linen, materials.

Hospital elevators are of two types: (i) for passenger traffic including doctors, nurses, personnel and visitors, and (ii) for vehicular traffic including beds, stretchers, wheelchairs, portable machines, food carts and the accompanying personnel.

At least one hospital-type elevator should be installed when up to 59 patient beds are located on any floor other than the main entrance floor; at least two hospital-type elevators are needed when 60 to 200 patient beds are located on floors other than the main entrance floor; and at least three hospital-type elevators are needed when 201 to 350 patient beds are on floors other than the entrance floor.

Some Specifications

- The hospital-type elevator cars should have inside dimensions that will accommodate a patient bed, attendants, and necessary equipment. It should be at least 1.52 m (5′ 0″) wide by 2.29 m (7′ 6″) deep. The car door should have a clear opening of not less than 1.22 m (4′ 0″) wide and 2.13 m (7′ 0″) high.
  Additional elevators meant for visitors, personnel, and materials handling can be of a smaller size.
- All elevators should be equipped with automatic levelling devices.
- Some hospitals equip the patient transporting elevators with a two-way special services switch to permit cars to bypass all landing button calls and be dispatched directly to any floor.
- Every elevator should be equipped with a telephone and an alarm for use in an emergency.
- If there is a bank of elevators, at least one elevator should have dual control to obviate the necessity for an operator during the night or the hours when traffic is light. This is particularly useful in smaller hospitals.
- Elevator call button and controls should be of the type that will not be activated by heat or smoke.
- As a rule, hospital elevators are slow. And there is a misconception that they should be so.

This is not true. There is no reason why they should not move as fast as those in hotels or in commercial buildings.

♦ Elevators are one place where people panic in case of power failure. To obviate this, electric service and switching facilities should be arranged to permit operation of elevators from alternate (emergency) power source in case of interruption in normal electrical service. Where it is not possible to connect all elevators to the alternate source of power, at least one elevator in each bank of elevators should be powered by alternate source.

♦ For their proper and efficient functioning, elevators require routine maintenance and inspection by skilled elevator mechanics with special training and experience in this field. Elevator inspection and maintenance service should be contracted to the manufacturer or his authorized agent, and not undertaken by the in-house personnel as it is normally considered beyond their capabilities.

Lift

An elevator which is primarily used to move materials one or two floors is called a lift. In India, however, the term lift is used synonymously with elevator. In this context, lifts (elevators) are used to move all kinds of traffic and materials.

Dumbwaiters

Dumbwaiters are small lifts or elevators which are used to deliver food trays, medicines and supplies. Before installing a dumbwaiter, the purpose for which it is going to be used should be decided. One purpose for which it is used is to deliver sterile supplies and instruments from the CSSD to the operating rooms when these two departments are located in different floors but directly one below the other so that the dumbwaiters open directly into the departments.

There are two types of dumbwaiters: the conventional waist-loading type and the floor-loading type. The latter permits a greater variety of use including transportation of small carts between upper and lower floors. The cart can be rolled directly onto the platform. This eliminates manual handling of items. This is not possible in the waist-loading type.

Dumbwaiters can be equipped with automatic loading and unloading devices. They are available in various sizes and capacities. One popular size is of 9 sq. feet of floor size and 4 feet in height.

Large dumbwaiters can be used to transport bulky equipment like food trucks or trolleys and laundry trucks.

When used for transporting sterile supplies from CSSD to the upper floor operating rooms, generally twin dumbwaiters are used: one to transport sterile instruments and supplies and the other to send dirty instruments and supplies to CSSD for reprocessing.

➤STAIRWAYS AND RAMPS

Stairways

Wherever there are elevators, stairways do not play a major role in handling normal traffic. However, they are required largely for use if the elevators break down and as a means of egress in case of fire. Therefore, they must be planned with considerable care. There should always be at least two stairways — in larger facilities, even more than two —. leading from the top floor to a ground level exit. It is also necessary to locate them in different areas of the building. The fire department may require, as a measure of protection against fire, that there should be a complete enclosure in the entry to the stairways with self-closing doors and lighted exit signs over the door in the corridors. A minimum width of 3 ft. 8 in. and wide landings are necessary for handling stretchers in an emergency as, for example, when patients have to be evacuated during a fire. Continuous railings on both sides at a height of approximately 0.9 m. (3 ft.) are necessary for ease as well as safety of patients and personnel. Treads with grooves should be provided to make the surface of the steps non-slippery.

In smaller hospitals with only ground and first floors, there may not be any elevators. In that case, stairway in combination with a ramp becomes the major means of handling traffic of all kinds with ramps being used for transporting patients on stretchers and wheelchairs.

Ramps

Ramps are a common feature in many hospitals. They are largely used for transporting stretcher patients. Some features of the ramp are: gradient is 1:10, width is 2.5 m., width at landing at the U turn is 3.0 m., concrete railings are at a height of 0.9 m. and top of the railings with M.S. pipe or wooden railings have a diameter of 75 mm.

The flooring of the ramp should have grooves in perpendicular direction to the slope to avoid skidding. The floor may be of tiles, stone slabs or ribbed vinyl. When ramps are located in the periphery of the building, as they sometimes are, they should be sheltered from weather.

CHAPTER 9

Appendices

9 CHAPTER

CONTENTS

APPENDIX A
CHECKLIST OF MINOR FACILITIES

Planners and architects generally are not found wanting in their attention to major items of facilities planning. But more often than not, myriads of small items escape their attention. We present below some of the generally forgotten items as a checklist for hospital planners. The list is not in any particular order.

- Gift shop, book shop and florist's shop.
- Coffee/snack bar.
- Pay telephones in the outpatient department, emergency department and the inpatient areas, particularly near labour-delivery suites, ICUs and CCU.
- Drinking water fountains.
- Assisted STD and ISD call facilities.
- Cashier's booth(s).
- Offices for night administrator and night nursing supervisor.
- Police and duty driver's posts near the emergency department.
- Doctors' lounge and medical staff facilities.
- Hospital mail sorting and delivery office.
- Medical library.
- Duty rooms, separate for male and female doctors, with sleeping accommodation and toilets and bathroom facilities for night duty residents and doctors, and for those who are on call duty, like surgeons, anaesthesiologists, obstetricians, etc.
- Doctors' chart completion and chart review room adjacent to the medical record department with dictating facilities.
- Medical transcription area in the medical records department.
- Facilities from where in-house video programmes are telecast and music is broadcast.
- Room for copier, mimeographing. Also provision for FAX, TELEX, etc.
- Staff and employees' health clinic.
- Employees' locker rooms — separate for male and female employees — with toilet facilities.
- Board room.
- Conference hall.
- Special dining room for VIPs and medical staff.

- Place(s) near the staff entrance where employees can punch time cards. If the staff is large, more than one area may be needed. A separate place for nurses is recommended.
- Name boards of doctors in the main lobby.
- Directory/floor plan of building and floors.
- Covered car park for officials and senior medical staff.
- Covered park for hospital vehicles.
- Adequate number of staff toilets and public toilets.
- Place for security chief and change room for guards.
- Alcoves for wheelchairs, stretchers, etc. at the main entrance and at emergency.
- Quiet room/prayer room/chapel.
- Ramps at the main entrance and at the emergency room unless the ground floor is at grade level.
- Fire-proof vault for important business records.
- Burglar-proof safe or locker for patients' money and valuables either in the admitting office or in the business office.
- Solarium/vista lounge/sun rooms/day rooms.
- Janitor's closet(s) on every floor.
- Protective guards at wall corners against knocking of wheelchairs, carts, etc. as well as skirting.
- Doorman's station.
- Magazine stands for promotional work.
- Retiring facilities for anxious or bereaved relatives. Quiet room may be used.
- Electrical outlets on corridors for use of cleaning and polishing equipment, and for mobile x-ray, spaced to reach every room without having to use unduly long extension cords.
- Recessed spaces for fire extinguishers and hoses.
- Sleeping accommodation for on-call duty personnel like pharmacists, x-ray and lab technicians.
- Sleeping/living accommodation for kitchen employees who have to be resident to start work early.

➤ESSENTIAL FACILITIES THAT SHOULD BE LOCATED CLOSE TO OUTPATIENT DEPARTMENT AND EMERGENCY SERVICES ON GROUND FLOOR

- Registration and enquiry or information desk.
- Medical records.
- Admitting office.
- Outpatient cashier.
- Pharmacy.
- Laboratory services.

♦ Radiology.

♦ Finance department, particularly billing and cashier.

These interrelated services are open round the clock and should be close to one another.

➤ OTHER FACILITIES THAT SHOULD BE IN THE VICINITY ON THE GROUND FLOOR

♦ Gift shop, book shop, and florist's shop.

♦ Coffee shop and snack bar, preferably accessible to emergency patients as it is most needed in the night.

♦ Pay telephones and water coolers drinking water fountains.

♦ Night administrator's room.

♦ Security post near emergency. It can also be used by the police summoned in connection with accident cases.

Note the following:

♦ Facilities should be organized in such a way that departments and areas that are not functioning in the night could remain closed at that time.

♦ In such areas and floors, there should not be a solitary section open or functioning, necessitating the entire facility to be kept open.

♦ Uneconomical use and waste of resources should be avoided. Unnecessary traffic will pose a security risk.

♦ The following areas must remain closed or kept locked allowing access only to personnel on duty and in emergency.

  ♦ Outpatient department and clinics except adjunct services like medical records, admission, etc.

  ♦ All inpatient areas after visiting hours. Gates will be opened only to physicians, duty personnel and for admission of patients.

  A single lockable gate should be provided for each ward for strict control.

➤ DEPARTMENTS AND FACILITIES FUNCTIONING IN THE NIGHT WHICH SHOULD REMAIN OPEN OR BE AVAILABLE

♦ Accident and emergency services

♦ Registration and enquiry

♦ Admitting

♦ Medical records.

* Pharmacy
* Radiology
* Laboratories
* Billing and night cashier
* Coffee or snack bar
* Pay telephones and STD or ISD telephone booth(s)
* Night administrator and supervisor's office
* Security post
* Ambulance and the duty driver

APPENDIX B
HOSPITAL SAFETY RULES

➤PATIENT CARE

1. Prevent patients from falling from bed. It occurs frequently as they attempt to get on or off the bed unaided. Many of them may be feeble, disoriented or under sedation.
2. Make infirm patients feel at ease. Make them understand that they need to get assistance.
3. Provide for patients' personal belongings to be kept within their easy reach. Ask them to use nurses' call bell to get bedpan or urinal.
4. Use bedside rails on both sides wherever provided, particularly for elderly or restless patients, those coming out of anaesthetic and whenever conditions warrant.
5. Check and double check medications regarding instructions, labels and patient identity.
6. Label all bottles and containers. Keep the medicine supply locked. Keep caution, warning signs against toxic substances, isolation, etc.
7. Lift patients correctly with your leg power keeping your back straight. Use mechanical aids where available.
8. Know proper techniques for:
* turning a patient toward you
* turning a patient from you
* turning a helpless patient
* lifting a patient up in bed, and
* lifting a helpless patient from sitting position or wheelchair.

Familiarize with and follow written procedures regarding proper techniques. These should be available in the nursing service department.

9. Return equipment and materials after use to the correct storage and containers.

➤ TRAFFIC

1. Secure wheelchair or stretcher in place by locking wheel brakes or by other means before loading or unloading a patient or when assisting a patient on or off the vehicle.
2. Always use safety belts or side rails on stretchers to protect patients from falling while transporting.
3. Push carts, wheelchairs and stretchers slowly. Watch your way ahead of you.
4. Push stretchers and beds from the end and not on the sides to avoid jamming your hand against something.
5. Control stretchers and wheelchairs from the lower side while going up or down a ramp. Get help if load or traffic is heavy.
6. Pull vehicles through swinging door. Do not ram through.
7. Before entering or leaving an elevator with wheelchair or stretchers, be sure floor is at level. Wheelchair is always back first.
8. Transport patient's feet first. Have assistant guide at front. Never leave the patient unattended.
9. When you have transferred the patient or have to wait with the patient, park wheelchair or stretcher out of traffic at one side of the corridor.

➤ TRIPS AND FALLS

1. Trips and falls can cause serious injury. Pick up the little things on the floor such as banana or plantain peelings, flower petals, pencils, broken glass, etc.
2. A liquid spill can be risky. Clean it up immediately. Block off the area until cleaned.
3. Keep drawers and cabinet doors closed, particularly the doors of wall mounted cabinets.
4. Never be too busy to look ahead when you are walking.
5. Beware of electric cords. You may easily trip. Place them out of the way. Remove them when not needed.
6. Take one stair at a time. Always use handrails when walking up or down the stairs.
7. Never use fingers to pick up broken glass. Use brush or pan instead.
8. Place objects carefully overhead. Carelessly kept objects may drop and hurt people.

►SAFETY IN ELECTRIC GOODS

1. Prevent dampness near switches, wiring and appliances. Keep hands dry when you handle them.
2. Protect cords. Heat, oil and abuse will damage electric insulation.
3. Inspect cords, plugs, switches, sockets and outlets frequently to ensure that they are not damaged.
4. Report electrical faults immediately. A "small shock", overheating, sparking or noise are urgent warnings.
5. Report defective wiring such as worn out cords, loose or broken plugs or receptacles, blown fuse, etc. to the maintenance department.
6. Do not use an electrical outlet when a plug does not fit smugly. Get the outlet changed.
7. Be sure the equipment is properly grounded. Three-wire "ground" plugs are a good protection.
8. When connecting and disconnecting an electrical equipment, turn the on-off switch to the "off" position.
9. Avoid using an adapter to fit a three-pinned plug in a two-pinned outlet.
10. Take particular care with electrical fittings in areas where it is difficult to keep the floor dry such as the laundry, kitchen, etc. because of spillage, steam condensation, melting ice, etc. All items of equipment and machinery should be grounded. No brass electric light sockets, handles, guards, etc. should be used. If they are currently being used, they should be replaced with non-conducting or rubber-covered-type material.
11. Never attach decorations of paper, cotton, cloth, etc. to electric light wires, fixtures etc. nor keep them within 3 ft. of any open light.
12. Never hang or fasten electric cords with nails, staples or other metal supports.
13. Keep wires, lamps, etc. free from contact with curtains, furniture, packing materials, etc.
14. Do not use any portable electric appliance until it has been checked by the engineering department for safety.

►HEALTH HAZARDS — TOXICITY

1. Each work place is different. Check procedures with your supervisor.
2. Certain chemical, physical and biological exposures can be hazardous to your health. Exposure may be through the eyes, ears, nose, mouth, skin contact, absorption and the nervous system. A hazardous exposure or its effect may be immediate or spread over a long period.
3. Comply with all safety procedures, exposure limits (of radiation, for example), and emergency aid.
4. Never store flammable liquids in your desk or cabinet.

➤GOOD POSTURE IS IMPORTANT

Poor posture not only looks bad, it leads to serious health problems like muscle tension, stiffness, fatigue, backache, neck pain, etc. and even a loss of self-confidence. Good standing posture and good posture in motion, sitting, sleeping and turning are important. Good posture in motion is the safest way to bend, reach and move throughout the day.

➤LOW BACK PAIN

Low back pain is one of mankind's most common ailments. It is estimated that eight out of ten persons have a back injury sometime during their lives. And yet, it can be prevented by learning good posture, correct techniques of lifting, etc. All of us tend to neglect our backs until one day we wince: "Oh, ouch, my aching back!" Poor posture is one of the greatest causes of back pain. Back injuries cost people and their employers a great deal of money, not to mention the loss on account of reduced production, absenteeism and high medical bills.

Follow these simple rules to protect your back.

1. Standing: Rest one foot on a low stool to support your lower back. Do not bend forward with straight legs or stand in one position for a long time. If you are using a lectern (high reading desk), raise or lower the work surface so that your shoulders and neck stay relaxed.

2. Sitting: It is said that 40 hours of sitting can put more strain on the back of a workaholic than 40 hours of standing or even lifting. So if you have to sit for a long time, minimize damage to your back by practising correct posture.

   The chair must be low enough for you to place both feet on the floor and the knees slightly higher than your hips. Cross your legs or put your feet on a stool or footrest if you wish. Always sit firmly against the back of your chair. You may support your lower back with a cushion.

3. Walking: Walk with a good posture. Walk with your head high, chin tucked in, pull in your abdominal muscles lightly to support your lower back.

   When properly performed, walking is the most healthy exercise (particularly recommended for those with high blood pressure). Keep the body loose. The proper position is to bend slightly forward with the toes pointed straight ahead. The knee should be lifted with every step which activates the knee and hip joints.

4. Sleeping: A good night's sleep is necessary. Sleep on a firm mattress which is good for your back. It will support the three natural curves of your backbone. The best way to sleep is on your back with a pillow under your knees.

   Sleeping on your side is all right with hips and knees drawn up and arms extending below the shoulder line. Use no more than one pillow under your head. Do not sleep on your stomach. That may strain your back.

5. Bending: Keep your back straight. The back and neck should be in line as you bend over at the hips. Tighten your abdominal muscles to protect your lower back.

6. Reaching: If you have to reach across the bed in your work with the patient, rest one knee on the bed to support your lower back. Then bend forward with your back straight. Keep your shoulders down; do not hunch over (thus making your back and shoulders into a rounded shape).

7. Lifting: Use your head before you use your back while lifting objects. Avoid painful and costly injury.
Use the power of your legs and not your back to lift correctly and safely. Your legs are strong but your back is weak.

Follow these simple rules.

• Evaluate the load before lifting — whether it is too heavy or too bulky and whether it is within your lifting capacity.
• Inspect the route over which the object is to be carried, the distance to be covered and obstacles on the way like door(s), traffic, etc.
• Stand close to the load or the object being lifted.
• Place the feet apart with one foot forward for a firm and balanced footing.
• Keep your back straight.
• Use your arm, thigh and leg muscles.
• Divide the weight between two hands.
• Take firm natural footing.
• Squat or bend the knees keeping your back in good alignment and erect.
• Extend your hands down to the object and grasp it firmly and in balance.
• Bring the object close to the body.
• Lift with your legs, straighten the ankles, knees, and hips to an upright position by applying a smooth steady pressure with the leg muscles.
• Avoid sudden jerking; do not twist while lifting.

While carrying the object
• Use leg muscles for all carrying movements.
• Keep weight of the load close to your body and at waist level.
• To counterbalance the load, shift part of your body in the opposite direction of the weight.
• Keep your back as straight as possible.
• Proceed by taking short steps.
• Keep a clear vision of the route.

Mistakes that cause back injuries
• Bending the back
• Reaching too far
• Lifting to one side
• Twisting with load

◆ Off-balance shifting
◆ Attempting too much

The above are general safety rules. In addition, there are rules specific to certain areas or departments such as housekeeping, laboratory, radiology, engineering and maintenance, food service, and laundry. Every department must formulate written rules specific to its area.

BIBLIOGRAPHY

1. *Hospital Architecture: Guidelines for Design and Renovation*, David R. Porter, Health Administration Press, Ann Arbor, Michigan, 1982.

2. *Hospital Management: A Guide to Departments*, Howard S. Rowland and Beatrice L. Rowland, An Aspen Publication, Rockville, Maryland, 1984.

3. *Hospital Departmental Profiles*, Edited by Alan J. Goldberg and Robert A DeNoble, American Hospital Association, Chicago, Illinois, 1986.

4. *Guidelines for Construction and Equipment of Hospital and Medical Facilities*, U.S. Department of Health and Human Services, Public Health Service, Health Resources and Service Administration, Washington, D.C. 1983–1984 Edition.

5. *Basic Guide to Hospital Public Relations*, American Society for Hospital Public Relations of the American Hospital Association, Chicago, Illinois, 1984.

6. *Hospitals and the News Media*: *A Guide to Good Media Relations*, Mary Laing Babich, American Hospital Association, Chicago, Illinois, 1985.

7. *Organization of Medical Record Department in Hospitals*, Margaret Flettre Skurka and Mary E. Converse, American Hospital Association, Chicago, Illinois, 1984.

8. *Medical Record Departments in Hospitals*: *Guide to Organization*, American Hospital Association, Chicago, Illinois, 1972.

9. *Hospital Engineering Handbook*, American Hospital Association, Chicago, Illinois, 1980.

10. *Hospital Materiel Management*, Charles E. Housley, An Aspen Publication, Germantown, Maryland, 1978.

11. *Physical Therapy Administration and Management*, Robert J. Hickok, The American Physical Therapy Association, Williams & Wilkins, Baltimore, Md. 1982.

12. *Understanding Computer Science*, Roger S. Walker, A Howard W. Sams and Co. Publication, Indianapolis, IN, 1986.

13. *Maintenance Management for Health Care Facilities*, American Hospital Association, Chicago, Illinois, 1984.

14. *Housekeeping Manual for Health Care Facilities*, American Hospital Association, Chicago, Illinois, 1984.

15. *How to Organize and Maintain an Efficient Hospital Housekeeping Department*, Charles B. Miller, American Hospital Association, Chicago, Illinois, 1981.

16. *Housekeeping Manual for Health Care Facilities*, American Hospital Association, Chicago, Illinois, 1966.

17. *Cafeteria Management for Hospitals*, Faisal A Kaud, R Paul Miller and Robert F. Underwood, American Hospital Association, Chicago, Illinois, 1982.

18. *Hospital Organization and Management*, Malcom T. MacEachern, Physicians' Record Company, Berwyn, Illinois, 1969.

19. *Principles of Hospital Administration*, John R. McGibony, G.P. Putnam's Sons, New York, 1969.